PHOTOMESIC AND PHOTONUCLEAR PROCESSES

FOTOMEZONNYE I FOTOYADERNYE PROTSESSY

ФОТОМЕЗОННЫЕ И ФОТОЯДЕРНЫЕ ПРОЦЕССЫ

The Lebedev Physics Institute Series

Editor: Academician D. V. Skobel'tsyn

Director, P. N. Lebedev Physics Institute, Academy of Sciences of the USSR

Proceedings (Trudy) of the P. N. Lebedev Physics Institute

Volume 34

PHOTOMESIC AND PHOTONUCLEAR PROCESSES

Edited by
Academican D. V. Skobel'tsyn
Director, P. N. Lebedev Physics Institute
Academy of Sciences of the USSR, Moscow

Translated from Russian

Springer Science+Business Media, LLC 1967

ISBN 978-1-4899-2719-4   ISBN 978-1-4899-2717-0 (eBook)
DOI 10.1007/978-1-4899-2717-0

The Russian text was published by Nauka Press in Moscow in 1966 for the
Academy of Sciences of the USSR as Volume XXXIV of the Proceedings
(Trudy) of the P. N. Lebedev Physics Institute.

Фотомезонные и фотоядерные процессы

Труды Физического института им. П. Н. Лебедева

Том XXXIV

Library of Congress Catalog Card Number 67-27904

CONTENTS

PHOTOPRODUCTION OF π^0-MESONS IN
DEUTERIUM NEAR THRESHOLD

A. I. Lebedev and E. I. Tamm

INTRODUCTION

Photoproduction and scattering of π-mesons on nucleons are among the simplest processes involving strongly interacting particles — the π-meson and nucleon. An investigation of the nature of strong interaction is a central problem in modern physics.

A complete study of π^0-meson photoproduction on nucleons entails an investigation of meson production both on protons and neutrons (we will not consider the prospects of an investigation of the reverse reactions). From this viewpoint, an investigation of meson photoproduction on the deuteron, which represents the most weakly bound system of nucleons, is of fundamental importance. For a complete analysis of the isotopic structure of the meson photoproduction amplitude on nucleons it would be sufficient to confine ourselves to an investigation of the reactions

$$\gamma + p \to n + \pi^+, \tag{1}$$

$$\gamma + p \to p + \pi^0, \tag{2}$$

$$\gamma + d \to p + p + \pi^-. \tag{3}$$

In this paper we consider meson photoproduction processes in the near-threshold region of γ-ray energies, where the meson momentum q in the center-of-mass system satisfies the condition $q \leq 1$.* This condition is fulfilled at γ-ray energies $\varkappa \lessgtr 230$ MeV.

An investigation of processes (1)-(3) in this energy region shows [1] that errors in the experimental data do not allow a definite isotopic analysis and a prediction of the amplitude for production of π^0-mesons on neutrons in the S state. The obtention of direct information regarding the process $\gamma + n \to n + \pi^0$ necessitates an investigation of the reactions

$$\gamma + d \to p + n + \pi^0, \tag{4}$$

$$\gamma + d \to d + \pi^0. \tag{5}$$

An investigation of the latter process can also provide information about the electromagnetic form factor of the deuteron.

An investigation of the isotopic structure of the photoproduction amplitude assumed particular value in connection with the study of the role of resonance $\pi-\pi$ interaction in meson photoproduction. When this inter-

---

*In this paper we use units with $\hbar = \mu_{\pi^0} = c = 1$.

1

action is associated with the two-meson intermediate state (ρ-meson), it makes a contribution only to the iso-calar part of the photoproduction amplitude F^0. When the interaction is associated with the three-meson inter-mediate state (ω- or φ-mesons) it makes a contribution to the isovector part of the amplitude F^+. These con-tributions are characterized by the constants of $\gamma\pi\rho$-, $\gamma\pi\omega$-, ρNN-, and ωNN-interactions, which are involved in a number of effects (lifetime of π^0-meson, electromagnetic form factors of nucleons and π-mesons, cross sections of processes $\gamma + N \rightarrow N + \pi + \pi$, $\pi + N \rightarrow N + \pi + \gamma$, etc. [2]). In this connection a determination of these constants is of great interest. As we will show below, the discussed reactions can give information about these constants. An investigation of these reactions can contribute toward a clarification of the question of the existence of the hypothetical π_0^0-meson [3].

For the description of meson photoproduction in deuterium the impulse approximation is used [4]. Its ap-plicability for the description of the reaction $\gamma + d \rightarrow p + p + \pi^-$ in the near-threshold energy region has been experimentally confirmed [5]. The case of π^0-mesons, however, requires an independent verification of the impulse approximation, since the cross section for the photoproduction of neutral mesons near threshold is much smaller than that for charged mesons and the corrections for multiple scattering of $\pi^\pm$ into π^0-mesons may be-come significant.

Thus, the problems which may be tackled by an investigation of π^0-meson photoproduction in deuterium in the region of low γ-ray energies make this process of exceptional interest for meson physics.

The aim of the present work is to give an account of the results of an experimental investigation of π^0-meson photoproduction in deuterium in the near-threshold region of γ-ray energies and to discuss these results from the viewpoint of phenomenological theory using the impulse approximation. We also give the results of careful calculations of the differential cross sections for the elastic ($\gamma + d \rightarrow d + \pi^0$) and inelastic ($\gamma + d \rightarrow p + n + \pi^0$) photoproduction of π^0-mesons on deuterons and suggestions regarding the use of these processes for a study of $\gamma\pi\rho$- and $\gamma\pi\omega$-interactions and for the investigation of the process $\gamma + n \rightarrow n + \pi^0$.

CHAPTER I: SHORT REVIEW OF THEORETICAL AND EXPERIMENTAL WORK
ON π^0-MESON PHOTOPRODUCTION

1. Meson Photoproduction on Free Nucleons

The operator for meson photoproduction on nucleons in the spin σ matrix representation has the form

$$T = e^{i2(\Delta r)} [i(\mathbf{K}\sigma) + L], \quad \Delta = \frac{1}{2}(\mathbf{k} - \mathbf{q}), \tag{1,1}$$

where the operators K and L in the meson−nucleon center-of-mass system can be written as

$$\mathbf{K} = \mathbf{e}(F_1 - F_2 \cos \theta) + \frac{\mathbf{k}(\mathbf{q}\mathbf{e})}{kq}(F_2 + F_3) + \frac{\mathbf{q}(\mathbf{q}\mathbf{e})}{q^2}F_4,$$

$$L = \frac{(\mathbf{e}[\mathbf{q}\mathbf{k}])}{qk}F_2. \tag{1,2}$$

Here, $\mathbf{k}$ and $\mathbf{e}$ are the momentum and polarization of the photon, $\mathbf{q}$ is the meson momentum, $\mathbf{r}$ is the radius vector of the nucleon, F_i (i = 1, 2, 3, 4) are functions of the momentum and angle of emission θ of the meson. Using operators (1.1) and (1.2) to determine the matrix element of the photoproduction process, we square it, sum, and average over the polarizations of the nucleon and γ-quantum and thus obtain the following expres-sion for the differential cross section of the process $\gamma + N \rightarrow N + \pi$:

$$d\sigma / d\Omega = q\sqrt{1 + q^2}\{|F_1|^2 + |F_2|^2 - 2\mathrm{Re}(F_1^*F_2)\cos\theta + \frac{1}{2}\sin^2\theta[|F_3|^2 +$$

$$+ |F_4|^2 + 2\mathrm{Re}(F_2^*F_3) + 2\mathrm{Re}(F_1^*F_4) + 2\mathrm{Re}(F_3^*F_4)\cos\theta]\}. \tag{1,3}$$

In the near-threshold region, where the mesons are produced mainly in the S and P states, when the function $F_i(q, \theta)$ is expanded in series in terms of multipole amplitudes, we can confine ourselves to a consideration of the electric dipole E_{0+} and quadrupole E_{1+} transitions $(E_j = l \pm \frac{1}{2})$ and magnetic dipole $M_{1\pm}$ transitions $(M_j = l \pm \frac{1}{2})$:

$$F_1(q, \theta) = E_{0+}(q) + 3\,[M_{1+}(q) + E_{1+}(q)]\cos\theta,$$

$$F_2(q, \theta) = 2M_{1+}(q) + M_{1-}(q),$$

$$F_3(q, \theta) = 3[E_{1+}(q) - M_{1+}(q)],$$

$$F_4(q, \theta) = 0. \tag{1.4}$$

The isotopic invariance of strong interactions in application to meson photoproduction processes allows the four meson-producing processes $(\gamma + p \to p + \pi^0,\ \gamma + p \to n + \pi^+,\ \gamma + n \to p + \pi^-,\ \text{and}\ \gamma + n \to n + \pi^0)$ to be described by three transition amplitudes in the isospin space of strongly interacting particles

$$\langle T_f = \tfrac{1}{2}\,|\,S\,|\,T_i = \tfrac{1}{2}\rangle,\quad \langle T_f = \tfrac{1}{2}\,|\,V_3\,|\,T_i = \tfrac{1}{2}\rangle,\quad \langle T_f = \tfrac{3}{2}\,|\,V_3\,|\,T_i = \tfrac{1}{2}\rangle,$$

where T is the isospin, and S and V_3 are the isoscalar and isovector parts, respectively, of the meson photoproduction operator. By combinations of these amplitudes

$$F^0 = \langle T_f = \tfrac{1}{2}\,|\,S\,|\,T_i = \tfrac{1}{2}\rangle,$$

$$F^+ = \tfrac{1}{3}\langle T_f = \tfrac{1}{2}\,|\,V_3\,|\,T_i = \tfrac{1}{2}\rangle + \tfrac{2}{3}\langle T_f = \tfrac{3}{2}\,|\,V_3\,|\,T_i = \tfrac{1}{2}\rangle, \tag{1.5}$$

$$F^- = \tfrac{1}{3}\langle T_f = \tfrac{1}{2}\,|\,V_3\,|\,T_i = \tfrac{1}{2}\rangle - \tfrac{1}{3}\langle T_f = \tfrac{3}{2}\,|\,V_3\,|\,T_i = \tfrac{1}{2}\rangle,$$

which possess particular crossing-symmetry properties, the amplitudes of meson production into particular charge states are expressed as follows:

$$F(\gamma + p \to p + \pi^0) = F^+ + F^0,\quad F(\gamma + p \to n + \pi^+) = \sqrt{2}\,(F^0 + F^-),$$

$$F(\gamma + n \to n + \pi^0) = F^+ - F^0,\quad F(\gamma + n \to p + \pi^-) = \sqrt{2}\,(F^0 - F^-). \tag{1.6}$$

Using the unitarity of the S matrix and its invariance to time reversal we can, for the near-threshold region, express the phases of the complex multipole amplitudes of meson production into states with a particular isospin of the $\pi - N$ system in terms of the corresponding phases of $\pi - N$ scattering $\delta^T_{j = l \pm \frac{1}{2}}$:

$$E^{T_f}_{l\pm}(M^{T_f}_{l\pm}) = \mathcal{E}^T_{l\pm}(\mathcal{M}^T_{l\pm})\,e^{i\delta^T_{l\pm}}. \tag{1.7}$$

The S phases of $\pi - N$ scattering are usually denoted by the symbol $\alpha_{2T} = \delta^T_{0+}$, and the P phases by the symbol $\alpha_{2T,2j} = \delta^T_{1\pm}$.

A phenomenological analysis of the differential cross sections of the process $\gamma + p \to p + \pi^0$ for several γ-ray energies, made on the assumption that the amplitude for $E^{\pi^0}_{1+}$ is small, gave the following values of the amplitudes $E^{\pi^0}_{0+}$, $M^{\pi^0}_{1-}$, and $M^{\pi^0}_{1+}$ [1]:

$$E^{\pi^0}_{0+} = -(0.24 \pm 0.04) \cdot 10^{-2},$$

$$M^{\pi^0}_{1+} \simeq M^+_{1+} \simeq \frac{2}{3}\,M^{\frac{3}{2}}_{1+} = \frac{a}{a_{33}}\frac{\sin\alpha_{33}}{q^2}\,e^{i\alpha_{33}},$$

$$M^{\pi^0}_{1-} = (0.0 \pm 0.01) \cdot 10^{-2}, \tag{1.8}$$

$$a = (0.65 \pm 0.05) \cdot 10^{-2},$$

$$a_{33} = \lim_{q \to 0} \frac{\alpha_{33}}{q^3}.$$

The results of a determination of the amplitudes $E_{0+}^{\pi+}$ and $E_{0+}^{\pi-}$ from the data for the differential cross section of the process $\gamma + p \rightarrow n + \pi^+$ and the ratio $\sigma^-(\gamma + n \rightarrow p + \pi^-)/\sigma^+(\gamma + p \rightarrow n + \pi^+)$ [6], together with the amplitude $E_{0+}^{\pi^0}$ give, according to relationships (1.6), the amplitudes E_{0+}^+, E_{0+}^-, and E_{0+}^0, and also the S-state π^0-meson photoproduction amplitude $E_{0+}^{n\pi^0}$ on neutrons:

$$E_{0+}^+ = (-0.10 \pm 0.03)\cdot 10^{-2}$$

$$E_{0+}^- = (+2.04 \pm 0.03)\cdot 10^{-2},$$

$$E_{0+}^0 = (-0.14 \pm 0.03)\cdot 10^{-2},$$ (1.9)

$$E_{0+}^{n\pi^0} = (+0.04 \pm 0.04)\cdot 10^{-2}.$$

The approximate equality of amplitudes E_{0+}^+ and E_{0+}^0 and the smallness of amplitude $E_{0+}^{n\pi^0}$ do not fit into the framework of the theory that $S \sim (\mu/M)V_3$ (M is the nucleon mass) [7]. Direct measurement of these amplitudes would help to clarify the parameters of the physics of slow π-mesons. It is from this viewpoint that the reactions $\gamma + d \rightarrow d + \pi^0$ and $\gamma + d \rightarrow p + n + \pi^0$ are of interest.

Owing to the relationship $E_{0+}^{\pi^0} \ll E_{0+}^{\pi\pm}$, an appreciable role in the process $\gamma + p \rightarrow p + \pi^0$ near threshold is played by charge transfer from the produced charged π-mesons to neutral mesons. It follows from the condition of unitarity of the S matrix, the threshold behavior of the cross sections, and the smallness of the amplitude $E_{0+}^{\pi^0}$ that the imaginary part of the π^0-meson photoproduction amplitude due to this charge exchange can be expressed in terms of E_{0+}^+ and the difference in S-state $\pi-N$ scattering lengths $a_1 - a_3$ ($a_{2T} = a_{2T}q$):

$$\text{Im } E_{0+}^+ = {}^2/_3 \ (a_1 - a_3)E_{0+}^- q_+,$$ (1.10)

where q_+ is the momentum of the charged meson produced by a photon with energy k. The imaginary parts of the other amplitudes (apart from the resonance amplitude $M_{1+}^{3/2}$) are negligibly small.

From the results of phenomenological analysis we can, by using relationships (1.3), (1.4), and (1.8), write the differential cross section of the process $\gamma + p \rightarrow p + \pi^0$ in the form

$$\frac{d\sigma}{d\Omega} = q\sqrt{1+q^2}\Big[|E_{0+}^{\pi^0}|^2 + 2\text{Re}(E_{0+}^{\pi^0}M_{1+}^{\pi^0*})\cos\theta + |M_{1+}^{\pi^0}|^2\frac{5-3\cos^2\theta}{2}\Big].$$ (1.11)

It follows from this relationship and (1.8) that for a forward angle of meson emission there is a γ-ray energy at which the real part of the amplitude becomes zero and the cross section is given by the imaginary part of the amplitude E_{0+}^+. This leads to the characteristic passage of the cross section $(d\sigma/d\Omega)|_{\theta=0}^{\pi^0}$ through a minimum and a maximum [8] (Fig. 1). The minimum value of the cross section can give valuable information regarding the difference $a_1 - a_3$ and the position of the minimum provides information about the magnitude of the amplitudes $E_{0+}^{\pi^0}$ and $M_{1+}^{3/2}$.

Investigations of the analytical properties of the amplitudes of processes involving strongly interacting particles have given the following dispersion relations [9, 10] for the amplitudes $F_i(q, \theta)$ (i = 1, 2, 3, 4):

$$\text{Re } F_i(q, \theta) = P + I(\Sigma\omega_{ij} \text{ Im } F_j) + \Delta F_i,$$ (1.12)

where the first term represents the photoproduction amplitudes in first-order perturbation theory for the $\pi-N$ interaction constant, the second term is the dispersion integral (ω_{ij} are coefficient functions), and the third term is the contribution of resonance $\pi-\pi$ interactions to the photoproduction amplitudes.

Dispersion relations are widely used for the analysis of experimental data on meson photoproduction on nucleons. There are two viewpoints on the choice of method for such an analysis. The first consists in conversion of the dispersion relations to approximate integral equations [10], the solutions of which are compared with the experimental data. Such an approach gives, for instance, the following solution for amplitude $M_{1+}^{3/2}$:

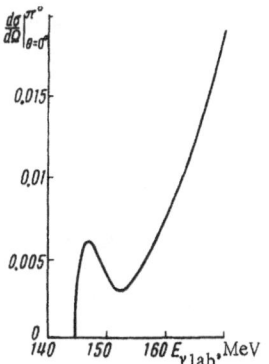

$$M_{1+}^{3/2} = \frac{(\mu_p - \mu_n)\,e}{4Mf}\,\frac{\sin\alpha_{33}}{q^2}\,e^{i\alpha_{33}} \qquad (1.13)$$

($\mu_p = 2.78$ and $\mu_n = -1.91$ are the magnetic moments of proton and neutron, f is the $\pi-N$ coupling constant, $f^2 = 0.08$, $e^2 = 1/137$), and satisfactorily describes the experimental data for meson photoproduction in the region of first resonance energies. In the near-threshold region, however, the solutions of the equations do not agree with the experimental data for the process $\gamma + p \to p + \pi^0$ [1] and, hence, cannot be used for a description of neutral meson photoproduction. Hence, the second viewpoint, according to which the experimental data for the imaginary parts of the amplitudes are substituted in the right sides of the dispersion relations, has been greatly developed. Such an approach [11, 12] has led to a much better description of experiments on π^0-meson photoproduction on protons. In [12], however, which gives the results of calculations of the amplitudes F_i, an error was made in the calculation of $F_1^+(q, \theta)$ and, hence, these amplitudes cannot be used for calculation of the cross section for π^0-meson photoproduction on deuterons.

Fig. 1. Differential cross section of process $\gamma + p \to p + \pi^0$ as a function of energy for angle $\theta = 0°$.

Discrepancies between the predictions of the dispersion relations and experimental data have recently been interpreted in terms of the contribution of resonance $\pi-\pi$ interactions to the photoproduction amplitudes [13]. It follows from the conservation of G-parity that the two-meson intermediate state (ρ-meson) makes a contribution only to amplitude F^0, and the three-meson state with isospin $T = 0$ (ω-meson) only to amplitude F^+. Ways of distinguishing these contributions in investigations of meson photoproduction on deuterons are discussed in Chapter II. Here we will merely point out that the separate determination of these contributions necessitates new theoretical calculations based on accurate dispersion relations for the amplitudes of π^0-meson photoproduction on nucleons.

2. Theory of π^0-Meson Photoproduction in Deuterium

Theories of π^0-meson photoproduction on nucleons bound on the deuteron are of a phenomenological nature. * They make use of several simplifying approximations, the most important of which is the impulse approximation [4]. The essential feature in this case is the assumption that the operator for photoproduction on the deuteron is the sum of the operators for photoproduction on the individual nucleons of the nucleus

$$T = T_p + T_n = e^{2i(\Delta \mathbf{r}_p)}\,[(\mathbf{K}_p\boldsymbol{\sigma}_p) + L_p] + e^{2i(\Delta \mathbf{r}_n)}\,[(\mathbf{K}_n\boldsymbol{\sigma}_n) + L_n], \qquad (1.14)$$

where r_i is the radius vector of the i-th nucleon. It is presumed here that the concept of the nucleon as an individual particle in the deuteron is still valid. Using the operator (1.14) we can obtain the matrix elements of the processes $\gamma + d \to d(p, n) + \pi^0$

$$M_{if} = \langle \psi_f,\, T\psi_i \rangle, \qquad (1.15)$$

where the wave function of the initial state ψ_i describes the deuteron

$$\psi_i = \chi_m^1 \varphi_d(\mathbf{r}), \quad \mathbf{r} = \mathbf{r}_p - \mathbf{r}_n, \qquad (1.16)$$

---

*This applies also to investigations carried out in the early stages of development of meson physics, where the calculations were based on model Hamiltonians [14]. Attempts to examine the problem of photoproduction in deuterium within the framework of the method of dispersion relations [15] represent an exception. However, no appreciable success has been obtained along this line.

and the wave function of the final state for the elastic process is (1.16) with allowance for motion of the recoil deuteron as a whole, and for the inelastic process:

$$\psi_f^{t,\,s} = \left(\frac{1}{2\pi}\right)^{3/2} e^{i(\mathbf{PR})} \varphi_{pn}^{t,\,s}(\mathbf{r},\,\mathbf{p}) \{\chi_m^1,\,\chi_0^0\},$$

$$\mathbf{R} = \frac{1}{2}\,(\mathbf{r}_p + \mathbf{r}_n),\ \ \mathbf{p} = \frac{1}{2}\,(\mathbf{p}_p - \mathbf{p}_n),\ \ \mathbf{P} = \mathbf{p}_p + \mathbf{p}_n,$$

(1.17)

where χ_m^1 and χ_0^0 are triplet and singlet spin functions of the two-nucleon system. Squaring the matrix element M, summing over the final, and averaging over the initial spin states, we obtain the following expressions for the differential cross sections of the processes $\gamma + d \to d + \pi^0$ and $\gamma + d \to p + n + \pi^0$:

$$d\sigma\,(\gamma + d \to d + \pi^0) = |\langle M_{if}\rangle_d|^2\, d\mathbf{q} d\mathbf{Q} \delta\,(\Sigma p_i)\,\delta\,(\Sigma E_i),$$

$$|\langle M_{if}\rangle_d|^2 = \left[\frac{2}{3}\,|\mathbf{K}_p + \mathbf{K}_n|^2 + |L_p + L_n|^2\right] I^2\,(\Delta);$$

(1.18)

where Q is the momentum of the recoil deuteron, $\Delta = \frac{1}{2}\,(k-q)$

$$d\sigma\,(\gamma + d \to p + n + \pi^0) = |\langle M_{if}\rangle_{pn}|^2\, d\mathbf{p}\, d\mathbf{P}\, d\mathbf{q} \delta\,(\Sigma p_i)\,\delta\,(\Delta E_i),$$

$$\langle M_{if}\rangle_{pn}|^2 = {}^2/_3\,|\mathbf{K}_p I_+^t + \mathbf{K}_n I_-^t|^2 + {}^1/_3\,|\mathbf{K}_p I_+^s - \mathbf{K}_n I_-^s|^2 + |L_p I_+^t + L_p I_-^t|^2.$$

(1.19)

In these expressions the deuteron form factor I(Δ) and functions $I_\pm^{t,\,s}(p,\Delta)$ have the form

$$I\,(\Delta) = \int d\mathbf{r} e^{i(\Delta \mathbf{r})}\,\varphi_d^2\,(\mathbf{r}),$$

(1.20)

$$I_\pm^{t,\,s}\,(\mathbf{p},\,\Delta) = \int d\mathbf{r} \varphi_{pn\,\mathrm{conv}}^{t,\,s*}\,e^{\pm i(\Delta \mathbf{r})}\varphi_d\,(\mathbf{r}).$$

(1.21)

It is usually assumed that the operators K and L depend weakly on the nucleon momentum. In some cases, however, this dependence may be significant. In view of this, Baldin devised a method of introducing corrections of the first order in v/c, where v is the nucleon velocity [4, 14].

The angular distributions of π^0-mesons are obtained by integrating expressions (1.18) and (1.19) with respect to all the variables, except $d\Omega_\pi$. However, integration of the cross section (1.19) with respect to dp requires specification of the nucleon wave functions and it is necessary to know the explicit form of the functions K and L. This difficulty is overcome by the substitution of infinity (∞) for the upper limit of integration p_{max}, the momentum determined by the laws of conservation of energy. The orthonormality of the functions φ_d and φ_{pn} in p space can be used and the following expression for the total cross section of processes of π^0-meson photoproduction on deuterons is obtained:

$$\frac{d\sigma}{d\Omega_\pi} = \frac{q^2}{\dfrac{q}{\sqrt{1+q^2}} + \dfrac{q - k\cos\theta}{2M}}\,\{|\,\mathbf{K}_p\,|^2 + |\,L_p\,|^2 + |\,\mathbf{K}_n\,|^2 + |\,L_n\,|^2 +$$

$$+\ 2\mathrm{Re}\,[{}^1/_3\,(\mathbf{K}_p^*\mathbf{K}_n) + L_p^* L_n]\,I\,(2\Delta)\}.$$

(1.22)

This approximation, called the approximation of the theorem of completeness, is often used for the analysis of experimental data. If it is assumed that the amplitudes of π^0-meson photoproduction on protons and neutrons are approximately equal, then it follows from expression (1.22) that the total cross section of π^0-meson photoproduction on deuterons differs from twice the cross section for the proton by a term proportional to the square

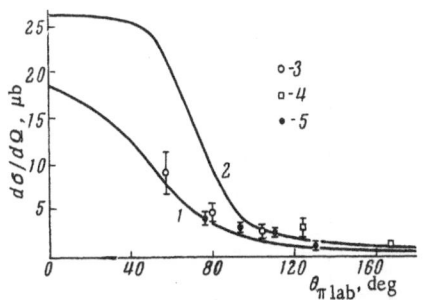

Fig. 2. Differential cross section of process $\gamma + d \rightarrow d + \pi^0$ for γ-ray energy 285 MeV. (1,2) Results of calculation of cross section with and without correction for multiple scattering of π^0-mesons [17]; (3, 4, 5) data of [22a], [24], and [22], respectively.

of the deuteron form factor.* Since Δ attains a high value in the case of backward angles of emission of mesons and the form factor I is close to zero, the cross section on the deuteron will be nearly twice the cross section for π^0-meson photoproduction on protons. At the same time, for $\theta_\pi \simeq 180°$ the cross section of the elastic process is small, since it is proportional to $I^2(\Delta)$. At angles $\theta_\pi \simeq 0°$, where the form factors $I(\Delta)$ and $I(2\Delta)$ are close to unity, the relationship $d\sigma_{tot} \approx d\sigma_{el} \approx 4d\sigma_p$ must hold. It is obvious, however, that in the considered near-threshold region of energies the approximation of the theorem of completeness is not valid, since p_{max} differs considerably from infinity (∞). In view of this, it was necessary to conduct careful calculations of the cross sections for π^0-meson photoproduction in deuterium with separate consideration of the elastic and inelastic processes.

The differential cross section of the elastic process was calculated in [16], where the solutions of the integral equations of [10] were used for the description of π^0-meson production on a nucleon. The differential cross section was obtained on the assumption that the resonance amplitude $M_{1+}^{3/2}$ plays the main role,

$$\frac{d\sigma}{d\Omega} = \frac{16}{27} \frac{e^2 k}{q^3 f^2} \left(\frac{\mu_p - \mu_n}{4M} \right)^2 \sin^2 \alpha_{33} (2 + 5 \sin^2 \theta) I^2(\Delta). \qquad (1.23)$$

However, the amplitudes obtained in [10] are not suitable for the description of the near-threshold energy region [1] and, hence, new calculations of the cross section of the process $\gamma + d \rightarrow d + \pi^0$, using the amplitudes (1.8) and (1.9), had to be made. In addition, in most theoretical calculations the calculation of the deuteron form factor has been based on model wave functions, whereas, in the near-threshold energy region it can be expressed in terms of the well-known parameters of N−N scattering: the effective triplet range of N−N interaction and the deuteron binding energy.

The question of the role of multiple scattering in the process $\gamma + d \rightarrow d + \pi^0$ was examined by Chappelear [17]. He showed that in the energy region of the first resonance in the π−N system multiple scattering reduces the cross section of the process by a factor of 1.5-2 (Fig. 2). Such calculations, however, have not yet been conducted for the near-threshold region nor for the process $\gamma + d \rightarrow p + n + \pi^0$. Hence, the question of the role of these corrections in the near-threshold region of γ-ray energies could be solved by a comparison of the experimental data for the processes $\gamma + d \rightarrow d(pn) + \pi^0$ with the results of calculations based on the impulse approximation.

3. Experimental Investigations

Investigations of π^0-meson photoproduction were begun in 1950 on the Berkeley accelerator [18]. The aim of the first experiments on nuclei, primarily in hydrogen and deuterium, was to find out if the amplitudes of π^0-meson photoproduction on protons and neutrons were the same. In [19], π^0-mesons were produced by bremsstrahlung with a maximum energy of 310 MeV in targets of H_2O, D_2O, heavy paraffin, and ordinary paraffin. One of the π^0-meson decay γ-quanta was detected. The ratio of the cross sections in deuterium and hydrogen averaged over various angles was $\sigma_d/\sigma_p \simeq 2$, and the corresponding ratio for carbon was $\sigma_C/\sigma_p \simeq 10$. However, the results were not accurate enough for a quantitative analysis (Table 1) and the only conclusion was that the cross sections for π^0-meson photoproduction on neutrons and protons were approximately the same.

---

*The difference in the value of the factors in front of the square of the matrix element is ignored. For forward angles of emission of mesons this difference is insignificant, but for $\theta_\pi = 90°$, it reaches a value of about 10%.

Table 1. Ratio of Yields of Decay Photons from π^0-Mesons Formed by Photoproduction in Deuterium and Hydrogen ($d\sigma_d/d\sigma_p$).
(Laboratory system of coordinates)

x, MeV	θ_γ, deg								Reference
	0	30	45	70	90	135	140	180	
310 273 237			1.76±0.20	2.03±0.17	1.91±0.09 2.11±0.20 1.88±0.37	1.53±0.17			[19]
500 400 300 500—400 400—300		1.98±0.04 1.92±0.06 2.50±0.22 2.10±0.28 1.72±0.12		1.76±0.04 1.82±0.04 1.72±0.08 1.58±0.22 1.88±0.10			1.82±0.04 1.66±0.08 2.00±0.10 3.40±1.20 1.36±0.12		[23]
Near threshold					4.2±1.3				[22, 22a] *
265±15	2.86±0.73				2.92±0.40			1.18±0.25	[25] *

*The data correspond directly to the ratio of the cross sections for π^0-meson photoproduction.

Table 2. Differential Cross Section of Process
$$\gamma + d \rightarrow d + \pi^0$$
(Laboratory system of coordinates)

$\varkappa$, MeV	θ_π, deg	$\frac{d\sigma}{d\Omega} \cdot 10^{30}$, cm^2/sr	Reference
250—300	90	$3.5 \pm {}^2_1$	[20]
250—300	76	4.2 ± 0.6	[22,22a]
	93	3.2 ± 0.5	
	110	2.5 ± 0.4	
	130	1.2 ± 0.3	
270 ± 14	124	3.2 ± 0.9	[24]
	168	1.3 ± 0.5	
229 ± 35	102	2.4 ± 0.9	[22, 22a]
246 ± 35	76	2.8 ± 1.0	
253 ± 35	104	2.7 ± 1.0	
279 ± 35	80	4.7 ± 1.2	
280 ± 35	56	9.0 ± 2.4	
	105	2.8 ± 0.8	

Table 3. Differential Cross Section of Process
$$\gamma + p \rightarrow p + \pi^0$$ and Total Cross Section
of Processes $\gamma + d \rightarrow d + \pi^0$, $\gamma + d \rightarrow p + n + \pi^0$
in Center-of-Mass System for γ-Ray Energy
$\varkappa = 265$ MeV

θ_π c.m., deg	$\left(\frac{d\sigma}{d\Omega}\right)_d \cdot 10^{30}$, cm^2/sr	$\left(\frac{d\sigma}{d\Omega}\right)_p \cdot 10^{30}$, cm^2/sr
0	15.2 ± 2.6	5.3 ± 1.0
20	26.8 ± 2.6	
40	29.4 ± 2.6	
60	48.0 ± 4.2	
90	40.0 ± 4.2	13.4 ± 1.2
135	21.6 ± 2.6	
180	13.4 ± 2.4	11.3 ± 1.2

Beginning in 1953, experiments with deuterium were conducted with the aim of verifying the hypothesis of isotopic invariance. A nontrivial consequence of this hypothesis is that the constants for coupling of the neutron (g_n) and proton (g_p) with the neutral meson field must have opposite signs. This was first experimentally confirmed in [20, 21], where the recoil deuterons and one of the π^0-meson decay γ-quanta were detected. In [20] the recoil deuterons and protons from the inelastic process were detected in two scintillation counters with NaI crystals. In the first counter, which had a thin crystal, pulses proportional to the energy loss dE/dx were produced. In the second counter, which had a thick crystal, the pulses were proportional to the nucleon energy E. The product of these pulses can be put in the form

$$E\frac{dE}{dx} \sim M^{0.8} Z^2 E^{0.2},$$

where M is the nucleon mass and Z is the charge. Thus, from an amplitude analysis of the signals obtained by multiplication of the pulses E and dE/dx, the deuterons and recoil protons could be distinguished.

In [21] the method of separating the recoil products from their time of flight in a semicircular orbit in a constant magnetic field was used. The protons and deuterons were focused simultaneously on a scintillation counter. The dimensions of this counter and the field geometry determined the range of emission angles, and the field strength determined the momentum of the detected particles. The results of the two investigations, if the impulse approximation is valid, show that the coupling constants g_p and g_n have different signs.

This result was later confirmed in [22, 22a, 23]. In [22, 22a] the differential cross section of the process $\gamma + d \rightarrow d + \pi^0$ at $E_\gamma \geq 230$ MeV and meson emission angles $\theta_\pi > 60°$ was measured (Table 2). The method used in these investigations was the same as that described in [20]. Figure 2 shows the results of a comparison of the measurements with Chappelear's calculations [17], which showed that scattering of the produced π^0-meson on the second nucleon of the deuteron leads to a reduction of the cross section for π^0-meson photoproduction on the deuteron. The figure also shows the results of [24], which were obtained by detection of the recoil deuterons on photographic plates. Unfortunately, the measurements were conducted only for meson emission angles of 124 and 168°.

The total cross section for π^0-meson photoproduction in deuterium (elastic and inelastic) for angles $\theta_\pi = 0$-$180°$ has been measured only in [25] for energies $\varkappa = 265 \pm 15$ MeV. The meson decay photons were detected by two γ-telescopes connected in coincidence. Measurements were made with liquid deuterium and hydrogen. No details of the experiment or treatment of the results are known. Hence, it is difficult to estimate the reliability of the data taken from the graph of [25] and given in Table 3. From these data the ratio of the cross sections for deuterium and hydrogen for angles $\theta_\pi = 0°$, 90°, and 180° can be determined (see Table 1).

Relatively complete measurements of the ratio of the π^0-meson yields in photoproduction in deuterium and hydrogen have been carried out only in 1956 [23] for maximum bremsstrahlung energies of 300, 400, and 500 MeV. Since the mesons were detected from one of the decay γ-quanta, the results of measurements are presented in the form of a ratio of the integral photon yields. The ratio of the differential yields was also determined by the "photon difference" method (see Table 1). In the discussion of the results of this work, it was suggested that the expressions under the integral sign, which connect the decay photon yield with the cross section for π^0-meson photoproduction, have the same form for hydrogen and deuterium. If this hypothesis is correct, the ratio of the photon yields from deuterium and hydrogen is equal to the ratio of the meson photoproduction cross sections. In view of this we can conclude that in the small-angle region ($\theta_\pi = 30°$) some reduction in the ratio $d\sigma_d/d\sigma_p$ in comparison with the value calculated by the impulse approximation may result from the correction to the elastic process in deuterium due to multiple scattering. For the large-angle region ($\theta_\pi = 73°$ and $140°$) at all energies the ratio is approximately constant and equal to 1.8. We can conclude from this that interference effects in deuterium in this region are negligible and the cross section for π^0-meson photoproduction in deuterium is equal to the sum of the cross sections on the neutron and proton.

We should also mention [26], where the ratio of the cross sections for π^0-meson photoproduction on bound and free protons was measured. Recoil protons produced in the processes $\gamma + p \to p + \pi^0$ and $\gamma + d \to p + n + \pi^0$ due to bremsstrahlung with maximum energy 500 MeV were detected. The ratio σ_d/σ_p, calculated from measurements of σ_p alone (the motion of the nucleons in deuterium was taken into account), was compared with the directly measured ratio. The results of this work for recoil proton emission angles $\theta_p > 70°$, and an analysis of the data obtained by other authors for maximum bremsstrahlung energy 310 MeV, show that the cross section for π^0-meson photoproduction on a nucleon bound in deuterium is 10% lower than in hydrogen. The statistical accuracy of these data is about 5%. This conclusion is not altered if the probability of reabsorption of the meson on the second nucleon is taken into consideration [27]. The quantitative results of [26] are not conclusive enough. However, the experimental assessment of the effect of the binding of the nucleons in deuterium on photoproduction is of interest from the viewpoint of isolation of meson photoproduction on the neutron.

In the experimental data given in Tables 1-3, only the statistical errors are indicated. It should be borne in mind that these results also contain an absolute error with an average value of about 25% due to calibration of the bremsstrahlung beam.

CHAPTER II: CALCULATION OF DIFFERENTIAL CROSS SECTIONS FOR π^0-MESON PHOTOPRODUCTION ON DEUTERONS

1. Approximations Employed

The processes $\gamma + d \to d(pn) + \pi^0$ are considered within the framework of the impulse approximation (1.14). For the description of meson photoproduction on nucleons the results of a phenomenological analysis (1.8)-(1.10) were used. The operators K and L will have the following form:

$$\mathbf{K}^+ = E_{0+}^+ \mathbf{e} + M_{1+}^+ \frac{[\mathbf{q}_p [\mathbf{ek}]]}{kq_p},$$

$$L^+ = 2M_{1+}^+ \frac{(\mathbf{q}_p [\mathbf{ke}])}{kq_p},$$

$$\mathbf{K}^0 = E_{0+}^0 \cdot \mathbf{e}, \qquad L^0 = 0.$$

(2.1)

The amplitudes M_{1+}^+ and $\mathrm{Im}E_{0+}^+$ strongly depend on the meson momentum q_p in the photon–nucleon center-of-mass system [system (1)]. In view of this, the calculation of the differential cross sections of the processes $\gamma + d \to d(pn) + \pi^0$ in the photon–deuteron center-of-mass system [system (2)] requires the expression of q_p in terms of the meson momentum q in system (2). For a given photon energy $\varkappa_{lab}$ in the laboratory system of coordinates, system (2) moves relative to system (1) with a velocity $u = 1/(3 + 2M/\varkappa)$ and the connection between q_p and q will be given by the Lorentz formulas for transformation from system (1) to system (2). Regarding the deuteron and nucleon as nonrelativistic particles, and expressing u in terms of the deuteron velocity v in the center-of-mass system:

$$v = \frac{1}{\sqrt{1 + (2M / \varkappa)^2}}, \quad u = \frac{v}{1 + v}, \quad (2.2)$$

from the energy transformation formula it is easy to obtain

$$q_p = \frac{1}{(3v + 1)(v + 1)} [q^2 (1 + 4v + 4v^2 + v^2 \cos \theta) - 2(1 + 2v) vq \sqrt{1 + q^2} \cos \theta + v^2], \quad (2.3)$$

where θ is the angle between $\varkappa$ and q.

In adopting this approach we regard the deuteron as a tightly bound system and neglect the velocity distribution of the nucleons in the deuteron. The assumption that the operators K and L are independent of the nucleon momenta is usually essential for the calculations.

We will describe the initial state of the two-nucleon system and the final state for elastic photoproduction (deuteron) outside the region of action of nuclear forces by a function of zero width, neglecting the contribution of the D state:

$$\varphi_d (r) = \sqrt{\frac{\alpha}{2\pi (1 - \alpha r_t)}} \frac{e^{-\alpha r}}{r}, \quad \alpha = \sqrt{M\varepsilon}, \quad (2.4)$$

where $\varepsilon = 0.0165$ is the deuteron binding energy. An estimate shows that the contribution of D waves to the cross sections of the processes under discussion in the near-threshold energy region is negligibly small [28].

In the inelastic production of mesons the final state of the two nucleons will be characterized by a plane wave distorted by the interaction of nucleons only in the S state:

$$\varphi_{pn \, conv}^{t, s}(r, p) = \left(\frac{1}{2\pi}\right)^{3/2} \left[e^{i(pr)} + \frac{1 - e^{-2i\delta^{t, s}}}{2i} \frac{e^{-ipr}}{pr} \right]. \quad (2.5)$$

We will describe the singlet and triplet nucleon−nucleon scattering phases δ^s and δ^t in an approximation which is independent of the form of the potential:

$$p \, ctg \, \delta^{t, s} = -\frac{1}{a_{t, s}} + \frac{1}{2} p^2 r_{t, s}, \quad (2.6)$$

where the scattering lengths $a_{t,s}$ and the effective ranges $r_{t,s}$ have the values [29]:

$$a_t = 3.687, \quad r_t = 1.160, \quad a_s = -16.27, \quad r_s = 1.826. \quad (2.7)$$

In the range of γ-ray energies considered, the relative momenta of the nucleons are such that a consideration of the interaction of nucleons only in the S state and the use of formulas (2.6) for the description of the N−N scattering phases are a satisfactory approximation.

We express the integrals of the wave functions of the two-nucleon system over the region of action of nuclear forces, which are encountered in the calculations of the matrix elements, in terms of the effective ranges of N−N interaction, as was done for the reaction $\gamma + d \rightarrow p + p + \pi^-$ in [4, 5].

2. Elastic Photoproduction of π^0 Mesons ($\gamma + d \rightarrow d + \pi^0$)

The kinematics of the process $\gamma + d \rightarrow d + \pi^0$ is characterized by two independent variables. For a fixed γ-ray energy there is a single-valued relationship, determined by the laws of conservation of energy and momentum, between the emission angle of the meson in the laboratory system and its energy:

$$\varkappa = q + P_d, \quad \varkappa = \sqrt{1 + q^2} + \frac{P_d^2}{4M}.$$

Here $\mathbf{p}_d$ is the deuteron recoil momentum. In the laboratory system there is a maximum angle of emission of the deuteron nucleus.

We consider the γ-ray—deuteron center-of-mass system. In this system we write the differential cross section of the process $\gamma + d \to d + \pi^0$ in a slightly different form from (1.18):

$$d\sigma_d = \frac{1}{1 + \dfrac{k}{2M}} \left\{ \frac{8}{3} |K^+|^2 + 4 |L^+|^2 \right\} I^2(\mathbf{\Delta}) \delta(\mathbf{q} + \mathbf{Q}) \delta\left(k + \frac{k^2}{4M} - \right.$$

$$\left. - \sqrt{1 + q^2} - \frac{Q^2}{4M} \right) d\mathbf{q} \, d\mathbf{Q}, \tag{2.8}$$

where $\mathbf{Q}$ is the deuteron momentum in the final state, the operators K^+ and L^+ are given by relationships (2.1), and the deuteron form factor $I(\mathbf{\Delta})$ can be put in the form

$$I(\mathbf{\Delta}) = \int dr e^{i(\mathbf{\Delta r})} \varphi_d^2(r) - \int dr e^{i(\mathbf{\Delta r})} [\varphi_d^2(r) - \varphi_{d\,\mathrm{tr}}^2(r)]. \tag{2.9}$$

Here $\varphi_{d\,\mathrm{tr}}(\mathbf{r})$ is the true wave function of the deuteron, which is the same as the model $\varphi_d(\mathbf{r})$ (2.4) outside the region of action of nuclear forces when $r \geq r_0$. Thus, the second integral in the expression (2.9) is actually taken with respect to r in the limits from zero to r_0. In the near-threshold energy region (Δr_0) $\lesssim 1$, and in the expansion of $e^{i(\mathbf{\Delta r})}$ in a series of powers of $(\mathbf{\Delta r})$ we can confine ourselves to the first two terms. Since $\varphi_{d\,\mathrm{tr}}(\mathbf{r}) = \varphi_d(\mathbf{r})$ for $r \geq r_0$, the parameter r_0 can tend to infinity. Then, proceeding from the definition of the effective triplet range of N—N interaction and performing the integration in the first term of formula (2.9), we obtain:

$$I(\mathbf{\Delta}) = \frac{1}{\Delta} \operatorname{arctg} \frac{\Delta}{2\alpha} - \frac{1}{2} r_t. \tag{2.10}$$

We note also that for the description of the deuteron form factor $I(\mathbf{\Delta})$ it would be possible in principle to use the results of experiments on electron scattering by deuterons.

Substituting in formula (2.8) the explicit form of the operators K^+ and L, averaging over the polarizations of the photons, and integrating with respect to $d\mathbf{Q}$ and $d\mathbf{q}$, we obtain the following expression for the differential cross section of the process $\gamma + d \to d + \pi^0$:

$$\frac{d\sigma_d}{d\Omega} = \frac{\left(1 + \dfrac{\sqrt{1 + q_p^2}}{M}\right) q_0 \sqrt{1 + q_0^2}}{\left(1 + \dfrac{k}{2M}\right)\left(1 + \dfrac{\sqrt{1 + q_0^2}}{2M}\right)} \frac{4}{3} \left[2|E_{0+}^+|^2 + 4\operatorname{Re}(M_{1+}^{+\,*} E_{0+}^+) \cos\theta + \right.$$

$$\left. + (7 - 5\cos^2\theta)|M_{1+}^+|^2 \right] \left[\frac{2\alpha}{1 - \alpha r_t} \left(\frac{1}{\Delta_0} \operatorname{arctg} \frac{\Delta_0}{2\alpha} - \frac{1}{2} r_t \right) \right]^2, \tag{2.11}$$

where $\mathbf{\Delta}_0 = (\mathbf{k} - \mathbf{q}_0)/2$, and q_0 is the root of the equation $k + (k^2/4M) - \sqrt{1 + q_0^2} - (q_0^2/4M) = 0$

The analysis of the experimental data on the process can conveniently be carried out by first dividing the differential cross section by the deuteron form factor. Then

$$\frac{d\sigma_d}{d\Omega} \Big/ I^2(\mathbf{\Delta}) = A' + B' \cos\theta + C' \cos^2\theta, \tag{2.12}$$

where the coefficients A', B', and C', together with the coefficients of the corresponding powers of $\cos\theta$ in the expansions of the differential cross section of the process (2) [see formula (3.8) below] give additional information (which can be obtained in polarization experiments) about the multipole and isotopic amplitudes of meson production on nucleons. In particular, coefficients A' and B' can provide information about the amplitude E_{0+}^+ and thus significantly increase the accuracy of the amplitudes $E_{0+}^{\pi^0}$, E_{0+}^0, and $E_{0+}^{n\pi^0}$.

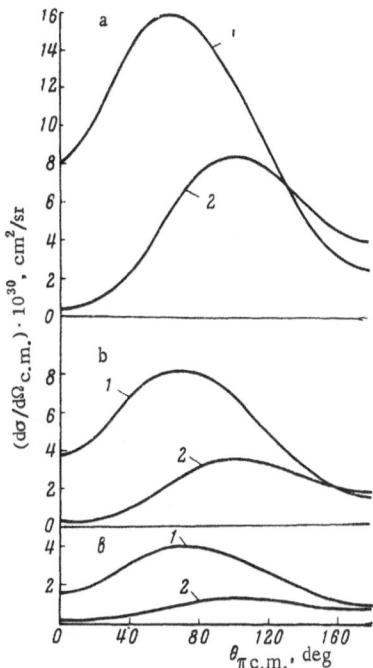

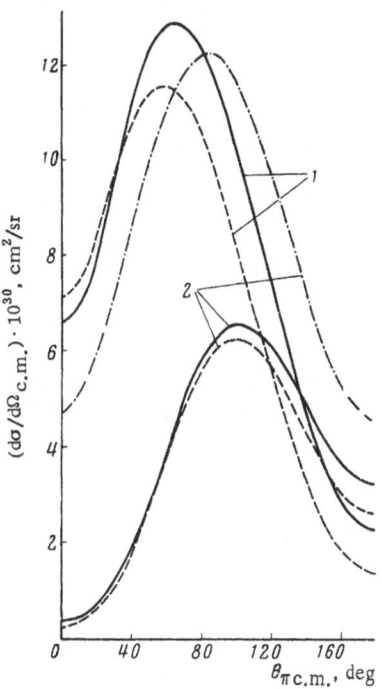

Fig. 3. Calculated differential cross sections of the processes $\gamma + d \rightarrow d + \pi^0$ and $\gamma + d \rightarrow p + n + \pi^0$ in photon–deuteron center-of-mass system. (a) $\varkappa_{lab} = 200$ MeV; (b) $\varkappa = 210$ MeV; (c) $\varkappa = 190$ MeV; (1) $\gamma(d, \pi^0)d$; (2) $\gamma(d, \pi^0)p, n$.

Fig. 4. Effect of amplitudes for meson production in S state and nucleon–nucleon interaction on the differential cross sections of reactions $\gamma + d \rightarrow d(pn) + \pi^0$ for $\varkappa_{lab} = 224$ MeV. Solid curves are calculated on the assumption that $E_{0+} \neq 0$ and broken lines on the assumption that $E_{0+} = 0$. The dot-dash curve is the result of calculation of the cross section of process $\gamma + d \rightarrow p + n + \pi^0$ without allowance for the interaction of nucleons in the final state. (1) $\gamma(d, \pi^0)d$; (2) $\gamma(d, \pi^0)p, n$.

In the approximation (2.1) the explicit form of coefficients A', B', and C' is given by formula (2.11). The results of calculations of the differential cross sections of the elastic process for three γ-ray energies in this approximation are given in Fig. 3. The reduction of the cross section for large meson emission angles is due to the deuteron form factor.

It follows from formula (2.11) that the cross section for the production of π^0-mesons at forward angles is sensitive to the amplitude E_{0+}^+. Figure 4 shows the results of calculation of the differential cross section of the process $\gamma + d \rightarrow d + \pi^0$ with and without consideration of the amplitude E_{0+}^+. The figure shows that consideration of the amplitude E_{0+}^+ leads to an appreciable reduction in the cross section for forward meson emission angles. Here we may have a situation similar to that for the process $\gamma + p \rightarrow p + \pi^0$, where the differential cross section $d\sigma/d\Omega$ of this process with $q^2 \simeq 0.18$ for $\theta = 0°$ passes through a maximum and a minimum (see Fig. 1). If the value of (1.9) is used for E_{0+}^+, the differential cross section of the process $\gamma + d \rightarrow d + \pi^0$ for $\theta = 0°$ will behave in this way when $q^2 \lesssim 0.1$. Hence, it follows that the ratio of the cross sections $d\sigma(\gamma + d \rightarrow d + \pi^0)/d\sigma(\gamma + p \rightarrow p + \pi^0)|_{\theta = 0°}$ at energies close to the threshold may change by a factor of tens (from 0.5 to 20), depending on the S-wave amplitudes. This circumstance can be used to determine these amplitudes. The ratio σ_d/σ_p, like the ratio σ^-/σ^+, gives information about the isoscalar part of the meson photoproduction amplitude and, thus, about the role of ρ-meson exchange in photoproduction processes. For angle $\theta = 0°$, where $I^2(\Delta) \simeq 1$ and $L^2 \simeq 0$, this ratio is

$$\left.\frac{d\sigma_d}{d\sigma_p}\right|_{\theta=0°} \simeq \frac{8}{3}\left|\frac{K^+}{K^+ + K^0}\right|^2. \qquad (2.13)$$

If mesons are produced equally on neutrons and protons (i.e., $K^0 \simeq 0$), then $d\sigma_d/d\sigma_p|_{\theta=0°} \simeq 8/3$.

3. Inelastic Photoproduction of π^0-Mesons ($\gamma + d \rightarrow p + n + \pi^0$)

The kinematics of the process $\gamma + d \rightarrow p + n + \pi^0$ for a given γ-ray energy is characterized by nine variables, in the capacity of which can be taken the nucleon momenta $\mathbf{p}_p$ and $\mathbf{p}_n$, or combinations of them

$$\mathbf{p} = \frac{1}{2}(\mathbf{p}_p - \mathbf{p}_n), \quad \mathbf{l} = \frac{1}{2}\mathbf{P} = \frac{1}{2}(\mathbf{p}_p + \mathbf{p}_n)$$

and the meson momentum $\mathbf{q}$. On the basis of the laws of conservation of energy and momentum we can distinguish five independent variables and an investigation of the differential distributions for these gives a complete picture of the investigated process. However, in view of the limited experimental possibilities the distributions are measured for a smaller number of variables. The conservation laws in this case define some region within the region of variation of the variables for which the distributions are investigated.

In the variables $\mathbf{p}$, $\mathbf{P}$, and $\mathbf{k}$ the conservation laws in the laboratory system have the form

$$\mathbf{k} = \mathbf{P} + \mathbf{q}, \quad k = \frac{p^2}{M} + \frac{P^2}{4M} + \sqrt{1 + q^2} + \varepsilon.$$

If the relationship between $\mathbf{p}$ and $\mathbf{P}$ is not fixed, then mesons with a particular momentum distribution will be emitted at a given angle θ_π between $\mathbf{k}$ and $\mathbf{q}$. This distribution will be determined by the phase factor, by the binding of the nucleons in the deuteron, and by the interaction of the nucleons in the final state. The variables $\mathbf{p}$ and $\mathbf{P}$ are suitable for writing the cross sections of the process $\gamma + d \rightarrow p + n + \pi^0$, since the transition matrix element depends only on the absolute values of p and P and the angle between them, whereas the number of independent variables is five. Using the laws of conservation of energy and momentum we find the allowed region of variation of p and P:

$$k = \frac{p^2}{M} + \frac{P^2}{4M} + \sqrt{1 + k^2 + P^2 - 2kP\cos\chi} + \varepsilon,$$

where χ is the angle between the vectors $\mathbf{k}$ and $\mathbf{P}$. From the condition $-1 \leq \cos\chi \leq 1$, we obtain

$$\frac{p^2}{M} \leqslant k - \varepsilon - \frac{P^2}{4M} - \sqrt{1 + (k - P)^2},$$

$$\frac{p^2}{M} \geqslant k - \varepsilon - \frac{P^2}{4M} - \sqrt{1 + (k + P)^2}.$$

Since the right side of the second inequality is always negative, the second condition simply means that $p \geq 0$. Along with the first condition this inequality gives

$$0 \leqslant p \leqslant \sqrt{M\left[k - \varepsilon - \frac{P^2}{4M} - \sqrt{1 + (k - P)^2}\right]}. \qquad (2.14)$$

Figure 5 shows the region of permissible values of p and q' $= \frac{1}{2}P$, determined by this relationship, for γ-ray energies up to 220 MeV. Diagrams of this type can be used to find the regions of integration when particular conditions are imposed on the variables. For example, if we detect protons with very low energy $p_p \approx 0$, and neutrons with energy which do not exceed a particular value $p_n \leq a$, then, because of the condition $p_p^2 + p_n^2 = 2p^2 + \frac{1}{2}P^2 \leq a^2$, the region of variation of p and q' will be given by the intersection of the regions (2.14) with a circle of radius $R = a/\sqrt{2}$.

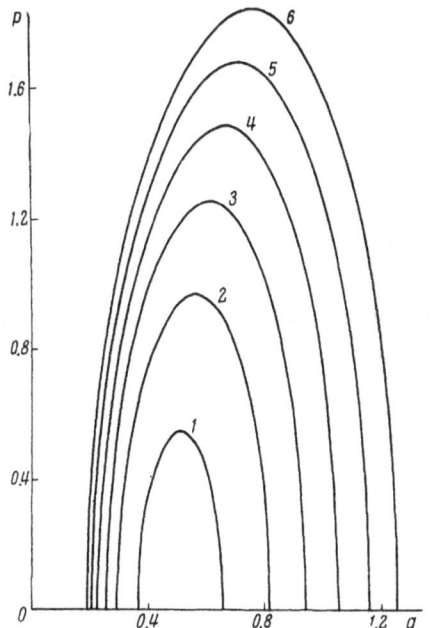

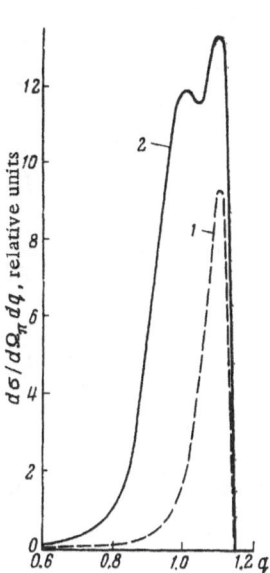

Fig. 5. Region of kinematically allowed values of p and q' for the process $\gamma + d \rightarrow p + n + \pi^0$. (1) $\varkappa = 1.1$; (2) $\varkappa = 1.2$; (3) $\varkappa = 1.3$; (4) $\varkappa = 1.4$; (5) $\varkappa = 1.5$; (6) $\varkappa = 1.6$.

Fig. 6. Spectra of π^0-mesons from the reaction $\gamma + d \rightarrow p + n + \pi^0$ for γ-ray energy $\varkappa = 200$ MeV, calculated from formula (2.28). (1) $\theta_\pi = 40°$; (2) $\theta_\pi = 100°$.

We will write the differential cross section of the process $\gamma + d \rightarrow p + n + \pi^0$ in the $\gamma - d$ center-of-mass system in a more suitable form than (1.19):

$$d\sigma_{pn} = \frac{1}{1 + \frac{k}{2M}} \left\{ \frac{2}{3} \,| \mathbf{K}^+ [I_+^t\,(\mathbf{p}, \boldsymbol{\Delta}) + I_-^t\,(\mathbf{p}, \boldsymbol{\Delta})] + \mathbf{K}^0\,[I_+^t\,(\mathbf{p}, \boldsymbol{\Delta}) -- I_-^t\,(\mathbf{p}, \boldsymbol{\Delta})]|^2 + \right.$$

$$+ \frac{1}{3}\,|\,\mathbf{K}^+\,[I_+^s\,(\mathbf{p}, \boldsymbol{\Delta}) - I_-^s\,(\mathbf{p}, \boldsymbol{\Delta})] + \mathbf{K}^0\,[I_+^s\,(\mathbf{p}, \boldsymbol{\Delta}) + I_-^s\,(\mathbf{p}, \boldsymbol{\Delta})]|^2 +$$

$$+ |\,\mathbf{L}^+\,[I_+^t\,(\mathbf{p}, \boldsymbol{\Delta}) + I_-^t\,(\mathbf{p}, \boldsymbol{\Delta})] + \mathbf{L}^0\,[I_+^t\,(\mathbf{p}, \boldsymbol{\Delta}) - I_-^t\,(\mathbf{p}, \boldsymbol{\Delta})]|^2 \left. \right\} \times$$

$$\times \delta\,(\mathbf{q} + \mathbf{P})\,\delta\,\left(k + \frac{k^2}{4M} - \varepsilon - \sqrt{1 + q^2} - \frac{P^2}{4M} - \frac{p^2}{M} \right) d\mathbf{q}\, d\mathbf{p}\, d\mathbf{P}. \qquad (2.15)$$

As in the case of the elastic process, we express the functions $I_\pm^{t, s}(\mathbf{p}, \boldsymbol{\Delta})$, which contain integration over the region of action of nuclear forces, in terms of the effective ranges of N−N interaction. We put the functions $I_\pm^{t, s}(\mathbf{p}, \boldsymbol{\Delta})$ in the form

$$I_\pm^{t, s}\,(\mathbf{p}, \boldsymbol{\Delta}) = \int d\mathbf{r} \varphi_{pn\,\text{conv}}^{t, s*}\,(\mathbf{r}, \,\mathbf{p})\,e^{\pm\,i(\Delta \mathbf{r})} \varphi_d\,(\mathbf{r}) -$$

$$- \int d\mathbf{r} e^{\pm\,i(\Delta \mathbf{r})}\,[\varphi_{pn\,\text{conv}}^{t, s*}\,(\mathbf{r}, \,\mathbf{p})\,\varphi_d\,(\mathbf{r}) - \varphi_{pn\,\text{conv, \,tr}}^{t, s*}\,(\mathbf{r}, \,\mathbf{p})\,\varphi_{d\,\text{tr}}\,(\mathbf{r})], \qquad (2.16)$$

where $\varphi_{pn}(\mathbf{r}, \mathbf{p})$ is the wave function of the unbound state of the two nucleons, which is the same as the model function (2.5) when $r \geq r_0$. Confining ourselves in the second term, as in the case of (2.9), to the lowest powers in the expansion of $e^{i(\Delta \mathbf{r})}$ in a series of powers of $(\Delta \mathbf{r})$ and using the explicit form of the wave functions $\varphi_{pn}^{t,s}(r,p)$ (2.5) and $\varphi_d(\mathbf{r})$ (2.4) we obtain, after an elementary but tedious integration,

$$I_{\pm}^{t,\,s}(\mathbf{p},\,\Delta) = \sqrt{\frac{\alpha}{1 - \alpha r_t}} \, \frac{1}{2\pi} \, [f_{1\pm}(p,\,\Delta,\,\widehat{\mathbf{p}\Delta}) + f_2^{t,\,s}(p,\,\Delta) + i f_3^{t,\,s}(p,\,\Delta)], \qquad (2.17)$$

$$f_{1\pm}(p,\,\Delta,\,\widehat{\mathbf{p}\Delta}) = \frac{2}{\alpha^2 + (\mathbf{p} \pm \Delta)^2}, \qquad (2.18)$$

$$f_2^{t,\,s}(p,\,\Delta) = \frac{1}{p\Delta} \, [\cos \delta^{t,\,s} \sin \delta^{t,\,s}(B - \Delta \rho_{t,\,s}) - \sin^2 \delta^{t,\,s} \, A], \qquad (2.19)$$

$$f_3^{t,\,s}(p,\,\Delta) = \frac{1}{p\Delta} \, [\sin^2 \delta^{t,\,s}(B - \Delta \rho_{t,\,s}) + \cos \delta^{t,\,s} \sin \delta^{t,\,s} \, A], \qquad (2.20)$$

$$B(p,\,\Delta) = \operatorname{arctg} \frac{2\alpha\Delta}{(\alpha^2 + p^2 - \Delta^2)}, \quad A(p,\,\Delta) = \frac{1}{2} \ln \frac{\alpha^2 + (p + \Delta)^2}{\alpha^2 + (p - \Delta)^2}. \qquad (2.21)$$

Here B varies from zero to π, ρ_s is the mixed effective range of N−N interaction [30], which is defined in terms of r_s and r_t:

$$\rho_t = r_t; \quad \rho_s \cong \frac{1}{2}(r_s + r_t). \qquad (2.22)$$

After squaring the individual terms of expression (2.15) and averaging the square of the matrix element over the polarization of the γ-rays, we can obtain an expression for the cross section, which in the general case (1.2) is rather unwieldy. Hence, here we give the cross section obtained on the assumption of (2.1): *

$$d\sigma_{pn} = \frac{1}{1 + \frac{k}{2M}} \left\{ \frac{2}{3} \, [\, |E_{0+}^+|^2 + \frac{1}{2}(1 + \cos^2\theta)\,|M_{1+}^+|^2 + 2\operatorname{Re}(E_{0+}^{+*}M_{1+}^+)\cos\theta] \times \right.$$

$$\times [\, |I_+^t|^2 + |I_-^t|^2 + 2\operatorname{Re}(I_+^t I_-^{t*})] + 2(1 - \cos^2\theta)\,|M_{1+}^+|^2 \, [\,|I_+^t|^2 + |I_-^t|^2 +$$

$$+ 2\operatorname{Re}(I_+^t I_-^{t*})] + \frac{1}{3}\,[\,|E_{0+}^+|^2 + \frac{1}{2}(1 + \cos^2\theta)\,|M_{1+}^+|^2 + 2\cos\theta\operatorname{Re}(E_{0+}^+ M_{1+}^{+*})] \times$$

$$\times [\,|I_+^s|^2 + |I_-^s|^2 - 2\operatorname{Re}(I_+^s I_-^{s*})] \Big\} \delta(\mathbf{q} + \mathbf{P}) \, \delta\left(k + \frac{k^2}{4M} - \varepsilon - \right.$$

$$\left. - \sqrt{1 + q^2} - \frac{p^2}{M} - \frac{P^2}{4M}\right) d\mathbf{q}\,d\mathbf{p}\,d\mathbf{P}. \qquad (2.23)$$

We write the expressions $|I_{\pm}^{t,s}|^2$ and $(I_+^{t,s}I_-^{t,s*})$ occurring in this formula in terms of the functions $F_2(p, \Delta)$ and $F_3(p, \Delta)$, which are determined from the relationships

$$F_2(p,\,\Delta) = 2f_2^{t,s}(p,\,\Delta), \qquad (2.24)$$

$$F_3(p,\,\Delta) = |f_2^{t,s}(p,\,\Delta)|^2 + |f_3^{t,s}(p,\,\Delta)|^2. \qquad (2.25)$$

Then

$$|I_{\pm}^{t,s}|^2 = \frac{\alpha}{1 - \alpha r_t}\left(\frac{1}{2\pi}\right)^2 [f_{1\pm}^2 + f_{1\pm} F_2^{t,s} + F_3^{t,s}], \qquad (2.26)$$

---

* Formula (2.23) drops the terms which give zero values after integration with respect to $d\theta_p$.

$$I_+^{t,s}I_-^{t,s*} = \frac{\alpha}{1-\alpha r_t}\left(\frac{1}{2\pi}\right)^2\left[f_{1+}f_{1-} + \frac{1}{2}\left(f_{1+}+f_{1-}\right)F_2^{t,s} + F_3^{t,s} + if_3^{t,s}\left(f_{1-}-f_{1+}\right)\right]. \qquad (2.27)$$

Functions $f_2^{t,s}$ and $f_3^{t,s}$ (or $F_2^{t,s}$ and $F_3^{t,s}$) allow for interaction of the nucleons in the final state. If this interaction is ignored (i.e., $\delta^{t,s}$ is put equal to zero), then it is clear from expression (2.18) for $f_{1\pm}$ that the regions $p \simeq \pm\Delta$ make the main contribution to the cross section.

Using formulas (2.23)-(2.27), we can find the required differential cross sections. Integrating expression (2.23) with respect to all the variables, except dq, we find the spectrum of π^0-mesons emitted at a particular angle θ_π (in the integration with respect to $d\Omega_p$ the angle between vectors Δ and $\mathbf{p}$ was chosen as θ_p):

$$\frac{d\sigma}{dq\,d\Omega_\pi} = \frac{1}{1+\frac{k}{2M}}\frac{\alpha Mq^2p_0}{3\pi(1-\alpha r_t)}\left(1+\frac{\sqrt{1+q_p^2}}{M}\right)\{[2\,|E_{0+}^+|^2 + 4\mathrm{Re}\,(M_{1+}^+E_{0+}^{+*})\cos\theta +$$

$$+ (7-5\cos^2\theta)\,|M_{1+}^+|^2]\,[F^t(p_0,\ \Delta) + E^t(p_0,\ \Delta)] +$$

$$+ \frac{1}{2}(1+\cos^2\theta)\,|M_{1+}^+|^2\,[F^s(p_0,\ \Delta) - E^s(p_0,\ \Delta)] + 2\,|E_{0+}^0|^2\,[F^t(p_0,\ \Delta) -$$

$$- E^t(p_0,\ \Delta)] + |E_{0+}^0|^2\,[F^s(p_0,\Delta)+E^s(p_0,\Delta)]\}. \qquad (2.28)$$

Here $p_0 = \sqrt{M[k + (k^2/4M) - \varepsilon - \sqrt{1+q^2} - (q^2/4M)}$, and the functions $F^{t,s}$ and $E^{t,s}$ have the following form:

$$F^{t,s}(p,\ \Delta) = \frac{4}{(\alpha^2+p^2+\Delta^2)^2-4p^2\Delta^2} + \Phi^{t,s}(p,\ \Delta), \qquad (2.29)$$

$$E^{t,s}(p,\ \Delta) = \frac{1}{p\Delta\,(\alpha^2+p^2+\Delta^2)}\,A\,(p,\ \Delta) + \Phi^{t,s}(p,\Delta), \qquad (2.30)$$

$$\Phi^{t,s}(p,\ \Delta) = \frac{\sin\delta^{t,s}}{p^2\Delta^2}\Big\{\cos\delta^{t,s}\,A\,(p,\ \Delta)\,[B\,(p,\ \Delta)-\Delta\rho_{t,s}] +$$

$$+ \sin\delta^{t,s}\Big[(B\,(p,\ \Delta) - \Delta\rho_{t,s})^2 - \frac{1}{4}\,A^2\,(p,\ \Delta)\Big]\Big\}. \qquad (2.31)$$

The meson spectra calculated from formula (2.28) for γ-ray energy $\varkappa = 200$ MeV and for meson emission angles $\theta_\pi = 40°$ and $100°$ are shown in Fig. 6. It should be noted that the meson spectrum is determined by the phase space and the interaction of the nucleons in the final state, which are responsible for the characteristic momentum distribution with a maximum. The halfwidth of the spectrum depends on the meson emission angle; its mean value is about 10 MeV.

By integrating expression (2.28) with respect to the meson momenta, we obtain the angular distribution of π^0-mesons in the reaction $\gamma+d \to p+n+\pi^0$. The results of integration for three γ-ray energies are shown in Fig. 3. A comparison with the corresponding distributions of π^0-mesons from the reaction $\gamma +d \to d +\pi^0$ shows that in the low-angle region the production of π^0-mesons on deuterons is due mainly to the elastic process. In the large-angle region the differential cross section of the process $\gamma +d \to p+n +\pi^0$ is comparable with the cross section of the elastic process. In view of this, the identification of inelastic photoproduction in experiments where only the angular distribution of the mesons is recorded is very problematic. It is obvious that one of the recoil nucleons (the neutron, for instance) will also have to be detected. The formulas required for the evaluation of this cross section can easily be obtained from the relationships (2.23)-(2.27).

It is of interest to consider the role of interaction of the nucleons in the final state of the considered reaction. Its contribution is significant only at small values of relative momentum of the nucleons. Figure 4 shows the results of calculation of the cross section of the process $\gamma +d \to p +n +\pi^0$ for γ-ray energy $\varkappa = 224$ MeV with and without allowance for nucleon interaction in the final state. Nucleon interaction reduces the differential cross section of the considered process. This is a result of the predominant role of the amplitude

M_{1+}^{+}, which is connected, as formula (2.28) clearly shows, with the triplet state of the nucleons in the final state. The overlap integrals in the functions $I_{\pm}^{t}(p, \Delta)$ which characterize the contribution of the triplet state to the cross section of the process $\gamma + d \rightarrow p + n + \pi^{0}$ are greatly reduced owing to the orthogonality of the wave functions $\varphi_{p,n}^{t}(\mathbf{r};\mathbf{p})$ and $\varphi_{d}(\mathbf{r})$. In the case of the reaction $\gamma + d \rightarrow p + p + \pi^{-}$, only the singlet state of nucleons is present in the final state and allowance for nucleon interaction in this reaction leads to an increase in the differential cross section of the process $\gamma + d \rightarrow p + p + \pi^{-}$ [5].

Figure 4 shows that nucleon interaction reduces the integral cross section of the process $\gamma + d \rightarrow p + n + \pi^{0}$ by a factor of several units and the differential cross section for angle $\theta_{\pi} = 0°$ by a factor of tens. This means that the data for inelastic π^{0}-meson photoproduction can be used for an investigation of nucleon–nucleon interaction. These data can also be used for an isotopic analysis of the photoproduction amplitude on the nucleon. In fact, expressions (2.15) and (2.28) show that the isoscalar K^{0}, L^{0} (isovector K^{+}, L^{+}) parts of the amplitudes appear as factors in terms which have, as regards relative momentum, poles corresponding to singlet virtual (triplet real) levels of the two-nucleon system. Thus, measurements of the differential relative momentum distributions of the nucleons and extrapolation of the results of measurements to the corresponding poles can give information about K^{0}, L^{0}, and K^{+}, L^{+} separately. This is particularly important for the separation of the contributions of resonance 2π and 3π states to meson photoproduction amplitudes. This method of measuring K^{0}, L^{0} involves less indeterminacy due to multiple scattering effects than the method of determining K^{0}, L^{0} from the data for the processes $\gamma + d \rightarrow d + \pi^{0}$ and $\gamma + d \rightarrow p + p + \pi^{-}$, the amplitudes of which may contain corrections to the impulse approximation equal to the amplitudes K^{0} and L^{0}.

A comparison of the results of calculation of the cross sections of the elastic and inelastic processes (see Fig. 2) shows that the elastic process predominates in the region of small π^{0}-meson emission angles. This can be used to distinguish the elastic process by means of the angular distribution of the mesons. The differential cross section of the elastic process for meson emission angles $\theta \approx 0°$ can give, as already mentioned, valuable information about the amplitude E_{0+}^{0} of meson production in the S state. The S-wave amplitudes E_{0+}^{+} and E_{0+}^{0} also affect the inelastic process and slightly increase its cross section for backward angles of meson emission (see Fig. 4).

CHAPTER III: EXPERIMENTAL INVESTIGATIONS OF π^{0}-MESON PHOTOPRODUCTION

This chapter deals with the setup and results of experimental investigations of the differential cross sections of the process $\gamma + d \rightarrow d(pn) + \pi^{0}$ and $\gamma + p \rightarrow p + \pi^{0}$ by the detection of one (Sec. A) or two (Sec. B) of the decay γ-quanta of π^{0}-mesons. The γ-quanta were detected by telescopes of scintillation counters connected up in coincidence and anticoincidence circuits. The investigation was conducted in the near-threshold energy region and at angles including $\theta \approx 0°$, where these processes have hardly been studied at all so far owing to the great experimental difficulties.

A. Measurement of Differential Cross Section by the Direction
of One Decay Photon of the π^{0}-Meson

1. Some Characteristics of π^{0}-Meson γ-Quanta

We will discuss the formulas which characterize the properties of π^{0}-meson decay γ-quanta. Many of the formulas given below are well known. A more detailed derivation of some of them can be found in [31].

In the laboratory system a π^{0}-meson with mass μ and momentum q decays into two γ-quanta with momenta $\bar{p}_{\gamma_{1}}$ and $\bar{p}_{\gamma_{2}}$. These γ-quanta are emitted at angles $\theta_{\gamma_{1}}$ and $\theta_{\gamma_{2}}$ to the direction of motion of the π^{0}-meson (Fig. 7). It can be shown by using the laws of conservation of energy and momentum that the energy of a decay γ-quantum and its angle of emission θ_{γ} in the laboratory system are connected by the expression

$$\varepsilon_{\gamma} = \frac{\mu}{2\gamma(1 - \beta_{\pi}\cos\theta_{\gamma})}, \tag{3.1}$$

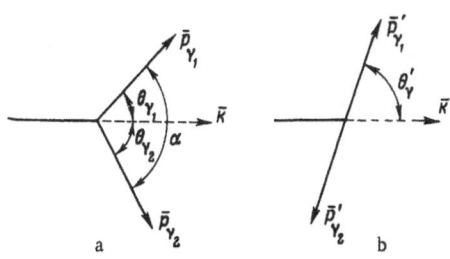

Fig. 7. Kinematics of π^0-meson decay to two γ-rays.
(a) Laboratory system; (b) π^0-meson rest system.

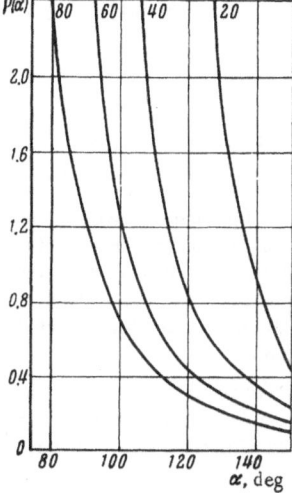

Fig. 8. Probability of separation of two π^0-meson decay γ-quanta at a given angle α for a given π^0-meson kinetic energy E_π. The figures beside the curves give the energy in MeV.

where $\gamma = (1 - \beta_\pi^2)^{-\frac{1}{2}}$, and β_π is the velocity of the π^0-meson. In the π^0-meson rest system all the decay γ-quanta have equal energies $\frac{1}{2}\mu$ and are emitted in opposite directions. The angular distribution of the decay γ-quanta in this system is isotropic. Hence, it follows that on conversion to the laboratory system we obtain a uniform momentum (p_γ) distribution for the decay γ-quanta in the range from $p_{\gamma max}$ to $p_{\gamma min}$. The limiting values of the momenta are given by formula (3.1) with θ_γ equal to 0 or 180°, respectively.

In the π^0-meson rest system the probability of a decay γ-quantum being emitted into a unit solid angle $d\Omega'_{\theta_\gamma}$ at angle θ'_γ to the direction of motion of the coordinate system can be written in the form

$$P(\theta'_\gamma)\, d\theta'_\gamma = \frac{1}{2\pi}\, d\Omega'_{\theta'_\gamma}. \qquad (3.2)$$

Using the formulas for relativistic conversion of solid angles, we obtain for the laboratory system

$$P(\theta_\gamma)\, d\Omega_\gamma = \frac{1 - \beta_\pi^2}{2\pi(1 - \beta_\pi \cos\theta_\gamma)^2} \sin\theta_\gamma\, d\theta_\gamma\, d\varphi. \quad (3.3)$$

This expression gives the probability of a decay γ-quantum being emitted into a unit solid angle $d\Omega_\gamma$ at an angle θ_γ to the direction of motion of a π^0-meson with velocity β_π.

By definition (see Fig. 7) the correlation angle between the π^0-meson decay photons in the laboratory system is $\alpha = \theta_{\gamma_1} + \theta_{\gamma_2}$ and

$$\cos\alpha = \cos\theta_{\gamma_1} \cos\theta_{\gamma_2} - \sin\theta_{\gamma_1} \sin\theta_{\gamma_2}. \quad (3.4)$$

Using the formulas for the relativistic conversion of angles and substituting them in (3.4), we find an expression connecting the angle α with the angle of emission θ'_γ of one of the γ-quanta in the π^0-meson rest system:

$$\sin\frac{\alpha}{2} = [\cos^2\theta'_\gamma + (1 - \beta_\pi^2)^{-1} \sin^2\theta'_\gamma]^{-\frac{1}{2}}. \qquad (3.5)$$

Hence, it is evident that when β_π is fixed, the minimum angle α_{min} between the decay γ-quanta corresponds to $\theta'_\gamma = \pi/2$. With this condition, symmetric decay occurs in the laboratory system when $\theta_{\gamma_1} = \theta_{\gamma_2}$, and it follows from (3.5) that

$$\sin\frac{\alpha_{min}}{2} = \sqrt{1 - \beta_\pi^2}, \quad \cos\frac{\alpha_{min}}{2} = \beta_\pi. \qquad (3.6)$$

We now determine the probability of a π^0-meson decay with a given correlation angle α. We write formula (3.2) in the form

$$P(\theta'_\gamma)\, d\theta'_\gamma = \frac{d\Omega_{\theta_\gamma'}}{2\pi} = \sin\theta'_\gamma d\theta'_\gamma = -d\cos\theta'_\gamma$$

and convert from the variable θ'_γ to the variable α. Then the transition Jacobian is $J = d\theta'_\gamma / d\alpha$ and

$$P(\alpha) = P(\theta'_\gamma)\frac{d\theta'_\gamma}{d\alpha} = -\frac{d\cos\theta'_\gamma}{d\alpha}.$$

Using formula (3.5) we find an expression for $\cos\theta'_\gamma$ and, on differentiating it, we obtain

$$P(\alpha) = \frac{\sin\alpha\, d\alpha}{\beta_\pi\gamma\,[\gamma^2(1-\cos\alpha)-2]^{1/2}(1-\cos\alpha)^{3/2}}, \tag{3.7}$$

The function $P(\alpha)$, illustrated in Fig. 8, gives the required distribution of the pairs of decay γ-quanta with respect to angle α for monoenergetic π^0-mesons. As the figure indicates, $P(\alpha)$ increases sharply when $\alpha \to \alpha_{min}$.

The maximum probability of a decay in which the γ-quanta separate at a given angle α is obtained when the π^0-mesons have the energy given by (3.6). Mesons of lower energy do not produce any γ-quanta with correlation angle α.

2. Method of Detecting π^0-Mesons from One of the Decay γ-Quanta

The mean lifetime of neutral π^0-mesons does not exceed $2 \cdot 10^{-16}$ sec. In other words, they decay practically instantaneously into two photons* and can be detected by detection of the decay products.

By the detection of one of the decay γ-quanta the cross sections of processes involving π^0-mesons can be rapidly measured with good statistical accuracy. This method is widely used for measurement of the total cross sections for π^0-meson production in photon–nucleon and nucleon–nucleon interactions. It was first used in [19]. When one γ-quantum is detected it is impossible to obtain the meson energy distribution directly, but the angular distributions can be accurately measured for mesons with energy greater than 50-100 MeV, when the decay photons are emitted at small angles to their direction of motion.

In the case of π^0-meson photoproduction on nucleons, there is a single-valued relationship between the angular distribution of the decay γ-quanta in the laboratory system and the angular distribution of the mesons in the center-of-mass system [19]. For the process $\gamma + p \to p + \pi^0$, the differential cross section for which in the center-of-mass system can be put in the form

$$d\sigma/d\Omega = A + B\cos\theta + C\cos^2\theta, \tag{3.8}$$

the distribution of the decay γ-quanta $f(\varepsilon_\gamma, \theta_\gamma)$ in the laboratory system for a given energy $\varkappa$ of incident γ-quantum can be written in the form [33]:

$$f(\varepsilon_\gamma, \theta_\gamma) = \frac{2}{\gamma_c q'(1-\beta_c\cos\theta_\gamma)}\left\{A + B\frac{\cos\theta_\gamma - \beta_c}{1-\beta_c\cos\theta_\gamma}\left(\frac{\sqrt{1+q^2}}{q'} - \right.\right.$$

$$\left.- \frac{1}{2q'\gamma_c\varepsilon_\gamma(1-\beta_c\cos\theta_\gamma)}\right) + C\left[\frac{\sin^2\theta_\gamma}{2\gamma_c^2(1-\beta_c\cos\theta_\gamma)^2} + \right.$$

$$\left.\left. + \frac{3}{2}\left(\frac{\sqrt{1+q'^2}}{q'} - \frac{1}{2q'\gamma_c\varepsilon_\gamma(1-\beta_c\cos\theta_\gamma)}\right)^2\left(\left(\frac{\cos\theta_\gamma - \beta_c}{1-\beta_c\cos\theta_\gamma}\right)^2 - \frac{1}{3}\right)\right]\right\}, \tag{3.9}$$

---

*The most accurate results of measurements of the mean lifetime of π^0-mesons give $\tau_{\pi^0} = (1.05 \pm 0.18)\cdot 10^{-16}$ sec [32]. Besides decay into two photons, π^0-mesons can decay according to the schemes $\pi^0 \to \gamma + e^+ + e^-$ and $\pi^0 \to 2e^+ + 2e^-$, which occur with a relative frequency of $\sim 1.2\cdot 10^{-2}$ and $\sim 3.5\cdot 10^{-5}$, respectively.

Here θ_γ is the angle between $\varkappa$ and the direction of the counter for the detection of the decay γ-quantum; q' is the meson momentum in the center-of-mass system; $\beta_c = \varkappa/(\varkappa + M)$ is the velocity of the center of mass; $\gamma_c = (1 - \beta_c^2)^{-\frac{1}{2}}$.

The number of detected γ-quanta per nucleus per unit primary beam intensity and per unit solid angle is expressed in terms of $f(\varepsilon_\gamma, \theta_\gamma)$ and the detection efficiency $\eta(\varepsilon_\gamma)$ of the scintillation telescope for photons with energy ε_γ

$$N(\theta_\gamma) = \int_{\varepsilon_\gamma^{\min}}^{\varepsilon_\gamma^{\max}} f(\varepsilon_\gamma, \theta_\gamma)\,\eta(\varepsilon_\gamma)\,d\varepsilon_\gamma. \tag{3.10}$$

Here,

$$\varepsilon_\gamma^{\max} = \frac{\sqrt{1+q'^2}\left(1 + \frac{q'}{\sqrt{1+q'^2}}\right)}{2\gamma_c(1 - \beta_c\cos\theta_\gamma)}, \qquad \varepsilon_\gamma^{\min} = \frac{\sqrt{1+q'^2}\left(1 - \frac{q'}{\sqrt{1+q'^2}}\right)}{2\gamma_c(1 - \beta_c\cos\theta_\gamma)}$$

are the maximum and minimum energies, respectively, in the spectrum of the decay γ-quanta. We use the expression (3.9) and rewrite (3.10) in the form $N(\theta_\gamma) = aA + bB + cC$, where A, B, and C are the coefficients in formula (3.8) and a, b, and c are given by

$$a(\theta_\gamma) = \frac{2}{\gamma q'(1 - \beta_c\cos\theta_\gamma)} \int_{\varepsilon_\gamma^{\min}}^{\varepsilon_\gamma^{\max}} \eta(\varepsilon_\gamma)\,d\varepsilon_\gamma,$$

$$b(\theta_\gamma) = \frac{2(\cos\theta_\gamma - \beta_c)}{\gamma q'(1 - \beta_c\cos\theta_\gamma)^2} \int_{\varepsilon_\gamma^{\min}}^{\varepsilon_\gamma^{\max}} \left[\frac{\sqrt{1+q'^2}}{q'} - \frac{1}{2q'\gamma\varepsilon_\gamma(1 - \beta_c\cos\theta_\gamma)}\right]\eta(\varepsilon_\gamma)\,d\varepsilon_\gamma, \tag{3.11}$$

$$c(\theta_\gamma) = \frac{\sin^2\theta_\gamma}{\gamma^3 q'(1 - \beta_c\cos\theta_\gamma)^3} \int_{\varepsilon_\gamma^{\min}}^{\varepsilon_\gamma^{\max}} \eta(\varepsilon_\gamma)\,d\varepsilon_\gamma + \frac{3}{\gamma q'(1 - \beta_c\cos\theta_\gamma)} \times$$

$$\times \left[\left(\frac{\cos\theta_\gamma - \beta_c}{1 - \beta_c\cos\theta_\gamma}\right)^2 - \frac{1}{3}\right]\left[\int_{\varepsilon_\gamma^{\min}}^{\varepsilon_\gamma^{\max}} \frac{\sqrt{1+q'^2}}{q'} - \frac{1}{2q'\gamma\varepsilon_\gamma(1 - \beta_c\cos\theta_\gamma)}\right]^2 \eta(\varepsilon_\gamma)\,d\varepsilon_\gamma.$$

Thus, to determine the coefficients A, B, and C it is sufficient to measure the photon yield $N(\theta_\gamma)$ for three angles θ_γ. This method of analyzing experimental data is suitable for the investigation of processes in which the angular distribution of the mesons is in the form (3.8). Another essential condition is the single-valued relationship between the energies of the meson and primary photon in the center-of-mass system.

However, there are π^0-meson production processes which do not satisfy these conditions. In this case there is another way of comparing the experimental data with theoretical predictions. The direct results of the measurements can be compared with the angular distributions of the decay photons calculated for the expected differential cross section for π^0-meson production. Such calculations require a knowledge of the angular resolution functions of the experimental apparatus, i.e., the probability of detection of a π^0-meson emitted in the photoproduction process at an angle θ_π by a telescope lying at an angle θ_γ (the angles are measured from the $\varkappa$ direction).

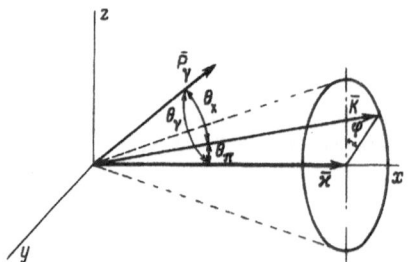

Fig. 9. Kinematics of decay of π^0-meson produced by a γ-quantum with momentum $\varkappa$.

Let $d\sigma/d\Omega$ be the differential cross section for meson formation in the laboratory system. Then the number of mesons emitted at angle θ_π into unit solid angle will be

$$n(\pi^0) = K \frac{d\sigma}{d\Omega} \sin \theta_\pi d\theta_\pi \, d\varphi, \qquad (3.12)$$

where K is a coefficient which depends on the number of nuclei in the target and the primary photon flux; we regard $\varkappa$ as fixed. We use the formula (3.3), which gives the probability that in the decay of a π^0-meson moving at velocity β_π one of the photons is emitted at an angle $\theta_\varkappa$ to β_π. Then the number of γ-quanta from the decay of π^0-mesons [see formula (3.12)] which move in the direction of the telescope will be

$$n(\gamma) = \frac{K}{2\pi} \frac{d\sigma}{d\Omega} \sin \theta_\pi d\theta_\pi \, d\varphi \frac{1 - \beta_\pi^2}{(1 - \beta_\pi \cos \theta_x)^2} . \qquad (3.13)$$

The meanings of angles θ_γ, θ_π, $\theta_\varkappa$, and φ are clear from Fig. 9. The telescope will receive γ-quanta from the decay of mesons emitted in the direction of the generatrices of a cone with an apical angle θ_π (angle $\theta_\varkappa$ is the variable). To include these γ-quanta, the expression (3.13) must be integrated with respect to φ. To include the decay γ-quanta of all the π^0-mesons, irrespective of their direction of emission, it is necessary to integrate (3.13) with respect to angle θ_π too. The final expression for the number of decay photons detected per unit intensity, per nucleus, and per steradian by a telescope placed at angle θ_γ to $\varkappa$ can be written as

$$N(\theta_\gamma) = \int_0^\pi \int_0^{2\pi} \frac{1}{2\pi} \frac{d\sigma}{d\Omega} \sin \theta_\pi \frac{\eta(\varepsilon_\gamma)(1 - \beta_\pi^2)}{(1 - \beta_\pi \cos \theta_x)^2} \, d\varphi \, d\theta_\pi. \qquad (3.14)$$

$N(\theta_\gamma)$ can be expressed in terms of the angular resolution function of the apparatus $W(\theta_\pi, \theta_\gamma)$. Then

$$N(\theta_\gamma) = \int_0^\pi \frac{d\sigma}{d\Omega} W(\theta_\pi \theta_\gamma) \, d\theta_\pi. \qquad (3.14a)$$

Comparing expressions (3.14) and (3.14a) we obtain

$$W(\theta_\pi, \theta_\gamma) = \frac{\sin \theta_\pi}{\pi} \int_0^\pi \frac{\eta(\varepsilon_\gamma)(1 - \beta_\pi^2) \, d\varphi}{[1 - \beta_\pi (\cos \theta_\pi \cos \theta_\gamma + \sin \theta_\pi \sin \theta_\gamma \cos \varphi)]} . \qquad (3.15)$$

Thus, for a given differential cross section for π^0-meson photoproduction $d\sigma/d\Omega$, knowing the angular resolution function $W(\theta_\pi, \theta_\gamma)$, we can calculate the angular distribution of the decay photons $N(\theta_\gamma)_{\text{theor}}$ and compare it with the experimentally measured $N(\theta_\gamma)_{\text{expt}}$.

The function $W(\theta_\pi, \theta_\gamma)$ can be accurately calculated only for the elastic process of π^0-meson photoproduction on deuterons, where there is a single-valued relationship between θ_π, β_π, and $\varkappa$. It was found (fuller account below) that the angular resolution function can be determined with sufficient accuracy for the process $\gamma + d \to p + n + \pi^0$.

3. Experimental Setup

The experiment consisted in measurement of the angular distribution of the decay photons of π^0-mesons

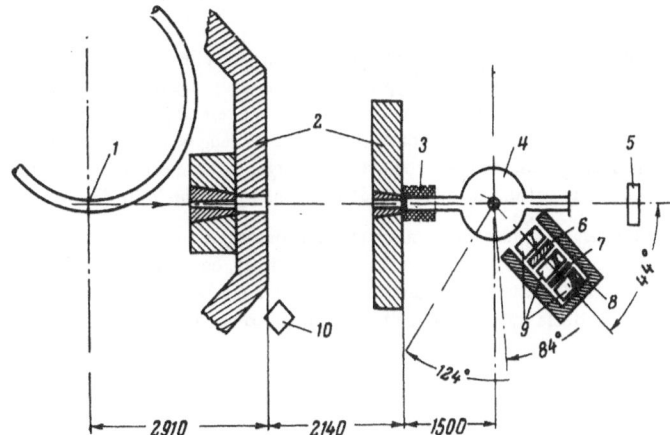

Fig. 10. Diagram of experimental apparatus for the investigation of π^0-meson photoproduction by detection of one of the decay quanta from the π^0-mesons. (1) Synchrotron; (2) Pb shield; (3) magnet; (4) target; (5) monitor (thin-walled ionization chamber); (6,7) lead and aluminum converters, respectively; (8) aluminum and lead shielding for γ-telescope; (9) scintillators; (10) differential monitor; (I, II, III) scintillation
counters.

produced by γ-quanta with energy $\varkappa \pm \Delta\varkappa$ in the processes

$$\gamma + d \to d + \pi^0, \tag{5}$$

$$\gamma + d \to p + n + \pi^0, \tag{4}$$

$$\gamma + p \to p + \pi^0. \tag{2}$$

The decay photons had to be identified against a background of a large number of charged particles which arrived at the counter, mostly from the target. The π^0-mesons were produced in liquid deuterium or hydrogen
by the bremsstrahlung (maximum energy 260 MeV) of the synchrotron of the FIAN (Physics Institute of the
Academy of Sciences). The decay photons emitted from the target at angles θ'_γ = 44°, 84°, and 124° were detected by the usual scintillation telescopes composed of three counters.

Figure 10 shows a diagram of the apparatus. Before striking the target, the collimated bremsstrahlung
beam passed through a scrubbing magnet 3. A thin-walled ionization chamber 5 was mounted behind the target for measurement of the integral intensity. The "instantaneous" intensity was measured by a differential
monitor 10 − a single scintillation counter. The γ-telescope was protected from the isotropic background by
enclosure in aluminum and lead shields 8 with a thickness of 2 and 20 cm, respectively. The electronic circuit
of the γ-telescope produced an output signal only when a particle was detected in counters II and III and not detected in counter I. Since a lead converter one t-unit thick was placed between counters I and II, this arrangement secured the detection of photons against the background of charged particles.

The output signal from the γ-telescope was delivered to a multichannel time analyzer. Each channel of
the analyzer corresponded to a change $\Delta\varkappa_{max}$ = 8 MeV in the maximum of the bremsstrahlung spectrum. The
decay γ-quantum yield was measured simultaneously for five intervals of primary photon energy between 170
and 210 MeV.

4. Target

In the experiments we used a cryogenic hydrogen—helium target designed in the Photomeson Laboratory of FIAN [34]. The working part of the target (appendix) was a cylinder made of brass foil 15 mg/cm^2 thick. The appendix was 100 mm high and 50 mm in diameter. With the selected collimation of the primary beam, the irradiated volume of the appendix was 53 cm^3. The bremsstrahlung beam entered and emerged through vacuum tubes communicating with the main volume of the target. At the entrance the beam passed through a 100-μ-thick aluminum foil and was then "scrubbed" in a constant magnetic field.

The design of the target provided for a minimum detection angle (relative to the primary beam) of 16° and a maximum of 156° on one side, and 24° and 164°, respectively, on the other side. The angle could be varied by steps of 20°. The deuterium was liquefied directly in the appendix, which was cooled by hydrogen. The mean consumption of liquid hydrogen in the target did not exceed 15 liters per 100 h of continuous operation in the beam.

5. Electronic Apparatus

A block diagram of the electronics is shown in Fig. 11.* FÉU-33 photomultipliers were used in the scintillation γ-telescopes. To increase the coincidence discrimination coefficient, the output stages of the photomultiplier operated in anode-clipping conditions. The coincidence selection element [35] consisted of D2-V diodes shunted by a capacitor $C = 8$-9 pF. Single pulses were suppressed by differentiation by an RC network with a small time constant.

The counter C_3 in the telescope operated simultaneously in two coincidence circuits — C_2-C_3 and C_1-C_3. The pulses of the C_1-C_3 circuit were applied to the anticoincidence channel, and those from the C_2-C_3 circuit to the coincidence channel. In the coincidence channel the pulses from the output of the selection element were fed to a broadband distributed amplifier (amplification factor $K \simeq 30$ with a settling time of $3 \cdot 10^{-9}$ sec). This provided for fast coincidences between the two telescopes (see Sec. B) or, where necessary, between the γ-telescope and a counter detecting the recoil nucleon.

A high-speed discriminator, connected up in a push-pull arrangement [36], was connected to the amplifier. After the discriminator the pulses from the coincidence channel were amplified and two signals were taken from the amplifier output. One went to a univibrator, which shaped the pulses for the control count, and the other went to an anticoincidence circuit incorporating two tubes with a common anode load. The second input of the anticoincidence circuit received the standard, previously shaped signals of the C_1-C_3 channel. Their length could vary from $1 \cdot 10^{-7}$ to $1 \cdot 10^{-6}$ sec, and was an order higher than the length of the coincidence channel signals. This ensured reliable and stable operation of the anticoincidence circuit.

After the coincidence circuit the pulses of channel C_2-C_3 were amplified and delivered to two outputs. One was used for a count of the total number of γ-telescope pulses and the other was used for the time analyzer.

The main characteristics of the apparatus were: resolving time of fast coincidence circuits $\tau_C = (4$-$5) \cdot 10^{-9}$ sec (a typical delayed coincidence curve is shown in Fig. 12); resolving time of anticoincidence circuit $\tau_{AC} = 2 \cdot 10^{-7}$ sec. For a check of the operation of the circuit there were three outputs for channels C_1-C_3, C_2-C_3, and the telescope, respectively. A pulse of length 10^{-6} sec was formed at the output for the time analyzer. In addition, there was an output (not shown on the diagram) giving "fast pulses" with a rise time of $(1$-$2) \cdot 10^{-8}$ sec and an amplitude of the order of 10 V.

The ten-channel time analyzer was similar in general features to that described in [37]. The number of analyzer channels was equal to the number of successively switched-on triggers.

In each channel trigger pulses of a specified length were applied to the pentode grids of two identical time discriminators (coincidence circuit) based on 6Zh4 tubes. The 6Zh4 tubes were conducting for a time equal to the length of these "time-setting" pulses. To the control grids of these tubes were applied the γ-

---

*Engineer P. N. Shareiko designed the electronic apparatus.

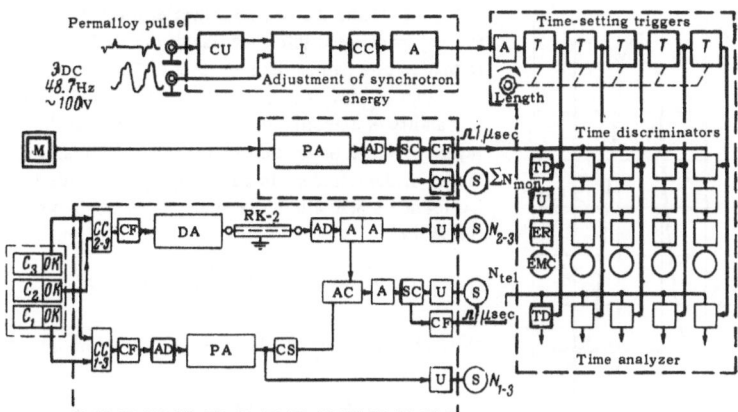

Fig. 11. Block diagram of apparatus. Symbols: $C_{1,2,3}$ – scintillation counters; CS – clipping stage; $CC_{1-3;2-3}$ – coincidence circuits; CF – cathode follower; DA – distributed amplifier; PA – pulse amplifier; RK-2 – time-delay cable; CU – controlled univibrator; I – integrator; AD – amplitude discriminator; A – amplifier; AC – anticoincidence circuit; CC – comparison circuit; SC – Schmitt circuit; U – univibrator; OT – output trigger; S – scaling circuit; TD – time discriminator; ER – electronic relay; M – monitor; EMC – electromechanical counter; ΣN_{mon} – total count of monitor counter; N_{tel} – total count of γ-telescope; N_{2-3} – count of coincidences from counters C_2 and C_3; N_{1-3} – count of coincidences from counters C_1 and C_3.

telescope pulses in one case (five channels) and the pulses from the monitor counter in the other case (five channels). The pulses from the telescope (and monitor counter) were delivered simultaneously to the time discriminators of all five channels and were registered in the channel for which the time discriminator was conducting at the time when the pulse appeared on the control grids of the 6Zh4 tubes. The "time-setting" triggers of the analyzer were operated by a special pulse, formed in the circuit described in [38], at the moment when the magnetic field on the accelerator orbit reached a value corresponding to the selected maximum energy of the bremsstrahlung spectrum. The "dead time" of the analyzer due to the length of the leading and trailing edges of the trigger pulses did not exceed 10% of the total length of all the channels.

The final adjustment of the whole electronic apparatus was carried out in working conditions. A carbon target was used to increase the count rate. The counting and time characteristics of the circuits were measured during the adjustment.

To obtain sufficient statistical accuracy we had to operate the apparatus for a long time and hence we had to have a reliable check on the sensitivity and adjustment of the whole apparatus. We carried out such a check regularly with the aid of a special device operating from a Co^{60} source.

6. Scintillation Gamma Telescope

In the experiments involving the detection of one decay γ-quantum of the π^0-meson we used the usual scintillation γ-telescope consisting of three counters with liquid scintillators. The efficiency of the telescope (probability of detection of a γ-quantum of particular energy) is its main characteristic. For an experimental determination of this characteristic of the telescope we used a monoenergetic electron beam. Yet this method involves additional, fairly laborious calculations. This is a drawback of the method, which is briefly described below.

We introduce the following symbols: $\eta(\varepsilon_\gamma)$ is the probability of detecting a photon with energy ε_γ (efficiency of γ-telescope); $W(\varepsilon_\gamma)$ is the probability of formation of an electron–positron pair by a photon with energy ε_γ in the Pb converter of the telescope; $\eta^+(E_c)$ and $\eta^-(E_c)$ are the probabilities of detecting pair components with energy E_c.

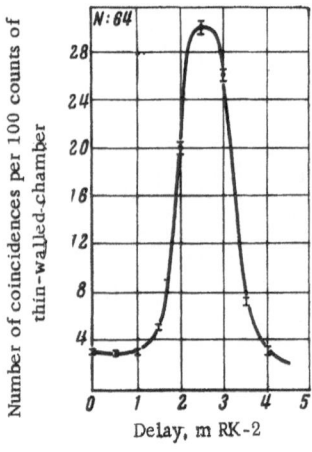

Fig. 12. Delayed coincidence curve
($\tau = 3.7 \cdot 10^{-9}$ sec).

We assume that the Pb converter is divided into m sufficiently thin layers of equal thickness and we denote by W_i the probability of formation of a pair in the i-th layer of the converter. We will assume that $W_1 = W_2 = \ldots W_i$. Then, if we denote by $\eta^{\pm}_{m-i}(E_c)$ the probability of detecting the components of a pair formed in the i-th layer and which passes through the $(m - i)$ following layers, the efficiency of the telescope in detecting a photon with energy ε_{γ} which has formed a pair with positron energy E_c can be written in the form (for m = 5)

$$\eta(\varepsilon_{\gamma}, E_c) = W_1(\varepsilon_{\gamma}) \times$$

$$\times \{\eta^+_4(E_c) + [1 - \eta^+_4(E_c)] \eta^-_4(\varepsilon_{\gamma} - E_c)\} + W_1(\varepsilon_{\gamma}) [1 - W_1(\varepsilon_{\gamma})] \times$$

$$\times \{\eta^+_3(E_c) + [1 - \eta^+_3(E_c)] \eta^-_3(\varepsilon_{\gamma} - E_c)\} + \cdots$$

$$\cdots + W_1(\varepsilon_{\gamma}) [1 - W_1(\varepsilon_{\gamma})]^4 \times$$

$$\times \{\eta^+_0(E_c) + [1 - \eta^+_0(E_c)] \eta^-_0(\varepsilon_{\gamma} - E_c)\}.$$

We assume that the energy spectrum of the pair components is rectangular in the range from zero to ε_{γ}. Then the probability of formation of a pair in the i-th layer of the converter with the energy of one of the components E_c is equal to $W_i / \varepsilon_{\gamma}$ and the probability of detecting a photon with energy ε_{γ} is

$$\eta(\varepsilon_{\gamma}) = \int_0^{\varepsilon_{\gamma}} \left(\frac{W_1(\varepsilon_{\gamma})}{\varepsilon_{\gamma}} \{\eta^+_4(E_c) + [1 - \eta^+_4(E_c)] \eta^-_4(\varepsilon_{\gamma} - E_c)\} + \cdots + \right.$$

$$\left. + \frac{W_1(\varepsilon_{\gamma})}{\varepsilon_{\gamma}} \left[1 - \frac{W_4(\varepsilon_{\gamma})}{\varepsilon_{\gamma}}\right]^4 \{\eta^+_0(E_c) + [1 - \eta^+_0(E_c)] \eta^-_0(\varepsilon_{\gamma} - E_c)\} \right) dE_c.$$

We now divide the whole energy interval from zero to ε_{γ} into small portions $\Delta\varepsilon$ and assume that the detection efficiency for electrons (positrons) with energy in the interval $\Delta\varepsilon$ is constant. Replacing integration by summation in the last formula, we finally obtain

$$\eta(\varepsilon_{\gamma}) = \Delta\varepsilon \sum_k \left(\frac{W_1(\varepsilon_{\gamma})}{\varepsilon_{\gamma}} \{\eta^-_4(E_c) + [1 - \eta^-_4(E_c)] \eta^+_4(\varepsilon_{\gamma} - E_c)\} + \cdots + \right.$$

$$\left. + \frac{W_1(\varepsilon_{\gamma})}{\varepsilon_{\gamma}} \left[1 - \frac{W_1(\varepsilon_{\gamma})}{\varepsilon_{\gamma}}\right]^4 \{\eta^-_0(E_c) + [1 - \eta^-_0(E_c)] \eta^+_0(\varepsilon_{\gamma} - E_c)\} \right).$$

To determine $\eta(\varepsilon_{\gamma})$ we experimentally measured $\eta^{\pm}_i(E_c)$. For this purpose a thin lead target was placed in the bremsstrahlung beam and electron−positron pairs were produced in this target. Using a traced magnetic field we determined the energy of the pair components entering the telescope. The electron energy was altered by a corresponding change in the magnetic field. The energy width of the magnetic analyzer was $\Delta E = 10\%$. The electron energy was varied in the range 30-110 MeV by steps of 10 MeV. For each electron energy we conducted measurements of $\eta^-(E_c)$ with a converter 0, 1, 2, 3, 4, and 5 mm thick.

To determine $W_i(\varepsilon_{\gamma})$ we used the experimental data of [39].

The telescope efficiency $\eta(\varepsilon_{\gamma})$ measured in this way had the following values:

ε_{γ}, MeV	30	40	50	60	70	80	90	100
$\eta(\varepsilon_{\gamma})$, %	1.1	8	14.5	20	24.5	28	32	35

The dependence of the telescope efficiency on the photon energy was satisfactorily described by the empirical formula

$$\eta(\varepsilon_\gamma) = \begin{cases} 0.0052 \ \varepsilon_\gamma - 0.12 & \text{for} \quad 110 > \varepsilon_\gamma > 25 \ \text{MeV,} \\ 0.42 & \text{for} \quad \varepsilon_\gamma > 110 \ \text{MeV.} \end{cases}$$

The maximum efficiency η = 0.42 corresponded to the probability of pair production in a Pb converter one t-unit thick by a photon with energy 1 BeV. An estimate was made for the cross section measured in [40].

In later experiments we used another method, which gave a direct determination of the photon detection efficiency. It is described in detail in Part 4 of Sec. B.

7. Monitoring of Bremsstrahlung Beam

The decay γ-quantum yield was measured simultaneously for five primary photon energy intervals in the range 170-210 MeV. Particular conditions were imposed on the operation of the accelerator: the length of each radiation pulse was artificially increased to approximately 1500 μsec, i.e., the "stretching" mode of operation was employed. This length corresponded to a change in the maximum of the bremsstrahlung spectrum from 160 to 220 MeV. The increase in the range of variation of the maximum energy removed the effect of instability of energy and intensity at the ends of the "stretching" region. Within the "stretch" the intensity distribution could vary. This redistribution had to be allowed for in each analyzer channel registering the intensity. During one radiation pulse each channel was open for approximately 200 μsec. Hence, the thin-walled ionization chamber usually employed for beam monitoring could not be used as a pulse detector for the time analyzer (a thin-walled chamber measured the integral intensity). As such a detector we used a differential monitor situated at the side of the bremsstrahlung beam to reduce the load [37].

We showed in specially conducted experiments that the differential monitor, despite its thorough shielding, was sensitive to induced activity and to the neutron background. In addition, the counting rate of such a monitor depends on the maximum energy in the bremsstrahlung spectrum. The data of these experiments were used to make the appropriate corrections to the calculations. To reduce the effect of these corrections on the final result, and to avoid additional calibrations, we used the differential monitor only to determine the relative distribution of intensity in the analyzer channels. The total intensity, measured by a thin-walled ionization chamber (TC) was distributed in the channels in accordance with this distribution. Thus, if during one series of measurements L_{TC} is the total count of the thin-walled ionization chamber and n_i is the count of the differential monitor, then

$$L_i = \frac{L_{TC} k_i}{\sum\limits_i k_i},$$

where L_i is the number of TC counts corresponding to the total bremsstrahlung intensity for the i-th channel. Here $k_i = \alpha_i \beta_i$, where $\beta_i = n_i / n_1$ is a coefficient allowing for the change in the number of counts of the differential monitor, corresponding to a constant number of TC counts, due to the change in the maximum of the bremsstrahlung spectrum.

In this part of the work the absolute calibration of the TC was carried out by two methods:

1. The TC readings were compared with the readings of a thick-walled graphite chamber, which measures the total flux of bremsstrahlung through the chamber in unit time.

2. The TC readings were compared with the intensity measured by means of the reaction $C^{12}(\gamma n)C^{11}$ [41].

The two methods of absolute calibration of the TC gave results which agreed to within 5-10%.

The induced activity method was used for a periodic check of the calibration. From the calibration we determined the "value" of a count of the thin-walled chamber in MeV for a given maximum of bremsstrahlung spectrum. We later used a much more convenient and reliable quantameter method of absolute monitoring of the beam.

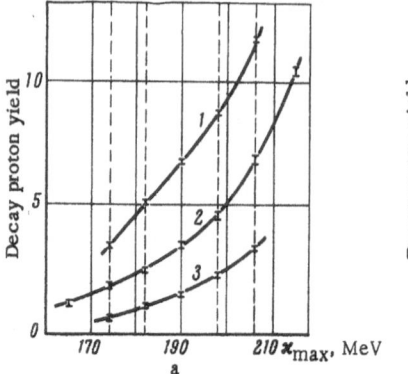 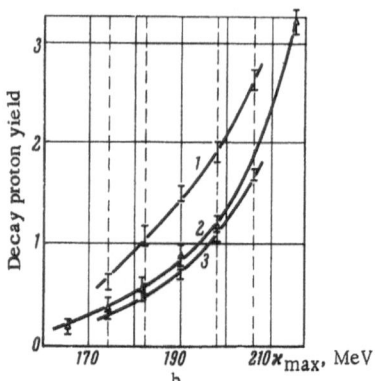

Fig. 13. Decay photon yield of π^0-mesons from reactions $\gamma + d \rightarrow d(pn) + \pi^0$ and $\gamma + p \rightarrow p + \pi^0$ as a function of maximum energy of bremsstrahlung. (a) For deuterium; (b) for hydrogen; $\theta_\gamma = 44°$ (1), 84° (2), and 124° (3) (Y per 10^2 counts of monitor).

Thus, the effect registered in each analyzer channel could be ascribed to a particular number of photons in the primary beam with energies from the photoproduction threshold to the maximum for the particular channel.

8. Measurements and Possible Errors

For angles θ_γ = 44 and 124° we measured the decay γ-quantum yield for mean values of maximum energy in the bremsstrahlung spectrum of 174, 182, 190, 198, and 206 MeV, and for an angle of 84° we measured the yield for energies 165, 174, 182, 190, 198, 206, and 215 MeV.

The constancy of the experimental conditions when the angle of the main γ-telescope was altered was checked by a "reference" telescope, the position of which remained fixed during the experiment. For each angle we conducted measurements with hydrogen, deuterium, and the empty target several times and in different sequence. The obtained energy dependences of the π^0-meson decay photon yield for photoproduction on hydrogen and deuterium are shown in Fig. 13. Here and henceforth only the statistical errors are indicated.

During the experiment we determined the background of random coincidences in the γ-telescope by the introduction of a time delay into one of the channels. In the most unfavorable case of an angle of 44° in the low-energy region, this background did not exceed 2-3% in any series. The background from the empty target varied from 30% (relative to count from deuterium) in the first channel (low energies) to 15% in the last. This resulted in relatively large statistical errors (Fig. 13), which varied from 1 to 5% for angles of 44 and 124° and from 2 to 7% for 84°.

The width of the time analyzer channels was periodically adjusted. The maximum drift of the width of any one channel did not exceed 5%. The relative accuracy of the determination of the maximum of the bremsstrahlung spectrum was within ± 2% [38].

One of the side processes which can contribute to the registered effect is the Compton effect on nucleons. The contribution of this process, however, was negligible, since control experiments showed that with the analyzer channels adjusted to an energy below the photoproduction threshold the count from filled and empty targets was the same. In addition, in the narrow energy region in which the experiment was conducted the cross section for the Compton effect on nucleons is small ($\sim 10^{-32}$ cm^2/sr) and does not depend greatly on the energy [43].

9. Treatment of Results. Control Experiment

The results of measurements for a given angle θ_γ give the integral γ-quantum yield due to π^0-meson decay. The π^0-mesons are produced by bremsstrahlung with energy from the photoproduction threshold to the maximum in the given channel (W_{max}^i).

To analyze the experimental data we had to determine the differential yield or the cross section for emission of decay γ-quanta for fixed values of $\varkappa$. For this purpose we treated the results of the measurements (Fig. 13) by a method based on [43].

The integral decay photon yield $Y(W)$ for a given angle θ_γ can be written in the form

$$Y(W) = \int_0^\infty f(W, \varkappa) N(\varkappa) d\varkappa, \qquad (3.16)$$

where $f(W, \varkappa)$ is the number of photons with energy $\varkappa \pm d\varkappa$ in the spectrum of bremsstrahlung from an electron with energy W [42, 44]; $N(\varkappa)$ is the number of detected decay γ-quanta in $cm^2/sr \cdot photon \cdot nucleus$ for a fixed primary photon energy, i.e., $N(\varkappa)$ is the differential yield; $Y(W)$ is the integral yield in $cm^2/sr \cdot nucleus$ referred to a bremsstrahlung intensity corresponding to one electron with energy W.

The yield measurements in our experiments were carried out so that the interval between the mean values of the maximum energy in the bremsstrahlung spectrum for neighboring analyzer channels was Δ MeV, and the maximum bremsstrahlung energy for the leading channel was W_m. In this case,

$$N(\varkappa_{\Delta, m}) = \frac{1}{\Delta} \sum_{i=a}^{m} B(W_m, \Delta, W_i) Y(W_i). \qquad (3.17)$$

Here the lower limit of summation is determined by the energy threshold of the process, i.e., $Y(W_i) = 0$ for $i < a$; $W_m > W_{m-1}$ and $W_i = W_m - \Delta(m - i)$; $\varkappa_{\Delta,m}$ is the mean energy for the m-th channel of width Δ; $B(W_m, \Delta, W_i)$ are coefficients which depend on the maximum energy in the leading channel, the maximum energy in the given channel, and the interval Δ.

Using the experimental values for the yields $Y(W_i)$ and the coefficients $B(W_m, \Delta, W_i)$ calculated for W_m and Δ, we could calculate the cross section of the investigated process, averaged over the energy interval of the leading channel. The whole procedure could then be repeated for all the intervals (channels), except the leading one. The new values of the coefficients $B(W_{m-1}, \Delta, W_i)$ gave the averaged value of the cross section $N(\varkappa_{\Delta, m-1})$, etc.

To clarify the physical sense of formula (3.17), we use the explicit form of the coefficients $B(W_m, \Delta, W_i)$. For simplicity we confine ourselves to the first two terms

$$N(\varkappa_{\Delta, m}) = \frac{1}{\Delta f(W_m, \varkappa_{\Delta, m})} \left[Y(W_m) - Y(W_{m-1}) \frac{f(W_m, \varkappa_{\Delta, m-1})}{f(W_{m-1}, \varkappa_{\Delta, m-1})} + \cdots \right].$$

We recall that $\varkappa_{\Delta,m}, \varkappa_{\Delta,m-1}$, etc., are the mean photon energies in the i-th interval of width Δ. The coefficient of the second and last terms in the square brackets allows for the difference in the number of photons "working" in each of the intervals (channels) in measurement of the yield $Y(W_m)$ and any of the yields $Y(W_{m-i})$, where $i > 0$. In other words, $f(W_m, \varkappa_{\Delta,i-k}) > f(W_{m-1}, \varkappa_{\Delta,i-k}) > \ldots$ (Fig. 14). The quantity $\Delta f(W_m, \varkappa_{\Delta,m})$ in front of the square brackets corresponds to the number of photons, averaged over energy, in the interval associated with the maximum energy — the leading interval.

In the treatment of the results we used the photon spectrum with allowance for the "stretching" of the radiation pulse $- f_{str}(W_i, \varkappa)$ [43]. The number of photons in the leading interval was determined not as the product $\Delta f(W_m, \varkappa_{\Delta,m})$, but as the integral

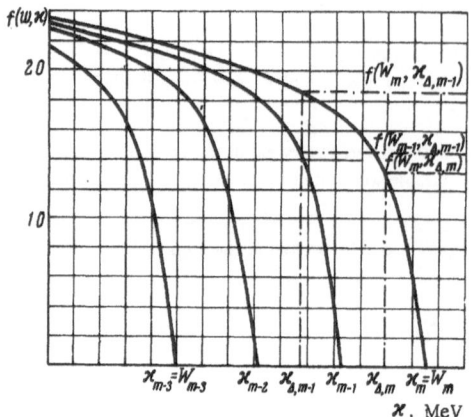

Fig. 14. Functions $f(W_m, \varkappa_{\Delta, i-k})$.

$$\int_{\varkappa_{\lim}}^{\varkappa = W_m} f_{\text{str}}(W_m, \varkappa)\, d\varkappa.$$

In this case the mean photon energy in the leading (and, by analogy, any) interval is given by the relationship

$$\widetilde{\varkappa}_{\Delta,\, m} = \dfrac{\displaystyle\int_{\varkappa_{\lim}}^{\varkappa = W_m} \varkappa f_{\text{str}}(W_m, \varkappa)\, d\varkappa}{\displaystyle\int_{\varkappa_{\lim}}^{\varkappa = W_m} f_{\text{str}}(W_m, \varkappa)\, d\varkappa},$$

where $\varkappa_{\lim}$ for the leading interval is determined by the halfwidth of the difference curve of the spectra $f_{\text{str}}(W_m, \varkappa)$ and $f_{\text{str}}(W_{m-i}, \varkappa)$ (Fig. 14). For other intervals, $\varkappa_{\lim}^i = \varkappa_{\lim}^m - (m - i)\Delta$.

The experimental data (see Fig. 13) were treated by the method described above. Table 4 gives the differential yield of decay γ-quanta formed in the processes $\gamma + d \to d(pn) + \pi^0$ and $\gamma + p \to p + \pi^0$.

The results of measurements and the data of Table 4 were used to determine the ratio of the yields and the ratio of the cross sections for the production of π^0-meson decay γ-quanta in deuterium and hydrogen. These ratios are given in Tables 5 and 6. The errors indicated here (as distinct from Table 4) do not include possible systematic errors due to absolute measurements of the intensity, telescope efficiency, etc.

Table 4. Differential Yield of Decay γ-Quanta in 10^{-30} cm^2/ sr · photon· nucleus
(Laboratory system of coordinates)

θ_γ, deg	$\varkappa \pm \Delta\varkappa$, MeV					
	167±5	176±4	184±4	192±4	200±4	209±5
Deuterium						
44	1.0±0.2	1.7 ±0.2	2.1 ±0.2	2.6 ±0.2	4.3 ±0.2	5.5±0.5
84		1.0 ±0.3	1.6 ±0.4	2.1 ±0.5	4.4 ±0.6	
124		0.47±0.07	0.64±0.08	0.91±0.08	1.64±0.09	
Hydrogen						
44	0.32±0.16	0.55±0.20	0.59±0.22	0.65±0.24	1.3±0.3	3.1±0.4
84		0.28±0.27	0.64±0.33	0.60±0.40	1.1±0.5	
124		0.24±0.09	0.26±0.09	0.61±0.10	1.1±0.1	

Table 5. Ratio of Integral Yields of Decay Photons Due to π^0-Meson Photoproduction on Deuterons and Protons $Y_i(d)/Y_i(p)$ in Relation to Maximum Energy of Bremsstrahlung Spectrum

θ_γ, deg	$\widetilde{\varkappa}$, MeV						
	167	174	182	190	198	206	215
44	5.1±2.0	4.6±0.64	4.1±0.40	3.9±0.23	3.9 ±0.21	3.8±0.15	2.8±0.13
84		3.4±0.8	4.0±0.77	3.5±0.43	3.5 ±0.45	3.6±0.27	
124		1.9±0.23	1.9±0.14	2.0±0.11	1.85±0.07	1.7±0.05	

Table 6. Ratio of Cross Sections $d\sigma_{id}/d\sigma_{ip}$ for Production of Decay γ Quanta Due to π^0-Meson Photoproduction on Deuterons and Protons for Mean Primary Photon Energies $\varkappa_{\Delta,i} \pm \Delta\varkappa$

θ_γ, deg	$\bar{\varkappa}$, MeV					
	167 ± 5	176 ± 4	184 ± 4	192 ± 4	200 ± 4	209 ± 5
44		3.1 ± 1.15	3.6 ± 1.4	4.1 ± 1.5	3.4 ± 0.8	
84	3.3 ± 1.7	3.6 ± 3.6	2.5 ± 1.4	3.5 ± 2.4	3.9 ± 1.7	1.8 ± 0.28
124		2.0 ± 0.8	2.5 ± 0.9	1.5 ± 0.28	1.5 ± 0.18	

The experimental apparatus, methods of calculation, and correctness of determining the main characteristics of the experiment are usually tested by measuring the parameters of well-investigated processes. In our experiment it was natural to use the part of the main results relating to the process $\gamma + p \to p + \pi^0$. For this purpose we treated the data obtained for the case of hydrogen by the method described in Part 2 of Sec. A. We determined the coefficients A, B, and C in the expression (3.8) for the angular distribution of π^0-mesons. In addition, we calculated the total cross section of the process $\gamma + p \to p + \pi^0$. The coefficients A, B, and C and the total cross section σ_t for different values of $\varkappa$ were compared with the data of [45, 46]. Although the accuracy of our results is low (they are only auxiliary), their agreement with the results of other investigations can be regarded as quite satisfactory.

B. Measurement of Differential Cross Sections at Small Angles
by the Detection of Two Decay γ-Quanta

1. Detection of π^0-Mesons from Two Decay γ-Quanta

In the investigation of the processes

$$\gamma + N \to N + \pi^0$$
$$ \lfloor\!\to \gamma + \gamma$$

on accelerators it is practically impossible to identify the π^0- mesons from one decay γ-quantum at angles $\theta_\pi \lesssim 40°$. Hence, at angles $\theta_\pi \simeq 0°$ we detected the two decay photons. This method is often employed in the investigation of processes of types (2), (4), and (5) when the nucleon or recoil deuteron is not detected. Detection of π^0-mesons from the two γ-quanta allows a determination, to a known degree of accuracy, of the probable angle of emission of the π^0-meson and the primary photon energy, i.e., measurement of the differential cross section of the process. The angular and energy resolution functions, which characterize this accuracy, depend on the geometry of the experiment, the characteristics of the bremsstrahlung beam, and the efficiency of the γ-ray counters. These functions determine the probability of detection by a pair of telescopes of a π^0-meson formed in the reaction $\gamma + N \to N + \pi^0$ by a photon with energy $\varkappa \pm \Delta\varkappa$ and emitted into a unit solid angle $\Delta\Omega_\pi$ at angle θ_π to $\varkappa$.

The relationship between the yield of the process $\gamma + N \to N + \pi^0$ and the differential cross section can be put in the form

$$Y(\alpha, \theta_1, \varkappa_{max}) = n \int_{\varkappa_{thr}}^{\varkappa_{max}} \int_{d\Omega} \frac{d\sigma}{d\Omega}(\varkappa, \theta_\pi) N(\varkappa, \Omega_\pi) f(\varkappa, W) d\Omega_\pi dx. \qquad (3.18)$$

Here, $Y(\alpha, \theta_1, \varkappa_{max})$ is the yield measured by a pair of telescopes, the position of which in relation to the bremsstrahlung beam is fixed by the angles α and θ_1; $\varkappa_{max}$ is the maximum energy in the bremsstrahlung spectrum; $(d\sigma/d\Omega)(\varkappa, \theta_\pi)$ is the differential cross section of the process; $f(\varkappa, W)$ is the bremsstrahlung spectrum; n is the number of nuclei per cm^2 of target; $N(\varkappa, \Omega_\pi)$ is the probability of detection of a meson formed by a photon with energy $\varkappa \pm \Delta\varkappa$ and emitted into a unit solid angle $\Delta\Omega_\pi$.

The energy and angular resolutions are given by the relationships

$$\frac{dY}{d\varkappa} = nf(\varkappa, W) \int_{\Omega_\pi} \frac{d\sigma}{d\Omega}(\varkappa, \theta_\pi) N(\varkappa, \Omega_\pi) d\Omega_\pi,$$

$$\frac{dY}{d\Omega_\pi}(\theta_\pi, \varphi_\pi) = n \int_{\varkappa_{thr}}^{\varkappa_{max}} f(\varkappa, W) \frac{d\sigma}{d\Omega_\pi}(\varkappa, \theta_\pi) N(\varkappa, \Omega_\pi) d\varkappa. \tag{3.19}$$

We introduce the symbols

$$J = \iint_{\varkappa \, \Omega} \frac{d\sigma}{d\Omega}(\varkappa, \theta_\pi) N(\varkappa, \Omega_\pi) f(\varkappa, W) d\Omega_\pi \, d\varkappa,$$

$$W(\varkappa) = \int_{\Omega} \frac{d\sigma}{d\Omega}(\varkappa, \theta_\pi) N(\varkappa, \Omega_\pi) d\Omega_\pi.$$

Then

$$J = \int W(\varkappa) f(\varkappa, W) d\varkappa$$

and expressions (3.18) and (3.19) can be rewritten as

$$Y(\alpha, \theta_1, \varkappa_{max}) = nJ, \tag{3.18a}$$

$$\frac{dY}{d\varkappa} = nf(\varkappa, W) W(\varkappa). \tag{3.19a}$$

Thus, measurement of the yield of the reaction $\gamma + N \to N + \pi^0$ allows a direct determination of the value of the integral J, which can then be compared with the value calculated for an assigned theoretical cross section $(d\sigma/d\Omega)_{theor}$. However, a direct experimental determination of the differential cross section is of great interest. It will be shown below that if the upper limit of the bremsstrahlung spectrum is correctly chosen the angular and energy resolutions have a distinct maximum and differ from zero in a relatively narrow region of angles and energies (Figs. 15 and 16). In a first approximation, assuming that the unknown differential cross section in expressions (3.18) and (3.19) is constant within the region of integration and is equal to the cross section for the mean value of angle and energy, we can write $Y(\alpha, \theta, \varkappa_{max})$ as follows:

$$Y(\alpha, \theta, \varkappa_{max}) = n \overline{\frac{d\sigma}{d\Omega}(\varkappa, \theta_\pi)} Q, \tag{3.20}$$

where

$$Q = \int_{\varkappa_{thr}}^{\varkappa_{max}} \int_{\Omega_\pi} N(\varkappa, \Omega_\pi) f(\varkappa, W) d\Omega_\pi \, d\varkappa. \tag{3.21}$$

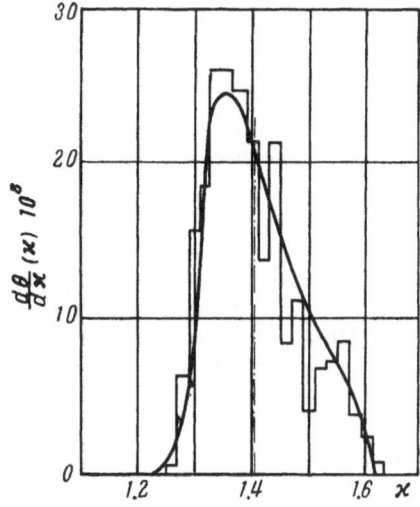

Fig. 15. Energy resolution function of apparatus. Histogram obtained by Monte Carlo method. The smooth curve is the calculation of [48]. Data: hydrogen, $\varkappa_{max} =$ 220 MeV, $\theta_\pi \simeq 0°$, $\overline{\varkappa} = 1.40$.

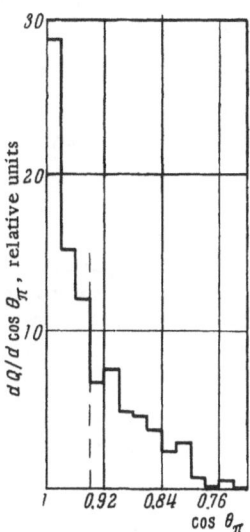

Fig. 16. Angular resolution function. Data: hydrogen, $\cos \theta_\pi = 0.938$.

In this case, the mean values are given by the equations

$$\widetilde{\varkappa} = \frac{\int\limits_{\Omega_\pi} \varkappa N (\varkappa, \Omega_\pi) \, d\Omega_\pi}{\int\limits_{\Omega_\pi} N (\varkappa, \Omega_\pi) \, d\Omega_\pi} \, , \qquad (3.22)$$

$$\widetilde{\cos} \, \theta_\pi = \frac{\int\limits_{\varkappa_{thr}}^{\varkappa_{max}} \cos \theta_\pi f (\varkappa, W) N (\varkappa, \Omega_\pi) \, d\varkappa}{\int\limits_{\varkappa_{thr}}^{\varkappa_{max}} f (\varkappa, W) N (\varkappa, \Omega_\pi) \, d\varkappa} \, . \qquad (3.23)$$

If the value of Q is known, expression (3.20) can be used to determine the differential cross section of the process $\gamma + N \rightarrow N + \pi^0$.

Calculation of the integral Q and the resolution functions (3.19a) with due regard to the bremsstrahlung spectrum, geometry, counter efficiency, etc., is a typical problem for the Monte Carlo method. In fact, all the events occurring in the experiment, beginning with the interaction of the primary particles and ending with the detection of the decay photons, can be regarded as a chain of random events occurring with a definite probability. Hence, the corresponding probability scheme can be constructed and repeated tests can be conducted on a high-speed electronic computer. The programming of this problem, however, is very laborious [47], and the calculation of one position with an accuracy of not better than 10% requires several machine hours.

Of interest from this viewpoint is the relatively simple method, given in [48], of calculating the probability W and the integral J. This calculation allows a thorough analysis of the possibility of detecting π^0-mesons from the two decay γ-quanta. It gives an integral representation of the function W, from which the values of the integrals J and Q, and also the energy resolution function of the apparatus, can be calculated. As Fig. 15

indicates, the results of such a calculation (smooth curve) agree satisfactorily with the predictions of the Monte Carlo method.

The angular resolution $dQ/d \cos\theta_\pi$, calculated by the Monte Carlo method, is shown in Fig. 16.

It should be noted that there is adequate justification for using formulas (3.20) and (3.21) to determine the differential cross section only in cases where it varies smoothly within the limits of the energy and angular resolutions.

2. Experimental Setup

It is clear from the previous section that by detecting the π^0-mesons from the two decay γ-quanta it is possible to measure the differential cross sections for photoproduction at any angles, including angles close to 0 and 180°. If the angular and energy resolutions functions of the apparatus for the processes $\gamma + p \rightarrow p + \pi^0$ and $\gamma + d \rightarrow d + \pi^0$ are identical, the ratio of the measured yields corresponds directly to the ratio of the cross sections of these processes. *

We measured the differential cross sections and the ratio of the cross sections of the processes $\gamma + d \rightarrow d(pn) + \pi^0$ and $\gamma + p \rightarrow p + \pi^0$ at angles $\theta_\pi \approx 0°$ for a maximum bremsstrahlung energy of 220 MeV (the correlation angle of the telescopes was chosen as $\alpha_0 = 100°$) and 260 MeV ($\alpha_0 = 80°$). For the given values of energy $\varkappa$ of the incident γ-quantum and angle θ_π the correlation angle of the telescopes is determined by the kinematics of the processes $\gamma + d \rightarrow d + \pi^0$ and $\gamma + p \rightarrow p + \pi^0$ and the relationship (3.6). The working range of energies $\varkappa$ and angles θ_π is determined by the dimensions of the telescopes and the selected maximum bremsstrahlung energy. The differential cross sections were measured for $\theta_\pi = 90°$ ($\alpha_0 = 100°$) and maximum bremsstrahlung energy $\varkappa_{max} = 260$ MeV, as well as for $\theta_\pi \simeq 0°$.

The choice of these experimental parameters was based on the desire to combine the results of this series of measurements with the data given in Sec. A. In addition, experiments for angle $\theta_\pi = 90°$ enabled us to verify our results by comparing the differential cross sections for hydrogen for $\theta_\pi = 90°$ with the well-known data of other investigations.

The only differences in the block diagram of the apparatus from that described in Part 3 of Sec. A were that two identical γ-telescopes were used and the signals from the output of the telescope circuits were not delivered to a time analyzer, but to an intertelescope coincidence circuit with a resolving time $\tau = 10^{-8}$ sec.

3. Electronic Apparatus

A block diagram of the electronics is shown in Fig. 17. FÉU-33 photomultipliers were used in the scintillation γ-telescopes. To increase the discrimination coefficient of the coincidence circuits, the photomultiplier output stages (6Zh11P tube) operated in anode-clipping conditions (Fig. 18). From the output stages the signals were transmitted through 50-m cables to the telescope circuit. The changes in this circuit consisted mainly in an increase in the anticoincidence efficiency. For this purpose, the suppressing pulses, which arrive at the pentode grid of the "anticoincidence" 6Zh2P tube (Fig. 19), were shaped by a low-threshold, high-speed trigger circuit. The latter provided a stable suppressing pulse with a base length of $3 \cdot 10^{-7}$ sec. A similar shaping stage at the output of the telescope circuit increased the counting rate from one telescope to $5 \cdot 10^6$ counts/sec. In addition, the parameters of the pulses applied to the intertelescope circuit ("fast" output) were improved. The pulses from this output after shaping on a short-circuited PK-2 cable 1 m long, secured a resolving time $\tau = 10^{-8}$ sec for intertelescope coincidences with almost 100% efficiency. This was achieved by the use of broadband 6Zh9P and 6Zh23P tubes in some units of the circuit.

In this version of the apparatus, the γ-telescope unit is simple to adjust and could be successfully used in other apparatuses. Its drawback is the absence of a common feedback loop. If there is a large number of

_____
* We recall that at angles $\theta_\pi \simeq 0°$ the cross section of the process $\gamma + d \rightarrow p + n + \pi^0$ is negligibly small in comparison with the cross section of the process $\gamma + d \rightarrow d + \pi^0$.

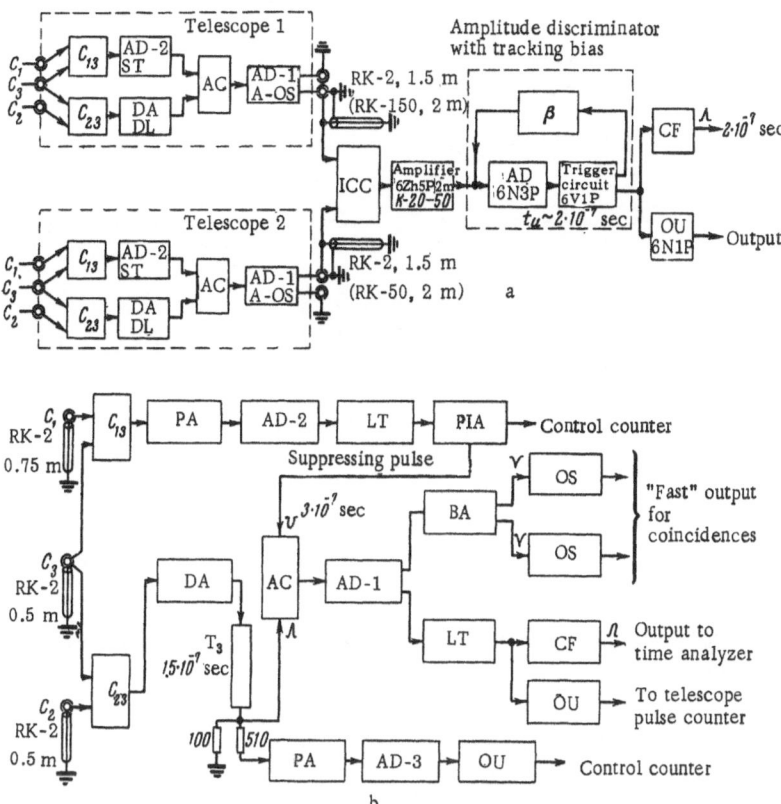

Fig. 17. Functional block diagram of intertelescope and telescope coincidences. (a) Intertelescope; (b) telescope coincidences. Symbols: ICC – intertelescope coincidence circuit ($\tau = 1 \cdot 10^{-8}$ sec); ST – shaping trigger circuit; DA – distributed amplifier; A-OS – amplifiers with output stages; C_1, C_2, C_3 – inputs; C_{13}, C_{23} – coincidence circuit ($\tau = 5 \cdot 10^{-9}$ sec); PA – preamplifiers; DL – delay line; AC – anticoincidence circuit; AD – amplitude discriminator; PIA – phase-inverting amplifier and pulse stretcher; BA – broadband amplifier; LT – low-threshold trigger; OS – output stage; CF – cathode follower; OU – output univibrator.

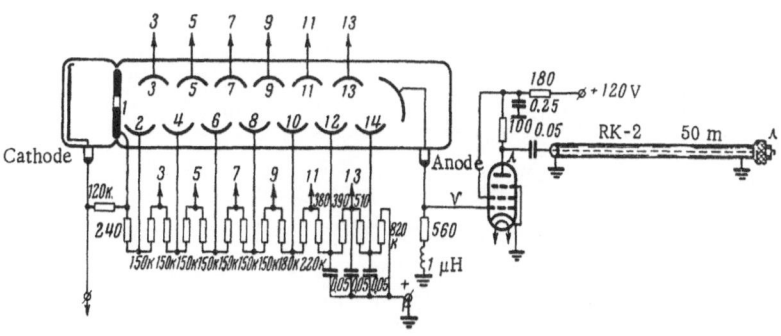

Fig. 18. Photomultiplier supply circuit.

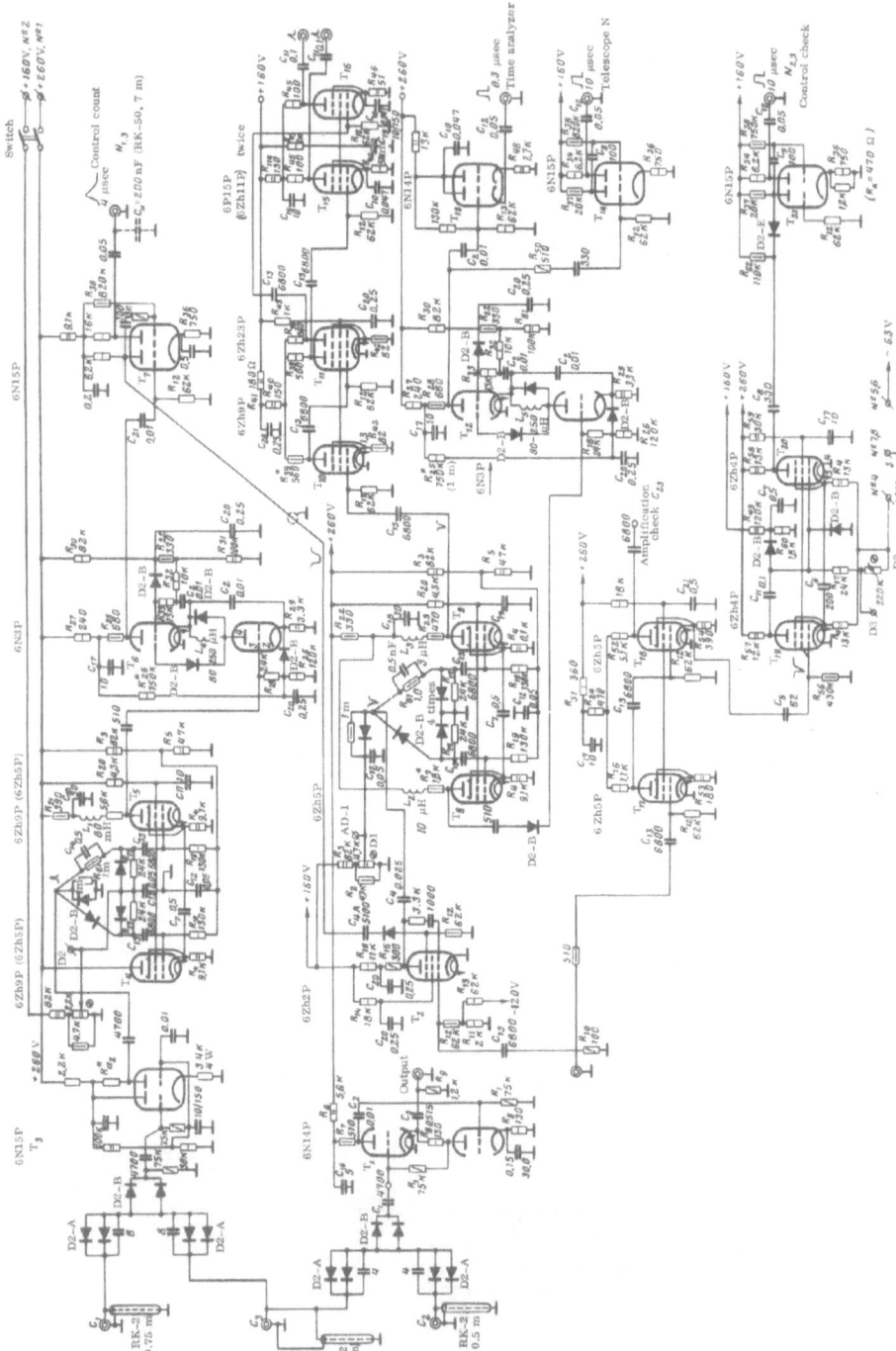

Fig. 19. Coincidence and anticoincidence circuit. The output from the cathode of T_1 goes to the distributed amplifier ($k \sim 15$). After the amplifier the signal arrives at the control grid of tube T_{1r}. In the anode of tube T_3, $R_{a_2}^* = 1.5$ kΩ.

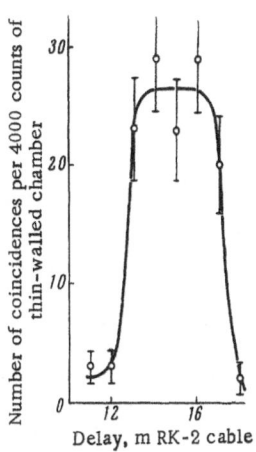

Fig. 20. Delayed coincidence curve
($\tau = 1.13 \cdot 10^{-8}$).

threshold devices it becomes fairly sensitive to aging of the tubes in prolonged operation.

The intertelescope circuit is mainly a fast coincidence circuit with the selection element described in Part 5 of Sec. A. The coincidence circuit operates on a cathode follower connected to an amplifier based on two 6Zh4 tubes. The amplifier is followed by the usual discriminator and output univibrator for shaping the pulses for the scaling circuit. A typical delayed coincidence curve for the intertelescope circuit, obtained during work on the accelerator, is shown in Fig. 20. A regular check of the sensitivity and adjustment of the apparatus, as in the first part of the work, was carried out with an apparatus operating from a Co^{60} source.

4. Scintillation γ-Telescope. Measurement of Efficiency

Each γ-telescope consisted of three scintillation counters, which differed from those described in Part 6 of Sec. A in their scintillators and the scintillator containers. Plastic scintillators, 120 mm in diameter and 25 mm thick, made from polystyrene + terphenyl + POPOP were mounted in cylindrical aluminum containers. The 25-mm-thick sides of the containers provided protection against soft radiation.

The efficiency of the γ-telescopes and electronic circuits in this part of the work was determined with the aid of monochromatic high-energy photons. For this purpose electrons with energy $E_1 \pm \Delta E_1$ were selected by a magnetic spectrometer from the electron flux produced by the synchrotron bremsstrahlung in a Pb target. The selected electron beam was slowed down in a second Pb target. The bremsstrahlung photons produced were detected by a γ-telescope connected up in coincidence with the counter for the "used" electrons. This counter could receive only "used" electrons with energy $E_2 \pm \Delta E_2$, which passed through a second magnetic spectrometer. The difference $E_1 - E_2$ gave the energy of the detected photons. This method is experimentally more complicated [49, 50] than that described in Chapter II. However, it allows direct measurement of the detection efficiency for photons of different energies, does not require absolute measurements, and has high energy resolution.

A diagram of the apparatus for efficiency determination is shown in Fig. 21.

The trajectory of electrons over the whole course from the first magnet to the counters for detecting the "used" electrons was traced with a single current-conducting wire. The accuracy of determination of the electron energy E_1 and E_2 in these experiments was about 0.5%, and the energy resolution was approximately 1.5%. The latter was determined by the ionization loss of the electrons in the second target. The electron trajectories were kindly traced for us by the authors of [49].

In the course of the measurements the "used" electrons were detected by a carefully collimated telescope composed of two scintillation counters. This practically eliminated any count due to charged particles which did not pass through the second spectrometer.

The distance between the working γ-telescope and the second target was chosen so that the $\gamma - e$ coincidence counting rate was unaltered when this distance was reduced. This meant that the bremsstrahlung photon emitted by each detected electron with energy $E_2 + \Delta E_2$ entered the γ-telescope. The $\gamma - e$ coincidence counting rate without the second target was approximately 10% of the counting rate from the target.

To determine the efficiency we simultaneously measured the $\gamma - e$ coincidence counting rate $N_{\gamma - e}$ and the counting rate of the electron telescope N_e for different values of primary electron energy E_1. The energy of the "used" electrons E_2 was always constant. For each value of energy E_1 we also conducted measurements without the second target. The detection efficiency for photons with energy $\varepsilon_\gamma = E_1 - E_2$ was determined as the ratio $\eta(\varepsilon_\gamma) = N'_{\gamma - e}/N'_e$, where $N'_{\gamma - e}$ and N'_e are the differences in the respective counting rates measured with

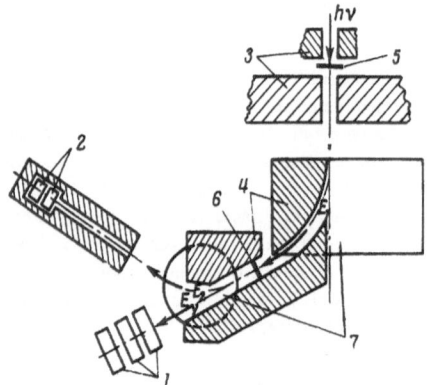

Fig. 21. Block diagram of apparatus for measuring γ-telescope efficiency. (1) γ-Telescope; (2) electron counters; (3) collimators; (4) Pb shield; (5) first target; (6) second target; (7) magnets.

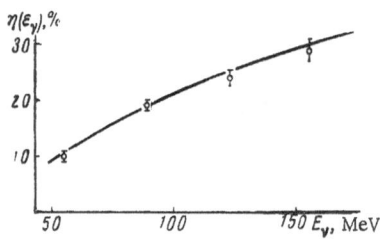

Fig. 22. Efficiency of γ-telescope.

$$\eta(E_\gamma) = 0.45 \left(1 - e^{-\frac{E_\gamma - 20}{123}}\right)$$

and without the second target. The results of measurement of the energy dependence of the efficiency for one of the γ-telescopes are shown in Fig. 22. The experimental points are approximated well by a relationship of the form

$$\eta(\varepsilon_\gamma) = \eta_\infty \left(1 - e^{-\frac{\varepsilon_\gamma - E_t}{E}}\right), \qquad (3.24)$$

where $\eta_\infty = 0.45$, $E_t = 20$ MeV, and $E = 123$ MeV. The value of E_t was chosen so that it corresponded to the energy threshold for electrons formed in the telescope converter. This value can easily be determined theoretically. η_∞ is the detection efficiency for photons of maximum energy. The value of η_∞ agrees well with the value $\eta_\infty = 0.42$ obtained on the basis of the theoretical estimates in Part 6 of Sec. A. The results of measurements of the efficiency of the second γ-telescope agreed with the limits of error with the data given in Fig. 22.

Since, in these experiments, we used the electronic part of the working apparatus we simultaneously chose the best conditions for operation of the circuits. In particular, we measured the efficiency of the intertelescope circuit. For this purpose we detected the "used" electrons with two telescopes (without the anticoincidence counters) arranged so that the two counters of the first telescope were between the two counters of the second. The ratio of the count of coincidences detected by the intertelescope circuit (on the plateau of the counting characteristic) to the counting rate of the second telescope determined the efficiency of the circuit, which was 98%.

5. Monitoring of Bremsstrahlung Beam

Relative measurements of the bremsstrahlung intensity in the experiments with two γ-telescopes were made with a thin-walled ionization chamber. Absolute measurements and calibration of the TC readings were carried out with the aid of a quantameter, instead of by the less accurate methods described in Part 7 of Sec. A. This method is now widely used in all laboratories engaged in work with high-energy bremsstrahlung beams. Hence, one of the advantages of this method, provided a standard quantameter is used, is the opportunity for a valid comparison with the results obtained by other groups.

The operation of a quantameter [51] is based on measurement of the cascade curve developing in copper and a simultaneous integration of this curve. In other words, a quantameter measures all the energy of the incident photons which is spent on ionization of the medium and neglects the energy spent on nuclear reactions.

The bremsstrahlung energy flux incident in unit time on a quantameter is proportional to the charge Q which accumulates in this time on the quantameter plates. The proportionality factor A (called the quantameter constant) can be calculated theoretically (in the quantameter which we used the parameters were chosen so that the constant A was the same as in [52]). The quantameter constant does not depend on the primary photon energy for $\varkappa > 21.5$ MeV. Yet, in the low-energy region, where the number of "working" gaps of the quantameter is small, the accuracy of integration of the cascade curve is greatly reduced.

According to the data of [51], the error of the absolute intensity measurements in our experiments was 5-6%. This includes the relative error in the measurement of Q, the relative error in the integration of the cascade curve, and the error in calculation of the constant A. The fact that the quantameter constant is independent of the maximum energy in the bremsstrahlung spectrum gives the quantameter an advantage over the thick-walled graphite chamber.

During the experiment we regularly checked the calibration of the TC with the quantameter and the π^0-meson yield obtained in each series of measurements could be referred directly to the bremsstrahlung energy flux.

6. Measurements and Results

The yield of the reactions $\gamma + d \rightarrow d(pn) + \pi^0$ and $\gamma + p \rightarrow p + \pi^0$ was measured in three different series of experiments. In the first series the measurements were made for angles $\theta_\pi \simeq 0°$ and a maximum bremsstrahlung energy of 220 MeV with the beam "stretched" from 216 to 220 MeV. The correlation angle α_0 between the γ-telescopes was chosen as 100°. The telescopes were arranged asymmetrically relative to the primary photon flux (this was due to the design of the target) and their position was determined by the angles $\theta_{1,0} = 56°$ and $\theta_{2,0} = 44°$. The angular aperture of each telescope was $\delta \simeq \pm 5°$. In the calculation of the energy and angular resolutions of the apparatus in work with deuterium we considered only elastic photoproduction.

The second series of measurements was conducted for angles $\theta_\pi \simeq 0°$ and maximum bremsstrahlung energy 260 MeV with "stretching" from 248 to 260 MeV. In this case the selected positions of the telescopes were as follows: $\alpha_0 = 80°$, $\theta_{1,0} = 36°$, $\theta_{2,0} = 44°$, and $\delta = \pm 5°$.

Finally, in the third series, the measurements were conducted for angles $\theta_\pi \simeq 90°$ and maximum energy 260 MeV with "stretching" from 248 to 260 MeV. The angles of the telescopes were $\alpha_0 = 100°$, $\theta_{1,0} = 44°$, $\theta_{2,0} = 144°$, and $\delta = \pm 5°$.

Although the reaction $\gamma + d \rightarrow p + n + \pi^0$ at angles $\theta_\pi \simeq 90°$ (as distinct from $\theta_\pi \simeq 0°$) makes a significant contribution to photoproduction on deuterium, the angular and energy resolutions, as in the first two cases, were calculated only for the elastic process. This did not introduce a large error, since the energy distribution of the π^0-mesons formed in the inelastic process has a distinct maximum in the region of maximum meson momenta (in the region $q \simeq 1$ with the selected experimental parameters) and a slight spread on the side of lower momenta (see Fig. 6). In this case the mean value of q was $q_0 = 0.94$. In the process of elastic photoproduction with the same experimental parameters $q \simeq 1$.

In each series of experiments measurements with hydrogen, deuterium, and the empty target were carried out several times and in different sequence. The coincidence counting rate with the empty target was approximately 4% of the counting rate with deuterium and 9% of the counting rate with hydrogen for a maximum energy of 260 MeV, and 15% and 40%, respectively, for a maximum energy of 220 MeV. In the control measurements we could not detect any chance coincidences between the telescopes. This means that the chance coincidence background did not exceed 1% of the counting rate with the filled target.

The yield measured in experiments with hydrogen and deuterium, after deduction of the contribution due to the empty target, was used to determine the differential cross sections [see formula (3.20)]. The latter were determined for mean values of energy $\widetilde{\varkappa}$ and angle $\widetilde{\cos} \theta_\pi$ by the following procedure:

1. In the experiment we measured the yield Y in units of 1/TC count, corresponding to a particular intensity, which was measured with a thin-walled ionization chamber.

2. From calibrations of the TC against the quantameter we determined the mean value k of a TC count for the series for a given maximum bremsstrahlung energy in units of MeV/count.

3. From the tables [43], with allowance for "stretching" of the beam, we determined the bremsstrahlung flux corresponding to a "hit" on the synchrotron target of one electron with energy

$$\varkappa_{max} - q = \int_0^{\varkappa_{max}} \varkappa f(\varkappa)\, d\varkappa.$$

Table 7. Measured Differential Cross Sections of Processes
$\gamma + p \rightarrow p + \pi^0$ and $\gamma + d \rightarrow d(pn) + \pi^0$
(Laboratory system of coordinates)

$\widetilde{\varkappa}$, MeV	$\left(\frac{d\sigma}{d\Omega}\right)_p \cdot 10^{30}$, cm^2/sr · photon · nucleus	$\left(\frac{d\sigma}{d\Omega}\right)_d \cdot 10^{30}$, cm^2/sr · photon · nucleus	$\overline{\cos \theta_\pi}$	$\Delta\theta_\pi$, deg
190	1.6±0.2	5.0±0.3	0.938	0±20
224	6.3±0.3	15.3 ±0.4	0.966	0±15
218	5.5±0.2	—	0.002	89±6
207	—	10.7 ±0.4	0.063	87±6

Table 8. Measured Ratio of Cross Sections $d\sigma_d/d\sigma_p$

$\widetilde{\varkappa}$, MeV	$d\sigma_d/d\sigma_p$	$\overline{\cos \theta_\pi}$	$\Delta\theta_\pi$, deg
190	3.12±0.40	0.938	0±20
225	2.43±0.13	0.966	0±15
250*	2.34±0.33	0.999	—

*Results of [53].

4. Analytically, and by the Monte Carlo method, we calculated the probability of detection Q [see formula (3.21)]. The value of Q was normalized to an intensity corresponding to a "hit" of one electron on the accelerator target.

5. Finally, we calculated the cross section from the formula

$$\frac{d\sigma}{d\Omega} = \frac{Y(\alpha, \theta_1, \varkappa_{max})q}{nkQ} ,$$

expressed in cm^2/sr · photon · nucleus (n is the number of nuclei per cm^2 of target).

The differential cross sections for π^0-meson photoproduction in hydrogen and deuterium, obtained in this way, are given in Table 7. The table indicates only the statistical errors. The error in the results due to the absolute measurements of the bremsstrahlung intensity, measurement of the γ-telescope efficiency, and calculation of Q is not given in the table. These sources of error can lead to an error of about 15-25% in the final result.

We mentioned above that as a control experiment we measured the cross section of the process $\gamma + p \rightarrow p + \pi^0$ for $\theta_\pi \simeq 90°$ and $\widetilde{\varkappa} = 218$ MeV. This result agreed within the limits of error with the results of [45, 46], where the corresponding results in the center-of-mass system were* $d\sigma/d\Omega = (4.5 \pm 0.3) \cdot 10^{-30}$ cm^2/sr · photon · nucleus and $d\sigma/d\Omega = 5.9 \cdot 10^{-30}$ cm^2/sr · photon · nucleus. For deuterium it is simpler to compare the experimental data with theory if they are expressed in the form of a ratio of cross sections for meson photoproduction in deuterium and hydrogen. The ratio of these cross sections for angles $\theta_\pi = 0°$ is given in Table 8. For comparison the table includes the results of [53].

Table 8 gives only the statistical errors. Other possible sources of error in the experiment do not affect this result, since the measurements with hydrogen and deuterium were conducted in the same conditions. For angle $\theta_\pi \simeq 90°$, the differential cross sections for meson photoproduction in deuterium and hydrogen were measured for mean energy $\widetilde{\varkappa} = 207$ and 218 MeV, respectively. To determine the ratio of the cross sections for $\widetilde{\varkappa} =$

---

*The slight disagreement with the data of [45] can possibly be attributed to the different methods of absolute intensity measurement. The results of [45] were converted to absolute values from the data of earlier investigations. Normalization of our results for this process for $\theta = 90°$ from the data of [45] requires multiplication of the differential cross sections given in Table 6 by a coefficient 0.76. Since the phenomenological analysis of the data on the process $\gamma + p \rightarrow p + \pi^0$ [1] involved the use of the results of [45], it is rational to carry out such a normalization for the comparison of our data with the predictions of phenomenological theory (see Table 8).

= 207 MeV we used the energy dependence, obtained in [45], for the cross section for hydrogen at θ_π = 90°. Converting, we find

$$\frac{d\sigma_d}{d\sigma_p}(\varkappa = 207 \text{ MeV})|_{\vartheta=90°} = 3.12 \pm 0.20.$$

This result agrees with the data, given in Tables 5 and 6, for energies $\tilde{\varkappa}$ = 206 MeV and $\bar{\varkappa}$ = 200 ± 4 MeV.

The results of the last series of measurements ($\theta_\pi \simeq$ 90°, $\theta_{1,0}$ = 44°) were also treated by the method described in Sec. A. This enabled us to determine the ratio of the cross sections $d\sigma_d/d\sigma_p$ for θ_γ = 44° and energy $\varkappa$ = 240 ± 20 MeV

$$\frac{d\sigma_d}{d\sigma_p}(\varkappa = 240 \pm 20 \text{ MeV})|_{\vartheta=44°} = 1.84 \pm 0.01.$$

We will make a brief comment on the question of the angular resolution in experiments with angle θ_π = 0°. For simplicity we consider a plane containing an axis from which the angles are measured. From the viewpoint of physical laws, the semiplanes on the right and left of the 0-180° axis are completely equivalent. On investigating the process for an angle of 90° or 45°, say, we can in principle obtain an angular resolution symmetrical relative to these angles, and thus determine the cross section for the range 90° ± δ or 45° ± δ. As distinct from this, angles 0° and 180° are always boundary values in the angular resolution, since the angles +δ and −δ are the same angle δ owing to symmetry. Hence, even if the probability of detection (angular resolution function) has a maximum at θ_π = 0°, we can never (having a finite resolution) measure the cross section for the mean angle $\tilde{\theta}_\pi$ = 0°.

The notation 0° ± δ frequently used in these cases must be understood as a characteristic of the apparatus indicating that the apparatus detects particles emitted at angle δ into the right and left semiplanes. This is the meaning of the column $\Delta\theta_\pi$ in Tables 7 and 8.

CHAPTER IV: COMPARISON OF OBTAINED EXPERIMENTAL DATA FOR PROCESSES $\gamma + d \rightarrow d(pn) + \pi^0$ WITH RESULTS OF THEORETICAL CALCULATIONS

As mentioned in Chapter III, the experimental part of the present work is a first stage in the investigation of the processes $\gamma + d \rightarrow d(pn) + \pi^0$ and is devoted mainly to the question of the applicability of the impulse approximation to this problem. However, from this investigation of π^0-meson photoproduction in deuterium and hydrogen we can also derive some conclusions regarding the amplitudes for photoproduction on free nucleons and the role of resonance $\pi - \pi$ interactions in meson photoproduction.

It is of interest to compare the results of the calculations carried out in Chapter II with the experimental data for the total cross section of the processes $\gamma + d \rightarrow d + \pi^0$ and $\gamma + d \rightarrow p + n + \pi^0$ and the relationship between them and the cross section of the process $\gamma + p \rightarrow p + \pi^0$: $d\sigma_d/d\sigma_p$. The ratio of these cross sections is a more reliable source of information on photoproduction amplitudes than the absolute values, since the ratio of the cross sections (this has been discussed earlier) contains fewer experimental errors. Table 9 compares the measured differential cross sections* for angles $\theta_\pi \simeq$ 0° and 90° with the cross sections calculated from formulas (2.11) and (2.28) with zero values and the values of (1.9) used for the amplitudes E_{0+}^+, E_{0+}^0, and $E_{0+}^{\pi^0}$. The table shows that within the limits of experimental error and theoretical uncertainties there is a qualitative agreement between the experimental data and the results of calculations. The large theoretical uncertainties in the ratio $d\sigma_d/d\sigma_p$ (θ = 0°) are due mainly to the inaccuracy of $d\sigma_p$ (θ = 0°)$_{\text{theor}}$. It might be falsely concluded from Table 8 that the case E_{0+} = 0 provides a better description of the data for meson production in both deuterium and in hydrogen. The experimentally established asymmetry in π^0-meson emission in the reaction $\gamma + p \rightarrow p + \pi^0$ [45] definitely indicates an appreciable value for the amplitude $E_{0+}^{\pi^0}$. But even in the extreme case E_{0+} = 0, no

---

*See previous footnote.

Table 9. Comparison of Measured Differential Cross Sections for π^0-Meson Photoproduction in Deuterium and Hydrogen and Their Ratios (in the Laboratory System) with the Predictions of Phenomenological Theory*

| | $x = 190$ MeV, $\theta_\pi = 20°$ | | | $x = 224$ MeV, $\theta_\pi = 15°$ | | | $x = 207$ MeV, $\theta_\pi = 90°$ | | |
| | Experiment | Theory | | Experiment | Theory | | Experiment | Theory | |
		$E_{0+}=0$	$E_{0+}\neq0$		$E_{0+}=0$	$E_{0+}\neq0$		$E_{0+}=0$	$E_{0+}\neq0$
$\left(\frac{d\sigma}{d\Omega}\right)_d \cdot 10^{30}$, cm²/sr ($\gamma + d \to d(pn) + \pi^0$)	3.8 ± 0.2	3.1 ± 0.4	2.9 ± 0.6	11.6 ± 0.3	10.8 ± 1.2	10.5 ± 1.8	8.1 ± 0.3	8.2 ± 0.9	9.1 ± 1.0
$\left(\frac{d\sigma}{d\Omega}\right)_p \cdot 10^{30}$, cm²/sr ($\gamma + p \to p + \pi^0$)	1.22 ± 0.15	0.9 ± 0.1	0.7 ± 0.2	4.8 ± 0.2	3.4 ± 0.4	2.7 ± 0.7	2.6 ± 0.1†	2.2 ± 0.3	2.9 ± 0.4
$d\sigma_d/d\sigma_p$	3.12 ± 0.40	3.3 ± 0.4	4.3 ± 1.3	2.43 ± 0.13	3.1 ± 0.4	3.9 ± 1.0	3.1 ± 0.2	3.4 ± 0.4	3.1 ± 0.4

*The experimental values of $d\sigma/d\Omega$ given here are reduced by a factor of 0.76 in comparison with the data of Table 7 (see footnote on p. 40).

†This value was obtained by extrapolation of the cross section measured for $x = 218$ MeV.

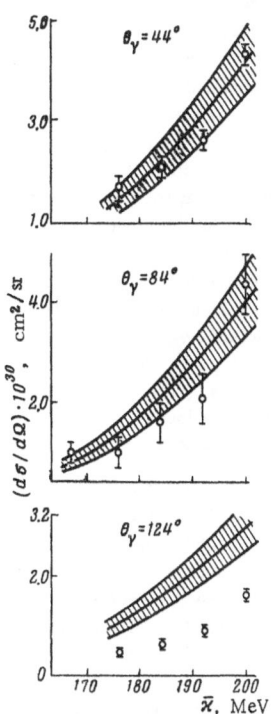

Fig. 23. Differential cross section for yield of decay photons from π^0-mesons formed in processes $\gamma + d \rightarrow d(pn) + \pi^0$. The hatched region indicates the inaccuracy in the determination of amplitude $M_{1+}^{3/2}$.

quantitative agreement for the cross section of the process $\gamma + p \rightarrow p + \pi^0$ and the ratio $d\sigma_d/d\sigma_p$ can be obtained. This fact, in conjunction with new results of investigation of the process $\gamma + p \rightarrow p + \pi^0$ [54], shows that in the phenomenological analysis of the experimental data other multipole amplitudes, besides the amplitudes E_{0+} and M_{1+}, must be taken into consideration. For instance, a consideration of the magnetic dipole transition to a state with total momentum $J = \frac{1}{2}$, if the sign of its amplitude M_{1-} is the opposite of that of M_{1+}, leads to better agreement with the experimental data. This amplitude can be determined by more thorough investigations of the processes $\gamma + d \rightarrow d + \pi^0$ and $\gamma + p \rightarrow p + \pi^0$ at small angles of emission.

The other group of experimental data, viz., the angular distributions of decay photons from π^0-mesons produced in processes $\gamma + d \rightarrow d(pn) + \pi^0$, contains some information, averaged over the angles, about the differential cross sections for meson production. In view of this, these data are not greatly affected by the contributions of small multipole amplitudes to the differential cross sections and can be used to verify the more general (than, for instance, the choice of multipole amplitudes) hypotheses, such as the impulse approximation, etc. For this purpose, the angular distributions of π^0-meson decay photons have to be obtained from the differential cross sections calculated in Chapter II (see Fig. 3). This can be done by using the angular resolution functions of the experimental apparatus $W(\theta_\pi, \theta_\gamma)$, which are given in Sec. A of Chapter III. It should be noted that $W(\theta_\pi, \theta_\gamma)$ cannot be calculated accurately for the inelastic process, since there is not a single-valued relationship between θ_π, θ_γ, and β_π (the π^0-meson velocity). However, since the momentum distribution of mesons from the reaction $\gamma + d \rightarrow p + n + \pi^0$ has a distinct maximum (see Fig. 6), $W(\theta_\pi, \theta_\gamma)$ can be calculated for a mean value of β_π. This gives a result which practically coincides with the result of calculations for the case where there are only two particles in the final state. Estimates show that the error in the differential cross section for decay photon production in the inelastic process when this calculation procedure is used does not exceed about 10%. We calculated $W(\theta_\pi, \theta_\gamma)$ separately for the elastic and inelastic processes and then, using graphic integration and the differential cross sections obtained in Chapter II, we calculated the angular distribution of decay photons for each of the processes $\gamma + d \rightarrow d + \pi^0$ and $\gamma + d \rightarrow p + n + \pi^0$.

The results of calculations of the differential cross sections for the production of π^0-meson decay photons are given in Fig. 23, where they are compared with the experimental data. Roughly speaking, the angles of emission of the γ-quanta correspond to the angles of emission of the π^0-mesons. The figure indicates that for forward meson angles of emission the experimental data agree with the predictions of the impulse approximation. The fact that this agreement holds through the whole considered energy range indicates the small role of multiple scattering of mesons when they are produced in deuterium. In fact, the multiple scattering of π^0-mesons, which is proportional to the resonance phase of $\pi - N$ scattering α_{33}, should be strongly dependent on the γ-ray energy. When $\varkappa = 280$ MeV, it reduces the cross section of the process $\gamma + d \rightarrow d + \pi^0$ by only 20% [17] for all meson emission angles and, thus, must be negligibly small at lower energies. Since, near threshold,

$$\sigma(\gamma + N \rightarrow N + \pi^\pm) \gg \sigma(\gamma + N \rightarrow N + \pi^0), \qquad (4.1)$$

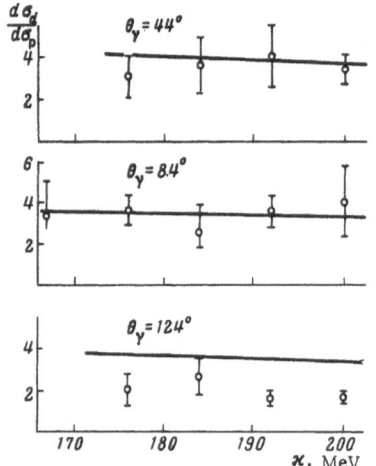

$\frac{d\sigma_d}{d\sigma_p}$

$\theta_\gamma = 44°$

$\theta_\gamma = 8.4°$

$\theta_\gamma = 124°$

$\varkappa$, MeV

Fig. 24. Energy dependence of ratio of differential yields of decay γ-quanta from π^0-mesons formed in processes $\gamma + d \rightarrow d(pn) + \pi^0$ and $\gamma + p \rightarrow p + \pi^0$. The solid line represents the results of calculations based on phenomenological theory.

scattering of charged π-mesons produced in deuterium to neutrals might play an appreciable role in the considered energy region. This effect will depend weakly on the meson emission angle (since near threshold charged mesons are produced and scattered mainly in the S state) and will depend strongly on the energy in the considered range 170-210 MeV, since for $\varkappa = 230$ MeV the inequality (4.1) is considerably weaker. As Fig. 23 shows, throughout this range of energies with $\theta_\pi < 90°$ there is no discrepancy between the experimental data and the predictions of the impulse approximation. This is confirmed also by the results of measurement of the ratio of the differential cross sections of the processes $\gamma + d \rightarrow d(pn) + \pi^0$ and $\gamma + p \rightarrow p + \pi^0$, which are given in Table 8 and in Fig. 24. Hence, charge exchange does not play any part either in the investigated processes.

Thus, the impulse approximation satisfactorily describes the data on π^0-meson photoproduction on deuterons in the angle region $\theta_\pi < 90°$. On the other hand, at angles $\theta_\pi \simeq 0°$, the production of mesons is due mainly to the process $\gamma + d \rightarrow d + \pi^0$, the amplitude of which is an isovector. Hence, further experimental investigations of π^0-meson photoproduction on protons and deuterons by the detection of only the π^0-mesons and a thorough analysis of the experimental data should lead to a separate determination of the isoscalar and isovector parts of the photoproduction amplitudes and the associated contributions of resonance 2π and 3π interactions. Thus, from the fact that near threshold $(d\sigma/d\Omega)(\gamma + d \rightarrow d + \pi^0) < (d\sigma/d\Omega)(\gamma + p \rightarrow n + \pi^+)$, and, on the other hand, the constants of $\gamma\pi\rho$ and $\gamma\pi\omega$ interactions satisfy the relationship $\Lambda_{\gamma\pi\rho} \ll \Lambda_{\gamma\pi\omega}$ [55], we can draw the important conclusion: $\Lambda_{\omega NN} \gg \Lambda_{\rho NN}$, where $\Lambda_{\rho NN}$ and $\Lambda_{\omega NN}$ are the coupling constants of ρ- and ω-mesons with nucleons.

Figures 23 and 24 show that for angles $\theta_\pi > 90°$ the cross section decreases more rapidly with increase in angle than the calculations predict.[*] For instance, for $\theta_\gamma = 124°$ the calculated cross sections are approximately twice the experimental ones. As the figure shows, this discrepancy depends strongly on the angle and weakly on the energy and, hence, cannot be due to multiple scattering effects. This discrepancy cannot be due to inaccurate allowance for N−N interaction in the final state of the reaction $\gamma + d \rightarrow p + n + \pi^0$, since the effect of this is insignificant for large π^0-meson emission angles (see Fig. 4). In addition, the contribution of the inelastic process to the total cross section of processes $\gamma + d \rightarrow d(pn) + \pi^0$ for large θ_π increases rapidly with energy increase (see Fig. 3), and the divergence is independent of energy. An estimate of the role of tensor forces, which we neglected in the description of the deuteron, shows [28] that their effect is negligible near threshold. Finally, a more accurate estimate of the range of action of nuclear forces in the integrals of the wave functions of two-nucleon systems [consideration of higher powers of expansion of function $e^{i(\Delta r)}$ in a series of powers of (Δr) in formulas (2.10) and (2.16)] increases the discrepancy under discussion, since it leads to an increase in the deuteron form factor for large transferred momenta.

[*] The discrepancy for angle $\theta_\gamma = 124°$ as regards $d\sigma_d/d\sigma_p$ (Fig. 24) is much less than in the differential cross sections (Fig. 23). This is due to the different behavior of the differential cross sections of the processes $\gamma + d \rightarrow d(pn) + \pi^0$ and $\gamma + p \rightarrow p + \pi^0$ within the region of angles bounded by the angular resolution functions of the experimental apparatus.

Thus, this discrepancy cannot be attributed to any of a whole series of effects. If we exclude systematic errors in the experimental data, the discrepancy implies the suppression of meson production on bound nucleons. With increase in photon energy this suppression extends into the region of smaller angles. For instance, for $\varkappa = 240$ MeV we obtained $d\sigma_d/d\sigma_p = 1.84 \pm 0.01$ for angle $\theta_\gamma = 44°$, whereas the calculations give $d\sigma_d/d\sigma_p \simeq 3$. Evidence of suppression of meson production has been obtained in several other investigations [56]. It is possibly due to competition between meson photoproduction on the bound nucleon and photodisintegration of the deuteron [57] due to the transfer of high momenta of the two-nucleon system.

CONCLUSION

The photoproduction of π^0 mesons in deuterium in the near-threshold region of γ-ray energies was investigated experimentally and theoretically. The main results are as follows:

1. In the impulse approximation expressions were obtained for the differential cross sections of the processes $\gamma + d \rightarrow d + \pi^0$ and $\gamma + d \rightarrow p + n + \pi^0$. These were independent of the explicit form of the wave functions of the two-nucleon system in the range of action of nuclear forces.

2. Calculations show that the relative role of elastic $\gamma + d \rightarrow d + \pi^0$ and inelastic $\gamma + d \rightarrow p + n + \pi^0$ processes of meson production on deuterons depends significantly on the meson emission angle. In the small-angle region the elastic process predominates. This can be used to determine the amplitudes for π^0-meson photoproduction on neutrons and to determine the role of resonance $\pi - \pi$ interactions in meson photoproduction.

3. Calculations showed that an important role in the process $\gamma + d \rightarrow p + n + \pi^0$ is played by the interaction of the nucleons in the final state. This reduces the total cross section of the process by a factor of units, and the differential cross section for $\theta_\pi \simeq 0°$ by a factor of tens. Small photoproduction amplitudes have an appreciable effect on the cross section of the process $\gamma + d \rightarrow d + \pi^0$ for forward and backward angles.

4. When the results of a phenomenological analysis of the data on the processes $\gamma + N \rightarrow N + \pi$ are used to calculate the cross sections for meson production on deuterons the difference in the $\gamma - p$ and $\gamma - d$ center-of-mass systems must be taken into account.

5. The differential cross sections for the production of π^0-meson decay γ-quanta in the processes $\gamma + d \rightarrow d(pn) + \pi^0$ were measured for bremsstrahlung energies of 170 to 210 MeV at intervals of $\Delta = 8$ MeV and γ-quantum emission angles $\theta_\gamma = 44°, 84°$, and $124°$, and for energy 240 MeV and $\theta_\gamma = 44°$. Similar measurements were made for the reaction $\gamma + p \rightarrow p + \pi^0$ and the ratios of the cross sections for π^0-meson photoproduction in deuterium and hydrogen were obtained.

6. Detection of the two π^0-meson decay photons was used to measure the differential cross sections for π^0-meson photoproduction in deuterium at angles $\theta_\pi \simeq 0°$ for γ-ray energies $\tilde{\varkappa} = 190$ and 224 MeV and at angle $\theta_\pi = 86°$ for $\tilde{\varkappa} = 207$ MeV. Similar measurements were carried out for the process $\gamma + p \rightarrow p + \pi^0$ for angle $\theta_\pi \simeq 0°$ and $\tilde{\varkappa} = 190$ and 224 MeV and for angle $\theta_\pi = 90°$ and energy $\tilde{\varkappa} = 218$ MeV. The ratios of the differential cross sections for π^0-meson photoproduction in deuterium and hydrogen were obtained for angles $\theta_\pi \approx 0°$ and energy $\tilde{\varkappa} = 190$ and 224 MeV and for angle $\theta_\pi = 90°$ and $\tilde{\varkappa} = 207$ MeV.

7. The obtained experimental data were compared with the predictions of phenomenological theory. The comparison led to the following conclusions:

(a) The impulse approximation satisfactorily describes the experimental data for the differential cross sections of the processes $\gamma + d \rightarrow d(pn) + \pi^0$ only for forward angles of meson emission;

(b) the measured differential cross sections for π^0-meson photoproduction on deuterons at angles $\theta_\pi > 90°$ decrease more rapidly with increase in angle than the calculations predict;

(c) the experimental data indicate that multiple scattering plays a minor role in π^0-meson photoproduction on deuterons; this agrees with the results of a qualitative assessment of this effect;

(d) the results of measurements of the differential cross sections of the processes $\gamma + d \to d(pn) + \pi^0$ and $\gamma + p \to p + \pi^0$ for angles $\theta_\pi \simeq 0°$ indicate that the photoproduction amplitudes obtained in [1] will have to be modified.

Thus, the investigation revealed a region of γ-ray energies (170-210 MeV) and π^0-meson emission angles ($\theta_\pi < 90°$) for which the impulse approximation satisfactorily describes π^0-meson photoproduction on deuterons (further investigation of these processes in this region may give valuable information on the role of resonance $\pi - \pi$ interactions in photoproduction processes). It is shown that the differences between the results of the theoretical calculations and the experimental data in the region of large meson emission angles imply a suppression of meson production on bound nucleons and cannot be attributed to any of the series of effects discussed above. This raises the problem of a quantitative examination of meson multiple scattering and the dependence of the operators K and L on the nucleon velocities [4]. From these viewpoints a further thorough (theoretical and experimental) investigation of π^0-meson photoproduction on deuterons and other few-nucleon systems is of interest.

In conclusion, the authors express their deep thanks to Corresponding Member of the Academy of Sciences of the USSR P. A. Cherenkov for his interest in the work, to Doctor of Physico-Mathematical Sciences A. M. Baldin for valuable discussion, and to Candidate of Physico-Mathematical Sciences A. S. Belousov, S. V. Rusakov, L. S. Tatarinskaya, and P. N. Shareiko, with whose collaboration we obtained and discussed the results given in this paper.

LITERATURE CITED

1. A. M. Baldin and B. B. Govorkov, Dokl. Akad. Nauk SSSR 127:998 (1959).
2. L. D. Solov'ev and Ch'en Ts'ung-mo, Zh. Eks. i Teor. Fiz. 42:527 (1962); M. Kawaguchi and H. Yokomi, Progr. Theoret. Phys. Suppl. 21:71 (1962).
3. A. M. Baldin and P. Kabir, Dokl. Akad. Nauk SSSR 122:361 (1958); A. M. Baldin and A. A. Komar, FIAN* Preprint A-37 (1964); Yu. D. Prokoshkin and V. I. Petrukhin, Joint Inst. Nuclear Research Preprint (1958), Dubna, 1964; A. M. Baldin, B. B. Govorkov, S. P. Denisov, and A. I. Lebedev, Yadernaya Fizika 1:92 (1965).
4. A. M. Baldin, Tr. Fiz. Inst. Akad. Nauk SSSR 19:3 (1963); G. F. Chew and H. W. Lewis, Phys. Rev. 84:772 (1952).
5. A. M. Baldin, Nuovo Cim. 8:569 (1958); M. I. Adamovich, V. G. Larionova, and S. P. Kharlamov, Tr. Fiz. Inst. Akad. Nauk SSSR 19:37 (1963).
6. E. H. Bellamy, Progr. Nucl. Phys. 8:237 (1960).
7. A. M. Baldin, Nuovo Cim. Suppl. 3:4 (1956).
8. A. M. Baldin and A. I. Lebedev, Zh. Eks. i Teor. Fiz. 41:1688 (1961).
9. A. A. Logunov, A. N. Tavkhelidse, and L. D. Soloviev, Nucl. Phys. 4:427 (1957).
10. G. F Chew, M. L. Goldberger, F. Low, and Y. Nambu, Phys. Rev. 106:1345 (1957).
11. A. M. Baldin and A. I. Lebedev, Nucl. Phys. 40:44 (1960).
12. J. S. Ball, Phys. Rev. 124:2014 (1961).
13. M. Gourdin, D. Lurie, and A. Martin, Nuovo Cim. 18:933 (1960).
14. A. M. Baldin, Dissertation, FIAN (1953).
15. V. Sachl, Nucl. Phys. 26:681 (1961).
16. A. Ramakrishnan, V. Devanathan, and G. Ramachandran, Nucl. Phys. 24:163 (1961).
17. J. Chappelear, Phys. Rev. 99:254 (1955).
18. E. M. McMillan and J. M. Peterson, Science 110:579 (1949); J. Steinberger, W. K. H. Panofsky, and J. S. Steller, Phys. Rev. 78:802 (1950).
19. G. Cocconi and A. Silverman, Phys. Rev. 92:520 (1953).
20. J. W. Dewire, A. Silverman, and B. Wolfe, Phys. Rev. 92:520 (1953).
21. A. S. Belousov, A. V. Kutsenko, and E. I. Tamm, Dokl. Akad. Nauk SSSR C120:921 (1955).
22. B. Wolfe, A. Silverman, and J. W. Dewire, Phys. Rev. 99:268 (1955).

---

*The abbreviation FIAN refers to the P. N. Lebedev Physics Institute of the Academy of Sciences of the USSR.

22a. J. W. Rosengren and N. Boron, Phys. Rev. 101 : 410 (1956).
23. J. C. Keck, A. V. Tollestrup, and H. H. Bingham, Phys. Rev. 103 : 1549 (1956).
24. H. L. Davis and D. R. Corson, Phys. Rev. 99 : 273 (1955).
25. R. Barringer, R. Mennier, and U. S. Osborne, Proc. CERN Sympos. (1956).
26. R. Smythe, R. M. Worlock, and A. V. Tollestrup, Phys. Rev. 109 : 518 (1958).
27. R. Wilson, Phys. Rev. 104 : 218 (1956).
28. F. T. Hadjioannon, Phys. Rev. 125 : 1414 (1962); J. I. Friedman and H. W. Kendal, Phys. Rev. 129 : 2802 (1963).
29. M. H. McGregor, M. J. Moravchik, and H. P. Stapp, Annual Rev. Nucl. Sci. 10 : 291 (1960).
30. H. Feshbach and J. Schwinger, Phys. Rev. 84 : 194 (1951).
31. C. F. Powell, U. Camerini, P. Fowler, et al., Usp. Fiz. Nauk 100 : 54 (1951); B. Rossi, High Energy Particles. Prentice-Hall, Englewood Cliffs, New Jersey (1952).
32. G. Von Dardel, D. Dekkers, R. Memod, J. D. Van Putten, M. Vivargent, G. Weber, and K. Winter, Phys. Letters 4 : 51 (1963).
33. L. S. Koester and F. E. Mills, Phys. Rev. 105 : 1900 (1957).
34. L. I. Slovokhotov, FIAN Report (1962).
35. A. A. Rudenko, Pribory i Tekhn. Eksp. 6 : 60 (1958).
36. M. F. Moody, G. J. Maclusky, and M. D. Deighton, Millimicrosecond Pulse Technique CREL (1950), p. 463.
37. R. G. Vasil'kov, B. B. Govorkov, and A. V. Kutsenko, Pribory i Tekhn. Eksp. 2 : 23 (1960).
38. V. S. Shirchenko, FIAN Report (1960).
39. J. W. Dewire, A. Ashkin, and L. A. Beach, Phys. Rev. 83 : 505 (1951).
40. E. Malamud, Phys. Rev. 115 : 687 (1959).
41. P. S. Baranov, Dissertation, FIAN (1959).
42. A. S. Penfold and G. E. Leiss, Analysis of Photo Cross Sections. Univ. Illinois (1958).
43. L. S. Hyman, Phys. Degree Thesis, Massachusetts Institute of Technology (1959).
44. L. I. Schiff, Phys. Rev. 83 : 252 (1951).
45. R. G. Vasil'kov, B. B. Govorkov, and V. I. Gol'danskii, Zh. Eks. i Teor. Foz. 37 : 11 (1959).
46. Y. Modesite, Ph.D. Thesis, Illinois Univ. (1958).
47. R. A. Schrack, Ph.D. Thesis, Maryland Univ. (1959).
48. E. I. Tamm, Dissertation, FIAN (1963); S. P. Denisov, FIAN Preprint A-154 (1962).
49. V. P. Agafonov, B. B. Govorkov, S. P. Denisov, and E. V. Minarik, Pribory i Tekhn. Eksp. 5 : 47 (1962).
50. A. S. Belousov, S. V. Rusakov, L. S. Tatarinskaya, and E. I. Tamm, Pribory i Tekhn. Eksp. 6 : 125 (1962).
51. Yu. M. Ado and V. V. Elyan, FIAN Report (1961).
52. R. Wilson, Nucl. Inst. 1 : 101 (1957).
53. G. Davidson, Ph.D. Thesis, Massachusetts Institute of Technology (1959).
54. B. B. Govorkov, S. P. Denisov, A. I. Lebedev, and E. V. Minarik, Zh. Eks. i Teor. Fiz. 44 : 163 (1963).
55. K. Kawarabayashi and A. Sato, Progr. Theoret. Phys. Suppl. 21 : 3 (1962).
56. J. C. Keck, A. V. Tollestrup, and H. H. Bingham, Phys. Rev. 103 : 1549 (1956); G. Cocconi and A. Silverman, Phys. Rev. 92 : 520 (1953); A. M. Baldin and A. I. Lebedev, Zh. Eks. i Teor. Fiz. 33 : 1221 (1957); R. Smythe, R. M. Worlock, and A. V. Tollestrup, Phys. Rev. 109 : 518 (1958); A. S. Belousov, B. B. Govorkov, and V. I. Gol'danskii, Zh. Eks. i Teor. Fiz. 36 : 244 (1959).
57. R. Wilson, Phys. Rev. 86 : 125 (1952).

PHOTOPRODUCTION OF π^+-MESONS ON PROTONS NEAR THRESHOLD

M. I. Adamovich, V. G. Larionova, A. I. Lebedev, S. P. Kharlamov, and F. R. Yagudina

INTRODUCTION

Among the physical processes which involve strongly interacting particles — pions and nucleons — one of the most accessible to experimental investigation is the photoproduction of pions on nucleons. Hence, the results of an experimental investigation of pion photoproduction provide an important measure of the correctness of attempts to create a theory of strong interactions.

The elementary processes of pion photoproduction on nucleons include

$$\gamma + p \rightarrow n + \pi^+, \tag{I}$$

$$\gamma + n \rightarrow p + \pi^-, \tag{II}$$

$$\gamma + p \rightarrow p + \pi^0, \tag{III}$$

$$\gamma + n \rightarrow n + \pi^0. \tag{IV}$$

The first process has been investigated over a long period and is one of the most reliable sources of our ideas in the field of π-meson physics. During the last few years, experiments on the photoproduction of positive π-mesons on protons in the γ-ray energy region close to the photoproduction threshold ($E_\gamma < 230$ MeV) have had the aim of increasing the accuracy of the experimental data and enlarging the investigated range of meson emission angles and γ-ray energies. This is linked up with the fact that quantitative measurements are required for the conduction of a phase shift analysis of photoproduction processes and, on the other hand, such measurements give information as to how well current theory predicts the experimental data.

The theoretical study of photoproduction processes has been conducted along two lines. Phenomenological theory, which employs general quantum conversation laws, has established some relationships between processes (I)-(IV), a relationship between these processes and pion scattering on nucleons, and a relationship between the dynamic quantities (squares of matrix elements) and the energy [1-3]. The other line has been the approach to the process $\gamma + N \rightarrow N + \pi$ on the basis of quantum field theory. Attempts have been made, initially within the framework of perturbation theory, for $\pi - N$ interaction and subsequently on the basis of extended source theory, to provide a qualitative explanation of the main features of the photoproduction process. The successful results of these theories have been substantiated by the method of dispersion relations [4, 5], which are now widely used for the analysis of experimental data.

These two approaches are particularly successful in the near-threshold γ-ray region, where there are several features which simplify the treatment. In addition, for the near-threshold energy region there are abundant experimental data obtained in foreign laboratories and in the Photomeson Laboratory of the Physical Institute of the Academy of Sciences of the USSR. In view of the simplicity and reliability of detection of charged mesons results of investigations of process (I) are of particular importance among these data. The accuracy of the measured differential cross sections of this process allows an assessment of the role of various effects in photoproduction processes, including the effect of resonance interaction of pions on photoproduction, from a comparison of experimental data with the results of careful theoretical calculations.

The aim of this paper is to give an account of the results of an experimental investigation of the process $\gamma + p \rightarrow n + \pi^+$ and new calculations of the cross sections of this process. Examination of the obtained data and a comparison of them with the predictions of dispersion relations has enabled us to determine some parameters of low-energy pion physics with greater accuracy than heretofore, and to give an upper limit for the $\gamma - 3\pi$ coupling constant.

1. Basic Ideas Regarding Pion Photoproduction on Nucleons

The assignment of amplitudes of pion photoproduction on nucleons requires, in addition to the spin and charge variables, the selection, from the three invariants written below, of two corresponding to the total energy and angle of emission of the pion in the center-of-mass system

$$s = -(K + P_1)^2 = -(Q + P_2)^2,$$
$$t = -(Q - K)^2 = -(P_1 - P_2)^2, \tag{1}$$
$$\bar{s} = -(K - P_2)^2 = -(Q - P_1)^2,$$

where $K = (k, \mathbf{k})$, $Q = (\omega, \mathbf{q})$, $P_1 = (E_1, \mathbf{p}_1)$, $P_2 = (E_2, \mathbf{p}_2)$ are the four-momenta of the photon, pion, and nucleon in the initial and final states, respectively; $E = \sqrt{M^2 + p^2}$, $\omega = \sqrt{1 + q^2}$, and M is the nucleon mass. * The three variables introduced are connected with one another by the relationship

$$s + t + \bar{s} = 2M^2 + 1. \tag{2}$$

Any two of the quantities s, t, and $\bar{s}$ can be selected as independent and then the third variable is given by relationship (2). In the photon—nucleon center-of-mass system (final state pion—nucleon) the variable s is expressed in terms of the total energy in the initial or final state: $W = k + \sqrt{M^2 + k^2} = \omega + \sqrt{M^2 + q^2}$, and the variable t is expressed in terms of the angle of emission θ of the pion:

$$s = W^2, \tag{3}$$

$$t = 1 - 2k(\omega - q\cos\theta).$$

The photoproduction amplitude for single pseudoscalar pions has the form

$$S_{fi} = \frac{i}{(2\pi)^2} \frac{M}{\sqrt{4E_1 E_2 \omega k}} \delta(P_1 + K - Q - P_2) \bar{u}(P_2) T u(P_1). \tag{4}$$

Here u are the nucleon 4-spinors, and the most general expression for the operator T, which satisfies the conditions of relativistic and calibration invariance, can be chosen in the following way:

$$T = A_1(s, t, \bar{s}) M_1 + A_2(s, t, \bar{s}) M_2 + A_3(s, t, \bar{s}) M_3 + A_4(s, t, \bar{s}) M_4, \tag{5}$$

where the invariant forms of M_i are expressed in terms of K, Q, P_1, P_2, the Dirac γ-matrix, and the 4-vector of photon polarization ε:

$$M_1 = i\gamma_5(\gamma\varepsilon)(\gamma K),$$
$$M_2 = i\gamma_5[((P_1 + P_2)\varepsilon)(QK) - ((P_1 + P_2)K)(Q\varepsilon)], \tag{6}$$

---

*In this paper we use the system of units in which $\hbar = c = \mu = 1$ (μ is the mass of the pion).

$$M_3 = \gamma_5 \left[(\gamma\varepsilon)(QK) - (\gamma K)(Q\varepsilon) \right],$$
$$M_4 = \gamma_5 \left[(\gamma\varepsilon)((P_1 + P_2)K) - (\gamma K)((P_1 + P_2)\varepsilon) - 2iM(\gamma\varepsilon)(\gamma K) \right]. \tag{6}$$

In the center-of-mass system for each of the four elementary processes (I)-(IV) the photoproduction amplitude is expressed in the following way [5]:

$$F = i(\boldsymbol{\sigma}\boldsymbol{\varepsilon}) F_1 + (\boldsymbol{\sigma} q_0)(\boldsymbol{\sigma} [k_0\varepsilon]) F_2 + i(\boldsymbol{\sigma} k_0)(q_0\varepsilon) F_3 + i(\boldsymbol{\sigma} q_0)(q_0\varepsilon) F_4, \tag{7}$$

where σ is the nucleon spin, $k_0 = k/|k|$ and $q_0 = q/|q|$ are the unit vectors of the photon and pion momenta, respectively. The scalar quantities F_i are complex functions of the total pion energy $\omega = \sqrt{1 + q^2}$, and the angle θ between the vectors of the photon ($\mathbf{k}$) and pion ($\mathbf{q}$) momenta.

Relationships (6)-(7) can provide the following formulas connecting the functions A_i and F_i with one another:

$$\frac{1}{N} F_1 = A_1 + (W - M) A_4 + \frac{t-1}{W-M}(A_3 - A_4),$$

$$\frac{M + E_2}{qN} F_2 = -A_1 + (W + M) A_4 + \frac{t-1}{W+M}(A_3 - A_4),$$

$$\frac{1}{qN} F_3 = (W - M) A_2 + A_3 - A_4, \tag{8}$$

$$\frac{M + E_2}{q^2 N} F_4 = -(W + M) A_2 + A_3 - A_4,$$

where

$$N = \frac{W - M}{8\pi W} \sqrt{(M + E_1)(M + E_2)}.$$

The angular dependence of the amplitudes F_i can be accurately expressed in the form of an expansion in derivatives of Legendre polynomials $P(\cos\theta)$:

$$F_1 = \sum_{l=0}^{\infty} [lM_{l+} + E_{l+}] P'_{l+1}(\cos\theta) + [(l+1)M_{l-} + E_{l-}] P'_{l-1}(\cos\theta),$$

$$F_2 = \sum_{l=1}^{\infty} [(l+1)M_{l+} + lM_{l-}] P'_l(\cos\theta),$$

$$F_3 = \sum_{l=1}^{\infty} [E_{l+} - M_{l+}] P''_{l+1}(\cos\theta) + [E_{l-} + M_{l-}] P''_{l-1}(\cos\theta), \tag{9}$$

$$F_4 = \sum_{l=1}^{\infty} [M_{l+} - E_{l+} - M_{l-} - E_{l-}] P''_l(\cos\theta).$$

Here, $M_{l\pm}$ and $E_{l\pm}$ are the multipole amplitudes describing magnetic and electron transitions to states of the pion—nucleon system with total angular momentum $j = l \pm \frac{1}{2}$, where l is the orbital momentum.

Using formula (7), we can find expressions for the differential cross section and polarization of the recoil nucleons for the case of interaction of nonpolarized and polarized photons with a nonpolarized and polarized proton target. For instance, the differential cross section for pion production from nonpolarized photons is written in the following way:

$$\frac{d\sigma}{d\Omega} = \frac{q}{k} \Big[|F_1|^2 + |F_2|^2 - 2\,\mathrm{Re}\,F_1^* F_2 \cos\theta + \frac{1}{2}\sin^2\theta \big(|F_3|^2 +$$

$$+ |F_4|^2 + 2\,\mathrm{Re}\,F_2^* F_3 + 2\,\mathrm{Re}\,F_1^* F_4 + 2\,\mathrm{Re}\,F_3^* F_4 \cos\theta \big) \Big]. \tag{10}$$

In the investigation of the photoproduction of charged pions on nucleons it is convenient to separate from the amplitudes F_3 and F_4 the terms corresponding to the photoeffect of pions on nucleons due to the interaction of photons with the pion current. Separating the terms due to direct interaction of photons with the pion current F_{3R} and F_{4R}, we can write

$$F_3 = F_3' + \frac{F_{3R}}{1 - \beta \cos \theta}, \quad F_4 = F_4' + \frac{F_{4R}}{1 - \beta \cos \theta}, \tag{11}$$

where β is the pion velocity in the center-of-mass system. We note that in the case of neutral pion photoproduction this interaction is absent and, thus, for π^0-mesons, $F_{3R} = 0$ and $F_{4R} = 0$. In the analysis of data on pion photoproduction the amplitudes F_1, F_2, F_3', and F_4' are usually expanded according to formulas (9), and the terms due to direct interaction of photons with the pion current are written in explicit form (11). This is due to the fact that expansion of F_{3R} and F_{4R} in multipoles will contain terms with as large values of l as desired.

Isotopic invariance in application to processes involving a γ-quantum allows us to describe the four elementary processes of π-meson photoproduction (I)-(IV) by three transition amplitudes in the isospin space of the system of strongly interacting particles [1]:

$$\left\langle \left(T_f = \tfrac{1}{2}\right) | S | \left(T_i = \tfrac{1}{2}\right) \right\rangle, \quad \left\langle \left(T_f = \tfrac{1}{2}\right) | V_3 | \left(T_i = \tfrac{1}{2}\right) \right\rangle,$$
$$\left\langle \left(T_f = \tfrac{3}{2}\right) | V_3 | \left(T_i = \tfrac{1}{2}\right) \right\rangle,$$

where T is the isospin in the final or initial state, S and V_3 are the isoscalar and isovector parts of the pion photoproduction operator. For example, for a magnetic multipole we can write

$$M_j^+ = \sqrt{2} \left(\tfrac{1}{3} M_{j,\,1/2} - \tfrac{1}{3} M_{j,\,3/2} + \delta M_{j,\,1/2} \right) \text{ for } \gamma + p \to n + \pi^+,$$
$$M_j^- = -\sqrt{2} \left(\tfrac{1}{3} M_{j,\,1/2} - \tfrac{1}{3} M_{j,\,3/2} - \delta M_{j,\,1/2} \right) \text{ for } \gamma + n \to p + \pi^-,$$
$$M_j^0 = \tfrac{1}{3} M_{j,\,1/2} + \tfrac{2}{3} M_{j,\,3/2} + \delta M_{j,\,1/2} \text{ for } \gamma + p \to p + \pi^0, \tag{12}$$
$$M_j^{n0} = \tfrac{1}{3} M_{j,\,1/2} + \tfrac{2}{3} M_{j,\,3/2} - \delta M_{j,\,1/2} \text{ for } \gamma + n \to n + \pi^0.$$

The first two quantities $M_{j,\,3/2}$ and $M_{j,\,1/2}$ describe transitions to the pure isotopic spin states $I = \tfrac{3}{2}$ and $I = \tfrac{1}{2}$ of the final state, consisting of the pion and nucleon. They behave in isotopic spin space on transformation of coordinates as a third vector component. The term $\delta M_{j,\,1/2}$ behaves in isospin space as a scalar quantity and permits only the state with isospin $I = \tfrac{1}{2}$ in the final state of the pion–nucleon system.

In the future we will find it more convenient to use other combinations of amplitudes which have definite crossing symmetry properties:

$$F\left(\gamma + p \to n + \pi^+\right) = \sqrt{2}\,(F^{(0)} + F^{(-)}),$$
$$F\left(\gamma + n \to p + \pi^-\right) = \sqrt{2}\,(F^{(0)} - F^{(-)}),$$
$$F\left(\gamma + p \to p + \pi^0\right) = F^{(+)} + F^{(0)}, \tag{13}$$
$$F\left(\gamma + n \to n + \pi^0\right) = F^{(+)} - F^{(0)},$$

where

$$F^{(0)} = \left\langle \tfrac{1}{2} | S | \tfrac{1}{2} \right\rangle,$$
$$F^{(+)} = \tfrac{1}{3} \left\langle \tfrac{1}{2} | V | \tfrac{1}{2} \right\rangle + \tfrac{2}{3} \left\langle \tfrac{3}{2} | V | \tfrac{1}{2} \right\rangle,$$

$$F^{(-)} = \frac{1}{3}\left\langle \frac{1}{2}|V|\frac{1}{2}\right\rangle - \frac{1}{3}\left\langle \frac{3}{2}|V|\frac{1}{2}\right\rangle.$$

The use of the unitarity of the S-matrix and its invariance to time reversal makes it possible to express the phases of the complex multipole amplitudes of photoproduction in a state with a particular isospin of the $\pi-N$ system in terms of the corresponding phases of $\pi-N$ scattering $\delta_{j,T}$ [1-3].

Thus, we can write

$$M_{j,\,3/_2} = e^{i\delta_{j,\,3/_2}} m_{j,\,3/_2},$$
$$M_{j,\,1/_2} = e^{i\delta_{j,\,1/_2}} m_{j,\,1/_2},$$
$$\delta M_{j,\,1/_2} = e^{i\delta_{j,\,1/_2}} \delta m_{j,\,1/_2},$$

(14)

where $\delta_{j,\,3/2}$ and $\delta_{j,\,1/2}$ are the phases of scattering of pions on nucleons in isospin states $I = \frac{3}{2}$ and $I = \frac{1}{2}$, respectively, and $m_{j,\,3/2}$, $m_{j,\,1/2}$, and $\delta m_{j,\,1/2}$ are real functions of the pion energy. It follows from the finiteness of the range of action of forces that at low pion energies the dependence of the multipole on the pion momentum is determined as q^l [3].

Thus, the process $\gamma + N \rightarrow N + \pi$ is described by twelve complex amplitudes $F_i^{(+),(-),(0)}(s, t, \bar{s})$ $(i = 1, 2, 3, 4)$, the determination of which for given values of s and t is the aim of a complete experiment for the meson photoproduction process. Experiments in the near-threshold energy region are of special interest.

The region of photon energies k is referred to as near threshold if the pion momentum in the center-of-mass system satisfies the condition $q \leq 1$. In the laboratory system this region of photon energies corresponds to the range from threshold to 230 MeV.

The condition $q \leq 1$ marks off a region in which the problem of analysis of the experimental data and theoretical calculations is greatly simplified. In this energy range the amplitudes (or multipoles) can be regarded as real, since the phases of scattering of pions on nucleons, which come into the expressions (14), are small and decrease with reduction of q (for instance, even the resonance phase shift does not exceed 0.23). Moreover, when the amplitudes are expanded (with the exception, of course, of the direct photoeffect terms) according to formulas (9) only small values of the pion orbital momentum, $l = 0$ and $l = 1$, need to be taken. In this case, the conduction of a complete experiment reduces to a determination of only four, angle-independent, real parameters E_{0+}, M_{1+}, M_{1-}, and E_{1+}. Finally, near-threshold expansions of the amplitudes in powers of the pion momentum q can be used and thus the threshold values of the photoproduction amplitudes can be obtained.

The condition $q \leq 1$ means that the differential cross section can be expressed in terms of real functions F_i and has the following form:

$$\frac{d\sigma}{d\Omega} = \frac{q}{k}\left[F_1^2 + F_2^2 - 2F_1F_2\cos\theta + \frac{1}{2}\sin^2\theta\,(F_3^2 + F_4^2 + 2F_2F_3 + 2F_1F_4 + 2F_3F_4\cos\theta)\right]. \quad (15)$$

This allows a determination of some combinations of amplitudes (multipoles). For instance, the differential cross sections at 0 and 180° include only F_1 and F_2

$$\frac{d\sigma}{d\Omega}(0°) = \frac{q}{k}[F_1(0°) - F_2(0°)]^2,$$

$$\frac{d\sigma}{d\Omega}(180°) = \frac{q}{k}[F_1(180°) + F_2(180°)]^2.$$

(16)

Substituting the experimental data in the left sides, using the expansions (9), and considering only l equal to zero or unity, the combination of multipoles can be determined

$$F_{10} = E_{0+} = \frac{1}{2} \sqrt{\frac{k}{q}} \left[\sqrt{\frac{d\sigma}{d\Omega}(0°)} + \sqrt{\frac{d\sigma}{d\Omega}(180°)} \right],$$

$$F_{11} - F_{20} = M_{1+} - M_{1-} - 3E_{1+} = \frac{1}{2} \sqrt{\frac{k}{q}} \left[\sqrt{\frac{d\sigma}{d\Omega}(180°)} - \sqrt{\frac{d\sigma}{d\Omega}(0°)} \right].$$

(17)

Restriction to the lowest states in orbital momentum ($l \leq 2$) allows the representation of the differential cross section of the process $\gamma + p \rightarrow n + \pi^+$ in the form

$$(1 - \beta \cos\theta)^2 \frac{d\sigma}{d\Omega} = \frac{q}{k}(a_0 + a_1 \cos\theta + a_2 \cos^2\theta + a_3 \cos^3\theta + a_4 \cos^4\theta + a_5 \cos^5\theta + a_6 \cos^6\theta). \quad (18)$$

By measuring the differential cross section for photoproduction at many angles, some multipoles or combinations of them can be determined.

An investigation of the energy dependence of the differential cross section enables obtaining the threshold value of the square of the amplitude $F_{10}^2 = E_{0+}^2$ by expansion of the cross section in powers of the pion momentum q and extrapolation of the experimental data to q = 0. As is known [6, 7], the threshold value of E_{0+}^2 is connected with such parameters of low-energy pion physics as the cross section for charge—exchange scattering, the ratio of the cross sections for photoproduction of π^-- and π^+-mesons on nucleons, and the Panofsky ratio P = $W(\pi^- + p \rightarrow n + \pi^0)/W(\pi^- + p \rightarrow n + \gamma)$, where W is the probability of transition of the $\pi^- + p$ system to one of the two final states. The probability $W(\pi^- + p \rightarrow n + \gamma)$ is connected with the probability of the reverse process $W(\gamma + n \rightarrow p + \pi^-)$ by the principle of detailed balancing of dynamic systems

$$W(\pi^- + p \rightarrow n + \gamma) = 2 \frac{k^2}{q^2} W(\gamma + n \rightarrow p + \pi^-). \quad (19)$$

In its turn,

$$W(\gamma + n \rightarrow p + \pi^-) = rW(\gamma + p \rightarrow n + \pi^+), \quad (20)$$

where r is equal to the ratio σ^-/σ^+ of the threshold cross sections for photoproduction of π^-- and π^+-mesons on free nucleons. The probability of charge—exchange scattering is proportional to $1/q^2 (\alpha_3 - \alpha_1)^2$, where α_3 and α_1 are the S-wave phase shifts of scattering of pions in isospin states I = $^3/_2$ and $^1/_2$, respectively.

Thus, the main parameters of low-energy pion physics are connected uniquely with one another. This connection is a connection of phenomenological parameters and does not depend on any specific meson theories. The comparison of the experimentally measured parameters has been the source of much discussion [8-14] in recent years and has stimulated theoretical and experimental investigations in low-energy pion physics.

2. Investigation of Meson Photoproduction on the Basis of Dispersion Relations

A consideration of the analytical properties of the invariant amplitudes $A_i(s, t)$ with complex values of the variable s has made it possible to write unidimensional dispersion relations for them [4, 5], which, with due regard to the properties of gradient invariance and crossing symmetry of the amplitudes $A_i(s, t)$ have the form

$$A_i(s, t) = R_i \left(\frac{1}{s - M^2} \pm \frac{1}{\bar{s} - M^2} \right) + \frac{1}{\pi} \int_{(M+1)^2}^{\infty} ds' \, \text{Im} \, A_i(s', t) \left(\frac{1}{s' - s} \pm \frac{1}{s' - \bar{s}} \right),$$

$$R_1 = \frac{e_r g_r}{2}, \quad R_2 = \frac{e_r g_r}{t - 1}(-1)^{\pm 1,0}, \quad R_3 = -\frac{1}{2} g_r \mu_r, \quad R_4 = -\frac{1}{2} g_r \mu_r, \quad (21)$$

$$e_r = \sqrt{4\pi} e, \quad g_r = \sqrt{4\pi} f, \quad e^2 = ^1/_{137}, \quad f^2 = 0.08,$$

where the upper sign corresponds to the isotopic indices (+) and (0) and the lower to (-); $\mu = \mu_p' - \mu_n$ for indices (+) and (-), $\mu = \mu_p' + \mu_n$ for (0). The dispersion relations (21) are a consequence of fundamental postulates of quantum field theory and the assumption that the amplitudes A(s, t) have at infinitely high energies a degree of growth not greater than zero.

Using relationship (8), we can write dispersion relations for amplitudes $F_i(s, t)$; they have a structure similar to (21):

$$\operatorname{Re} F_i(s, t) = \text{polar term} + \frac{1}{\pi} P \int_{M+1}^{\infty} dW' \sum_j \Phi_{ij}(s, s', t) \operatorname{Im} F_j(s', t), \qquad (22)$$

where the coefficient functions $\Phi_{ij}(s, s', t)$ contain singularities, and the dispersion integral is taken in the sense of the principal value.

Dispersion relations are widely used for the analysis of experimental data on meson photoproduction on nucleons. There are two views on the choice of methods for such an analysis. The first consists in conversion of the dispersion relations to approximate integral equations [5], the solutions of which are compared with the experimental data. However, the comparison of the equations and, in particular, their solutions, require several simplifying assumptions and hypotheses of a special order – restrictions to the lowest powers in the expansions of the amplitudes in terms of $1/M$, $(q/M)\cos\theta$, etc.

A direct comparison of the amplitudes obtained in this way [5] with data on the process $\gamma + p \to p + \pi^0$ in the near-threshold region of γ-ray energy [15, 16] reveals a considerable disagreement [17, 18]. In addition, the solutions of the equations are not unique [19]. This has led to a considerable development of the viewpoint which asserts that quantitative conclusions regarding meson photoproduction processes must be based on a direct comparison of rigorous dispersion relations with experiment by substitution of the experimental data in their left and right sides. The usual procedure in such a comparison is to limit the consideration of the dispersion integral to the imaginary part of the resonance amplitude $M_{1+}^{3/2}$. It is described by using the effective length formulas [20, 21], which are sometimes replaced by polar approximation formulas [22, 23]:

$$\operatorname{Im} M_{1+}^{3/2} = \frac{4.69}{3M} efkq\pi\delta (W - W_r); \qquad W_r = 8.8. \qquad (23)$$

The various ways of choosing the imaginary parts of the photoproduction amplitudes are discussed in Sec.6. Here we will merely mention that although the results of such approaches provide a more or less satisfactory description of the experimental data in the energy region up to 450 MeV, the calculations involve many special assumptions, such as restriction to the lowest multipoles, neglect of the S-wave and P-wave scattering phases, and the use of approximation (23). In view of this, it was desirable to conduct calculations of the differential cross sections with the minimum number of assumptions. The difficulties in the approach to the process of meson photoproduction on the basis of dispersion relations are associated with (apart from doubts regarding the fundamental postulates of quantum field theory) the evaluation of the dispersion integrals. These difficulties may be due either to the inadequacy of our information about the photoproduction amplitudes, or to possible contributions of the high-energy region and the nonphysical region to the dispersion integrals or to the effect of $\pi - \pi$ interaction. The contribution of the latter to meson photoproduction amplitudes is of course taken into account within the framework of double dispersion relations for the photoproduction process [21, 22].

In the majority of present-day discussions of the effect of $\pi - \pi$ interaction on the meson photoproduction process, the treatment is confined to a consideration of the two-meson resonance intermediate state (ρ-meson) [24], which makes a contribution only to the isoscalar part of the photoproduction amplitude. It follows from the procedure of approximate reduction of double dispersion relations to unidimensional relations [25], which is applicable near threshold, that this contribution is closely connected with the contribution of the high-energy region and appears in the form of an additional term to the unidimensional dispersion relations (21):

$$\Delta T = \Lambda \left[\frac{\xi_2}{t - t_R} (tM_1 - M_2) - \frac{\xi_1}{t - t_R} M_4 \right], \qquad (24)$$

where $t_R = m_\rho^2 \simeq 30$, $\xi_2 = 0.275$, $\xi_1 = 1$, and the constant Λ characterizes the strength of the $\gamma - \pi - \rho$ interaction. This interaction affects not only the photoproduction of single π-mesons on nucleons, but also ρ-meson decay ($\rho \to \pi + \gamma$), the lifetime of the π^0-meson, the photoproduction of π-meson pairs, etc. [26, 27]. Hence, Λ is an important parameter in the physics of elementary particles.

Table 1. Differential Cross Sections (in 10^{-30} cm^2/ sr) of π^+-Meson Photoproduction in Hydrogen (in center-of-mass system)

E_γ, MeV	θ, deg									
	40	59	75	93	107	123	135	148	159	165
170	—	5.7 ±0.44	—	5.94±0.44	—	—	—	—	—	—
175	—	6.7 ±0.77	—	7.04±0.44	—	7.04±0.77	—	—	—	—
180	—	7.15±0.55	—	8.25±0.44	—	8.36±0.55	—	—	7.26±0.99	—
185	6.6±1.54	6.93±1.65	—	7.59±0.88	—	8.14±0.66	—	9.35±1.43	9.02±0.88	—
190	—	7.92±0.55	—	10.0 ±0.55	—	9.56±0.66	—	—	8.25±0.66	—
195	8.7±1.2	—	10.5±1.4	9.9 ±0.77	9.8±0.99	12.0 ±1.1	—	10.9 ±1.54	9.9 ±1.32	—
200	7.7±1.2	9.35±0.55	—	10.8 ±0.66	11.1±0.88	11.5 ±0.66	13.1±2.09	—	10.5 ±0.77	—
210	—	11.0±0.77	—	13.4 ±0.77	13.7±1.43	11.8 ±0.77	—	15.1 ±1.54	11.8 ±0.88	—
220	—	10.7±0.99	—	14.7 ±0.77	14.0±1.1	13.3 ±0.66	—	15.6 ±1.54	13.3 ±0.88	13.2±1.32
230	—	—	—	—	—	16.9 ±0.99	—	—	14.3 ±1.1	—

3. Review of Experimental Work

The first results of experiments on the photoproduction of positive pions on protons were published in 1950 [28]. This was followed by a series of experiments in various laboratories. In these experiments various techniques were used and developed. The aim of the experiments was to obtain qualitative information about π^+-meson photoproduction on protons. In Berkeley, Steinberger and Bishop [29], using fast and delayed coincidences from counters, measured the excitation curve at 90° in the laboratory system in the photon energy range 170-320 MeV.

White et al. [30] repeated and extended the experiment by measuring π^+-mesons in the photon energy range 200-300 MeV using thick nuclear emulsions and counters connected to a fast coincidence circuit.

At the Massachusetts Institute of Technology, Feld et al. [31] used nuclear emulsions to measure the excitation curves at 26° and 90° in the laboratory system in the energy range 210-305 MeV. James and Kraushaar [32] measured the excitation curve of reaction (I) at 90° in the laboratory system in the photon energy range 165-280 MeV. At Cornell University, Jenkins et al. [33] measured the differential cross section at six angles in the range 40-180° at photon energies 200, 235, and 265 MeV by means of thin-walled proportional counters connected to a coincidence circuit.

At Illinois, Bernardini and Goldwasser [34] used thick nuclear emulsions to measure the differential cross section of the reaction at several angles for photon energies 175, 185, and 195 MeV. Of great interest from the methodological viewpoint is the work of Liess et al. [35], who attempted to measure the total cross section for photoproduction near threshold by detecting positrons from the decay of the μ-mesons produced in their turn by the decay of the pions in an absorber surrounding a liquid hydrogen target. In addition, Liess and Robinson [36] measured the differential cross section in the range 195-300 MeV at 90°, 135°, and 159° in the laboratory system with a telescope of counters.

At the California Institute of Technology, Tollestrup et al. [37] investigated reaction (I) in the photon energy range 200-400 MeV with a telescope of scintillation counters. They measured the ionization in relation to the residual range of pions. Walker et al. [38] measured the same cross sections in the same photon energy range by determining the energy and angle of emission of the pions by means of a magnetic spectrometer.

A fault of almost all these investigations is the scrappiness of the experimental data, their low statistical accuracy (in most investigations it was about 20%) and the uncoordinated nature of the experimental data, which were obtained at different times and in different laboratories.

The most accurate results in the photon energy range 170-220 MeV were obtained at four angles at Illinois by Beneventano et al. [39] in 1956. They are given in Table 1 along with the old data of Bernardini and Goldwasser [34] and the data obtained at Cornell University [33]. The values of the cross sections are increased by 10% in view of the change in calibration of the photon flux intensity, which Goldwasser reported at the Rochester conference [40].

The above experimental data were analyzed by means of phenomenological theory based on general principles of quantum mechanics and conservation laws. The bases of this phenomenological analysis were discussed in Sec. 1.

By analyzing the energy dependence of the cross section at low energies, Bernardini and Goldwasser [41] evaluated the pion—nucleon coupling constant and found that near threshold the main role is due to the photoproduction of pions in the S state.

Watson et al. [42] conducted a phenomenological analysis of the angular distributions of positive pion photoproduction in hydrogen in the photon energy range 200-470 MeV. Beneventano et al. [39] carried out a similar analysis in the range 170-265 MeV.

The result of the analysis showed the presence of a large S wave in the photoproduction of positive pions in hydrogen near threshold and a P-wave resonance at higher energies. In the analysis of the angular distributions, however, the terms due to direct interaction of the electromagnetic wave with the pion current (the so-called delay terms) were ignored. The analysis was based on the following form of angular distribution:

$$\frac{d\sigma}{d\Omega}(\theta) = A_0 + B\cos\theta + C\cos^2\theta, \tag{25}$$

but if the delay terms had been taken into account it should have been based on formula (18).

Unfortunately, an analysis of the angular distributions in formula (18) for low photon energies has been impossible until now owing to the absence of accurate data in a wide range of angles.

The next stage in the development of experiments on charged pion photoproduction involved theoretical investigations based on dispersion relations.

As expression (18) shows, the direct photoeffect for charged pions should be most pronounced at low emission angles. However, an investigation of the cross sections for pion photoproduction at small angles is very difficult owing to the high background of electrons and photons. Nevertheless, in 1958, Malmberg and Robinson [43] at Illinois, and Lazarus et al. [44] st Stanford University, managed to overcome the difficulties and established the existence of a direct photoeffect for charged pions. These investigations were the first in which the results were compared with calculations based on dispersion relations.

Using a magnetic spectrometer Malmberg and Robinson measured the angular distribution of positive pions produced in hydrogen by 225-MeV protons in the angle range 0-90°. The shape of the angular distribution was in good agreement with that predicted from dispersion relations [5]. No absolute measurements were made in this experiment.

Lazarus et al. used a magnetic spectrometer to investigate π^+-meson photoproduction in the angle range $7° \le \theta_{c.m.} \le 27°$ and in the photon energy range 220-390 MeV. The experimental results were compared with calculations based on the dispersion theory of Chew et al. [5]. The experiment agreed satisfactorily with the calculations and the agreement was better when the "delay term" was taken into account.

Direct interaction of the photons with the pion current affects the energy dependence of the square of the photoproduction matrix element near the photoproduction threshold. This is of great importance for the obtention of the threshold values of the square of the amplitude by extrapolation of the energy dependence to the threshold. In view of this, experiments aimed at measurement of the photoproduction cross sections near threshold have been carried out in many laboratories. Experiments near threshold are very laborious owing to the smallness of the cross sections, the high energy loss of the particles in the target itself, and the high background. The general effect [39, 45-47] established a satisfactory agreement between the theoretical and experimental cross sections for photoproduction at an angle close to 90°, although the statistical errors were large. A similar result was obtained in Glasgow [48] for an angle of 58° in the laboratory system. The dependence of the differential cross section on the photon energy near threshold was investigated in FIAN (Physics Institute of the Academy of Sciences) [49]. The results will be given in the later sections. McPherson and Kenney [50] investigated the dependence of the total cross section for π^+-meson photoproduction in hydrogen on photon energy with a hydrogen bubble chamber. The results of their investigation are given in Table 2, which also gives our data for comparison. It should be noted that McPherson and Kenney's data for photon energies 172, 177, and 182 MeV may be underestimates, since at these energies the detection efficiency of the bubble chamber for pions at certain angles is lower.

A comparison of old experimental data for the angular distributions [39, 43] and new data for energies 260 and 290 MeV [51] with the results of dispersion theory showed that the agreement between experiment and theory was not so good as might have been expected. For angles from 0 to 90° there was practically no disagreement between theory and experiment. For angles close to 180° the experimental points did not agree with the theoretical results. A similar situation has been observed for π^0-mesons [17]. As Baldin [18] pointed out, this might be due to the fact that the nonphysical region and the region of very high energies make a contribution to the dispersion integral. At his suggestion a series of experiments was carried out to measure the energy dependence of the cross section at angles satisfying the relationship

$$k\omega - kq\cos\theta = k_{thr} = 0.93, \tag{26}$$

Table 2. Cross Sections for π^+-Meson Photoproduction Near Threshold

E_γ, MeV	$\sigma/4\pi \cdot 10^{30}$, cm^2/sr *	E_γ, MeV	$\frac{d\sigma}{d\Omega_{c.m.}} \cdot 10^{30}$, cm^2/sr $(\theta_{lab} = 30°)$	$\frac{d\sigma}{d\Omega} \cdot 10^{30}$, cm^2/sr $(\theta_{lab} = 120°)$
154	2.71±0.76	153.4	3.28±0.55	—
157	3.48±0.12	155.7	4.0 ±0.76	3.65±0.65
162	4.62±0.14	157.6	4.4 ±0.79	—
167	6.07±0.22	159.3	4.1 ±0.78	4.5 ±1.8
172	6.47±0.31	160.8	4.0 ±0.99	—
177	5.74±0.43			
182	6.37±0.71			

Results of [50].

Table 3. Cross Sections $d\sigma/d\Omega$ for Fixed Value $k\omega - kq\cos\theta = 0.93$

θ_{lab}, deg	E_γ, MeV	E_π, MeV	$d\sigma/d\Omega \cdot 10^{30}$, cm^2/sr
43.53	187.4	37.2	6.93±0.41
44.97	182.1	31.9	6.75±0.40
46.68	175.9	25.7	6.41±0.38
48.55	168.9	18.7	5.82±0.35
49.55	164.0	13.9	5.20±0.31

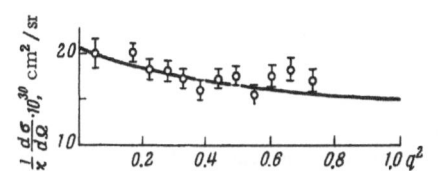

Fig. 1. $(1/\chi)(d\sigma/d\Omega)$ as a function of q^2 for angle $\theta =$ arccos$[(0.93 - k\omega)/kq]$.

where the contribution of the nonphysical region is zero and the contribution due to the high-energy region is constant.

In 1961, the results of experiments carried out in the P. N. Lebedev Physical Institute were published [52]. The differential cross section for π^+-meson photoproduction was obtained in the photon range 165-210 MeV and at angles satisfying relationship (26).

Agreement between the experimental data and theory was established and by extrapolation of the empirical dependence on the pion momentum to threshold the threshold value of the photoproduction amplitude was obtained. The results of this work are given in Fig. 1.

In 1962, Walker and Burk [53] conducted similar measurements with counters in the range 164-187 MeV and arrived at similar conclusions. The results of their measurements are given in Table 3.

In 1963, Bazin and Pine [54], at Stanford University, used a magnetic pulse analyzer and a telescope of scintillation counters to measure the cross section for π^+-meson photoproduction in the range 162-184 MeV at angles satisfying relationship (26). The authors of this work also found agreement between the experimental results and theoretical predictions [5, 55]. The same paper reported the measurement of the angular distributions of pions at 184 MeV. All the results obtained in this work are given in Table 4.

In 1963, in Bonn, Althoff [56] obtained the differential cross section along the line (26) in the range 200-260 MeV with a magnetic spectrometer and found good agreement with the theoretical predictions of Chew et al. [5] and Robinson [55].

Finally, in Glasgow, Patrick et al. [57] measured the π^+-meson yields in hydrogen at angles satisfying relationship (26) in the energy range 165-220 MeV.

Solov'ev et al. [23] showed that the contribution of the nonphysical region is insignificant in the energy range from the threshold to 250 MeV, but at high energies this contribution may reach 10-20%. Hence, more

Table 4. Differential Cross Sections for π^+-Meson Photoproduction in Hydrogen

E_γ, MeV	$\theta_{c.m.}$, deg	$d\sigma/d\Omega \cdot 10^{30}$, cm²/sr	Absolute error, %	Relative error, %	E_γ, MeV	$\theta_{c.m.}$, deg	$d\sigma/d\Omega \cdot 10^{30}$, cm²/sr	Absolute error, %	Relative error, %
162	70	5.16	6.6	5.1	180	105	7.85	5.2	3.1
165	90	5.81	6.9	5.5	180	130	8.43	6.3	4.7
165	67	5.86	6.6	5.1	180	140	8.45	5.9	4.1
167	90	6.25	6.6	5.1	180	150	7.83	6.2	4.6
170	65	6.32	6.4	4.8	184	58	6.82	5.7	3.9
170	90	6.32	7.4	6.1	194	90	8.71	5.6	3.7
170	110	6.78	6.9	5.5	194	120	9.64	6.5	5.0
180	35	6.78	6.5	50	194	150	9.80	6.3	4.7
180	60	6.84	5.7	3.9	225	150	12.35	5.5	3.8
180	90	7.41	7.0	5.6					

Table 5. Differential Cross Sections (in 10^{-30} cm²/sr) for π^+-Meson Photoproduction at E_γ = 185 and 187 MeV

θ, deg	E_γ=185 MeV	E_γ=185MeV [53]	E_γ=187 MeV [48]	θ, deg	E_γ=185 MeV	E_γ=185MeV [53]	E_γ=187 MeV [48]
15	7.54±0.41			95		8.18±0.15	
34	7.0±0.43			111.6			8.36±0.61
43	7.20±0.32			115	9.36±0.43		
45		6.53±0.83		125		8.17±0.95	
51			7.49±0.47	134	9.90±0.5		
69	7.44±0.40			146.5			9.54±0.61
70		7.43±0.85		150	10.17±0.56		
83.5			8.10±0.57	155		10.45±1.12	
93	8.24±0.38			165	10.97±0.58		

careful measurement of the angular distributions at low photon energies were carried out to assess the divergence from theory and to discover the nature of this divergence.

The angular distribution of π^+-mesons at photon energy 185 MeV has been measured in FIAN [58], Glasgow [59], Stanford [53], and Illinois [39]. The results are in good agreement with one another. They are shown in Table 5 and Fig. 2, which also shows the curve calculated from dispersion relations. Comparison with theory shows that at angles close to 180° the experimental points lie above the theoretical curve.

The following hypotheses on the nature of this disagreement can be put forward: It is either due to $\gamma - \pi - \rho$ interaction on charged pion photoproduction, or to the neglected contributions of small nonresonance multipoles and the effect of the contributions of multipoles for which the phases are not determined well by experiment, or to the effect of experimental corrections which may be a function of the pion emission angle.

If the disagreement is due to the effect of $\gamma - \pi - \rho$ interaction it is possible in principle to determine the coupling constant for this interaction. Such a program was carried out in FIAN [58] from coarse estimates of the coupling constant $\Lambda \simeq e\bar{f}$. This was obtained, however, by comparison with Ball's calculations [21], which contained some inaccuracies and several neglected multipoles. Hence, the estimate of Λ requires more systematic calculations to exclude the second hypothesis to explain the divergence.

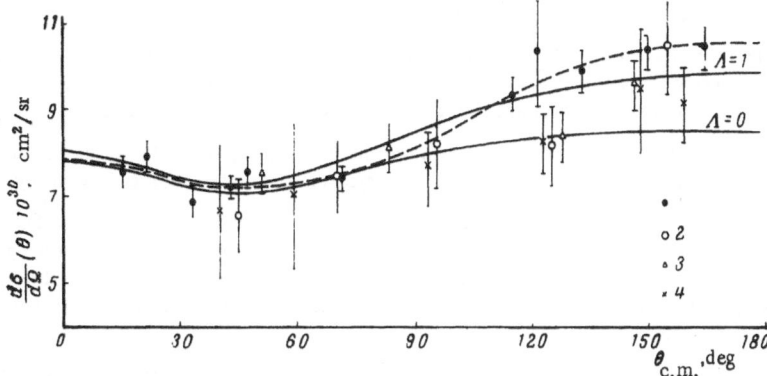

Fig. 2. Angular distribution of π^0-mesons at $E_\gamma = 185$ MeV. Data of: (1) [58]; (2) [59]; (3) [53]; (4) [39].

Table 6. Differential Cross Sections (in 10^{-30} cm^2/sr) for Fixed Meson Energy [61]

E_π, MeV	θ, deg							
	29.5	39.8	54.8	71.3	88.8	106.9	125.3	146.1
33.8	7.7 ±0.23	7.67 ±0.23	7.33 ±0.22	7.76 ±0.23	8.81 ±0.26	9.81 ±0.29	11.44 ±0.34	12.36 ±0.37

Table 7. Differential Cross Sections (in 10^{-30} cm^2/sr) for Fixed Meson Energies [62]

E_π, MeV	θ, deg				
	45	65	90	110	135
21.3	6.73±0.34	7.30±0.31	8.41±0.48	9.56±0.43	9.11±0.53
33.8	7.46±0.66	7.63±0.43	9.38±0.88	10.40±1.09	11.38±1.32
68.5	7.25±0.77	11.29±0.64	15.49±1.33	18.10+1.77	18.38±1.44

To rule out the possibility of experimental errors, Robinson et al. [60] at Illinois, conducted an experiment on pions of the same energy (44.3 MeV) at different angles. This excluded errors due to nuclear interaction of the pions. The pion momentum was isolated by a magnetic spectrometer. Unfortunately, the authors did not give absolute measurements. The experimental data were normalized to the theoretical at 60°. A comparison with McKinley's calculations gave the following result: $\Lambda = -1.2$ ef. This startling result led to another two similar experiments.

At Purdue University, Walker et al. [61] conducted an experiment with a magnetic spectrometer, which isolated pions of energy 33.8 MeV. Measurements were made at eight angles from 30 to 146°.

At Glasgow, Leith et al. [62] conducted measurements of the differential cross sections for three fixed pion energies — 21.3, 33.8, and 68.5 MeV.

The results of the two experiments, given in Tables 6 and 7, agree with one another, but the coupling constant depends on what theoretical calculations the experimental data are compared with.

In view of this, it was necessary to conduct systematic theoretical calculations and to obtain the experimental value of the differential cross sections for all possible angles and energies near threshold. This was the aim of our work, the results of which are given in the following sections.

4. Experimental Setup and Method

Photoproduction of π^+-Mesons in the Photon Energy Range 165-230 MeV. To investigate π^+-meson production in hydrogen in the photon energy range 165-230 MeV, we used liquid hydrogen as a target. The working part of the target was a brass cylinder 50 mm in diameter installed in a vacuum chamber.

The vacuum chamber had special aluminum windows 235 μ thick, which allowed the detection of mesons at sixteen angles to the direction of the photon flux. A diagram of the experiment is shown in Fig. 3.

The beam of photons with maximum energy 265 MeV from the FIAN synchrotron passed through a system of lead collimators and a scrubbing magnet and fell on the hydrogen target. The intensity of the photon beam was measured with a quantameter. Outside the vacuum chamber stacks of NIKFI BK-400 emulsions were mounted on special holders in such a way that the mesons entered them from the end. The emulsions were shielded from the scattered background by lead blocks.

The mesons were detected at eleven angles. The dimensions of the emulsions were chosen so that the whole meson energy range of interest was covered at each angle. The stacks consisted on the average of 20 emulsion pellicles, which, after marking and cementing to glass, were developed in the usual way. The emulsions were examined under MBI-1 and MBI-2 microscopes at a magnification of 20 × 7 × 1.5.

All cases of $\pi - \mu$ decays and for an additional check, tracks of μ- and π^+-mesons close to possible stoppages, were recorded. The tracks of μ- and π^+-mesons were followed into the neighboring pellicles and we were thus able to identify the $\pi - \mu$ decays definitely, to exclude the change π^+-meson background, and to check the examination of adjacent pellicles.

We also recorded the π^--meson tracks terminating in a star with one prong or more, so that we could deduct the background due to π^+-mesons formed in the target walls by using the known π^-/π^+ ratio for compound nuclei.

For each of the selected π^+-mesons we determined the energy from the residual range in the emulsion. The inaccuracy in the determination of the meson energy due to scattering did not exceed 0.1 MeV. We allowed for energy losses in the liquid hydrogen, in the walls of the brass container, and in the aluminum windows of the vacuum chamber. They amounted to 0.1-1 MeV, depending on the meson energy.

The intervals of kinetic energy of the pions were chosen so that we could obtain the meson photoproduction cross sections at photon energy intervals of 5 MeV. For each interval we introduced corrections for interaction of mesons with the emulsion nuclei and for meson decay in flight. As Fig. 4 shows, the total correction led from 6 to 19%, depending on the meson energy.

Loss of mesons due to multiple scattering was excluded by placing the stacks between two emulsion blocks during irradiation. The result of this was that the number of mesons leaving the working stack was balanced by the mesons scattered in the emulsion blocks. The background due to π^+-mesons produced in the target walls was 1.5%. The statistical accuracy in determining the cross section for meson photoproduction in each interval was 5% on the average.

The photon energy flux was measured by the cascade curve method with a quantameter. A quantameter is a set of copper plates separated by gas gaps of different width. Measurement of the current between the plates due to ionization of the gas in the gaps allows a determination of the total intensity of the photon flux. The main inaccuracy in this method, due to determination of the instrument constant, was 4.5%. The current was measured with an accuracy of 1%. Thus, the total inaccuracy in absolute measurement of the photon beam intensity was 4.6%.

Meson Photoproduction at Photon Energies Less Than 165 MeV [46, 49]. The photoproduction of mesons near threshold is of special interest. The obtention of reliable experimental data with good statistics is very difficult. In this energy region the photoproduction cross section and, hence, the meson yield, are very small and, in addition, for low-energy pions there are significant losses in the target and in its walls. Hence, liquid-hydrogen and gaseous targets cannot be used.

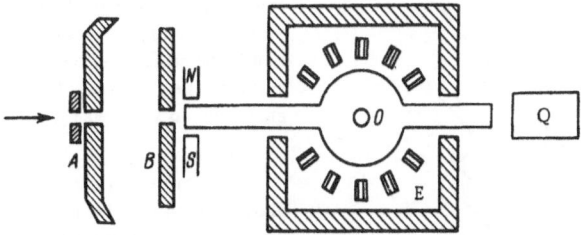

Fig. 3. Diagram of apparatus with liquid-hydrogen target. A,B — collimators; NS — scrubbing magnet; O — target; E — emulsion stacks; Q — quantameter.

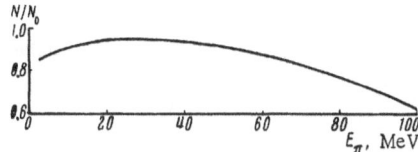

Fig. 4. Reduction of number of π^+-mesons due to nuclear interaction with emulsion and decay in flight.

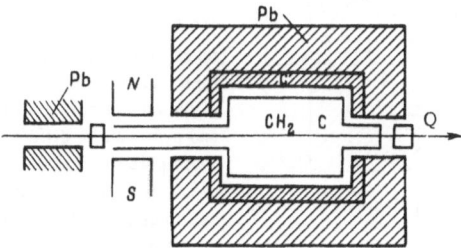

Fig. 5. Diagram of apparatus with CH_2 and C targets.

The use of thin polyethylene films removes the difficulties due to loss of mesons in the target, but the need to subtract the effects due to CH_2 and C in the case of a large pion yield from carbon greatly reduces the statistical accuracy of the results.

However, the Coulomb field of the nucleus greatly reduces the probability of production of low-energy π^+-mesons on carbon and the background from carbon can be small.

We carried out measurements of the cross section for meson photoproduction in hydrogen using a polyethylene target for two photon energy intervals: 152.9-158.3 and 158.3-161 MeV. A diagram of the experiment is shown in Fig. 5.

Mesons produced in polyethylene and carbon targets by the photon beam from the FIAN synchrotron with maximum energy 265 MeV were detected by mesons of NIKFI-K photographic plates. The geometry of the experiment was such that mesons with energy from 0.5 to 6 MeV making angles of 60-120° with the photon beam were detected in the plates. The thickness of the targets was chosen so that the meson energy loss could be neglected. The developed plates were scanned twice. Owing to the high grain background in the emulsions, which made it difficult to identify decays, we selected only $\pi - \mu$ decays with a μ-meson track ending in the

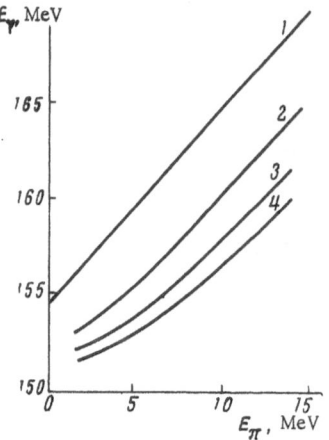

Fig. 6. Kinematics of π^+-meson photoproduction in carbon (1) and hydrogen (curves 2, 3, and 4 are for angles 50°, 30°, and 10°, respectively).

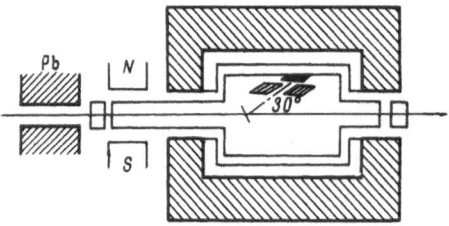

Fig. 7. Diagram of apparatus with polyethylene target.

emulsion. A geometrical correction was introduced for the remaining π^+-mesons. To find the cross section in the center-of-mass system we grouped all the events in energy and angular intervals. The background due to π^+-mesons from carbon was 26% on the average.

The low statistical accuracy of the experimental data was largely due to the use of the subtraction method. However, there are experimental conditions in which the π^+-meson yield from carbon can be practically suppressed. The method is based on the difference in the kinematic characteristics of photoproduction in hydrogen and carbon, including the difference in the thresholds of these reactions. The lowest thresholds for π^+-meson photoproduction in carbon are found in the following reactions:

$$\gamma + C_6^{12} \rightarrow \pi^+ + B_5^{12}, \qquad (27a)$$

$$\gamma + C_6^{12} \rightarrow \pi^+ + n + B_5^{11}, \qquad (27b)$$

$$\gamma + C_6^{12} \rightarrow \pi^+ + n + n + B_5^{10}. \qquad (27c)$$

The thresholds of these reactions are 154.5, 157.7, and 169.1 MeV, respectively. The threshold for π^+-meson photoproduction in hydrogen is 151.3 MeV. Figure 6 shows the photon energy as a function of the energy of mesons produced in carbon (curve 1) and hydrogen for different meson emission angles. The figure shows that for a fixed energy and small meson emission angle it is possible to select a maximum energy of the photon spectrum at which photomeson production in carbon will either not take place at all or will make up an insignificant part of the measured effect. The choice of experimental conditions is facilitated by the fact that meson photoproduction is a one-nucleon process and, hence, the reaction (27a) is a special and unlikely variant of reaction (27b). This allows an increase in the maximum energy of the spectrum so that a sufficient yield of mesons from hydrogen and a low yield of mesons from carbon can be obtained. For measurement of the cross section for photoproduction at an angle of 30° in the energy range 152-162 MeV, the vacuum chamber was placed in a beam of photons with maximum energy E_γ = 175 MeV. The chamber contained a polyethylene target, 200 μ thick, mounted in a special holder. Mesons with energy 1.5-16 MeV, emitted at 30° to the direction of the photons, were detected. The meson detector was a stack of NIKFI-K 400 emulsions. The photon flux was measured with a thick-walled graphite chamber calibrated against the quantameter. The geometry of the experiment (Fig. 7) was such that the stack detected mesons in the angle range 25-36° limited by a lead channel. Since the target was perpendicular to the direction of motion of the mesons, the energy loss for mesons with energy E_π = 3 MeV was 0.1 MeV. The energy intervals of the mesons detected by each layer of emulsion are shown in Fig. 8, where the heavy lines indicate the boundaries of the energy and angle intervals cut off by the plates.

In the determination of the pion energy we allowed for the energy loss in the target and the black paper in which the stacks were wrapped.

Table 8. Differential Cross Sections (in 10^{-30} cm²/sr) for Photoproduction of Positive Pions in Hydrogen in Center-of-Mass System for Various Directions of Emission of Pions in Laboratory System

E_γ, MeV	θ, deg										
	16	24	36	44	56	76	96	104	116	136	156
165	6.65±0.54	6.58±0.45	7.04±0.31		5.71±0.31	5.46±0.26	6.44±0.62				
170	6.05±0.29	5.81±0.39	6.55±0.32	5.92±0.22	6.71±0.35	6.89±0.24	6.82±0.40				
175	7.75±0.34	6.64±0.45	6.89±0.34	5.84±0.27	7.02±0.37	7.94±0.32	6.02±0.28	6.69±0.94	8.64±0.64		
180	7.80±0.39	7.46±0.47	7.33±0.36	6.62±0.30	7.57±0.40	7.33±0.31	7.44±0.42	10.63±1.19	8.88±0.65	8.51±0.39	
185	7.87±0.38	6.58±0.48	7.59±0.39	6.46±0.31	7.33±0.40	8.52±0.34	8.15±0.45	9.14±1.12	8.85±0.65	10.00±0.63	9.74±0.65
190	6.79±0.37	7.24±0.52	7.58±0.40	7.61±0.34	8.38±0.46	8.81±0.36	9.86±0.51	9.91±1.19	8.82±0.66	10.40±0.62	10.04±0.66
195	7.78±0.42	7.07±0.51	7.56±0.42	7.52±0.35	7.62±0.44	9.41±0.38	9.15±0.51	9.99±1.21	10.77±0.74	9.20±0.58	10.06±0.66
200	7.86±0.44	7.70±0.58	8.01±0.45	7.25±0.36	8.62±0.49	10.06±0.40	9.38±0.51	11.56±1.33	11.03±0.81	10.26±0.62	10.98±0.70
205	7.29±0.44	7.09±0.58	8.16±0.47	8.14±0.40	8.77±0.51	10.53±0.42	10.42±0.55	10.98±1.32	9.99±0.73	11.09±0.66	9.68±0.67
210	6.13±0.43	7.08±0.59	6.32±0.44	7.87±0.41	9.13±0.54	11.07±0.45	11.31±0.59	10.96±1.35	11.28±0.81	10.89±0.66	9.86±0.68
215	7.28±0.49	6.91±0.61	7.74±0.51	8.54±0.45	9.65±0.58	10.80±0.45	11.03±0.59	11.96±1.45	12.48±0.86	11.40±0.69	10.64±0.72
220	5.91±0.38	6.49±0.65	7.94±0.54	8.43±0.44	10.80±0.64	11.84±0.50	13.10±0.67	10.18±1.37	11.57±0.84	11.39±0.73	11.93±0.78
225		6.72±0.67	8.02±0.53	7.49±0.84	9.17±0.61	12.04±0.96	13.37±0.70	9.85±1.38	10.45±0.82	12.88±0.74	11.75±0.78
230		6.09±1.37	8.13±0.52	10.18±1.04	8.80±0.63		14.31±0.74	11.80±1.55	12.80±0.93	11.51±0.74	11.56±0.79

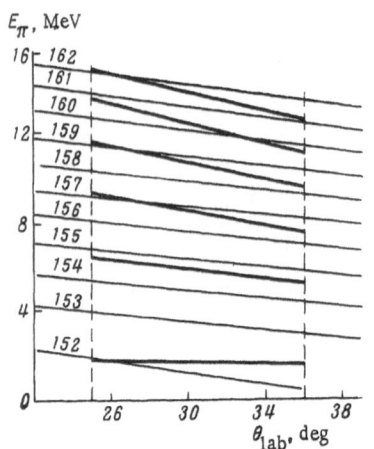

Fig. 8. Intervals of energies and angles cut off by emulsion stacks. The figures beside the curves give the photon energies in MeV.

The scanning and recording of the mesons were done in the usual way. The working area of emulsion was divided into parts, for each of which the permissible interval of angles of the detected mesons was determined. Each discovered meson was traced to its point of entry into the emulsion. At the point of discovery we measured the glancing angle, equal to the angle between the direction of the meson and the photon beam. The correspondence between the measured angle and the permissible interval of angles for the particular part was then established. This procedure excluded the contribution of pions scattered on the walls of the lead channel. The contribution of π^+-mesons from carbon was estimated from the number of π^- mesons detected and from the measured value of π^-/π^+ for carbon.

5. Experimental Results

The experimental determinations of the differential cross sections in the center-of-mass system in relation to the photon energy E_γ and the direction of emission of the pions in the laboratory system relative to the photon beam are given in Table 8.

Table 9. Values of χ^2 for Various Values of Constant Λ of $\gamma - \pi - \rho$ Interaction and Photon Beam Calibration Coefficient K

K \ Λ/ef	0	0.2	0.4	0.6
1.00	211	177	150	175
0.95	174	178	205	262
0.90	310	389	—	—

The results of determinations of the cross sections in experiments with the polyethylene target are given in Table 2. Only the statistical errors are indicated in the two tables. The absolute uncertainty was 5.5%. It was the same for all angles and energies.

A comparison of our tabulated data with the data of other authors [39, 50, 54], given in Tables 1, 2, 4, and 5, shows that the results of various laboratories agree well with one another within the indicated range of errors. An exception is the data obtained by Kirk and McElroy in Glasgow [63]. For small angles the cross section found by these authors is much smaller than the cross sections obtained in other laboratories.

The angular distributions of pions in the center-of-mass system are shown in Figs. 9 and 9a. The solid lines represent the results of numerical calculations from dispersion relations in the bipion approximation with various assumptions regarding the coupling constant Λ of $\gamma - \pi - \rho$ interaction. The following procedure was used to evaluate this constant. The deviation of each experimental point in the angular distribution from the theoretical curves calculated with different constants Λ was determined. This deviation was expressed in units of standard error at the given point and then from all the points the minimum value of χ^2 was determined. The values of χ^2 in relation to the constant Λ and the photon beam calibration coefficient K are given in Table 9. The calibration coefficient K varied within a range of twice the absolute uncertainty in calibration of the photon energy flux.

Figure 10 shows the dependence of $\varphi = \frac{1}{2}\sqrt{k/q}[\sqrt{(d\sigma/d\Omega)(163°)} - \sqrt{(d\sigma/d\Omega)(20°)}]$ on q^2. This value corresponds to the difference in the two photoproduction amplitudes for pions in the P state, and at angles 0 and 180° is exactly equal to the difference $F_{11} - F_{20}$. As the figure shows, the experimental points lie above the curve with $\Lambda = 0$. This discrepancy cannot be due to inaccuracy in the calibration of the photon flux, since the removal of the discrepancy would require alteration of the calibration by a factor of 1.5, which contradicts the accuracy of calibration. The same figure shows the curves with various values of Λ and a dashed curve obtained

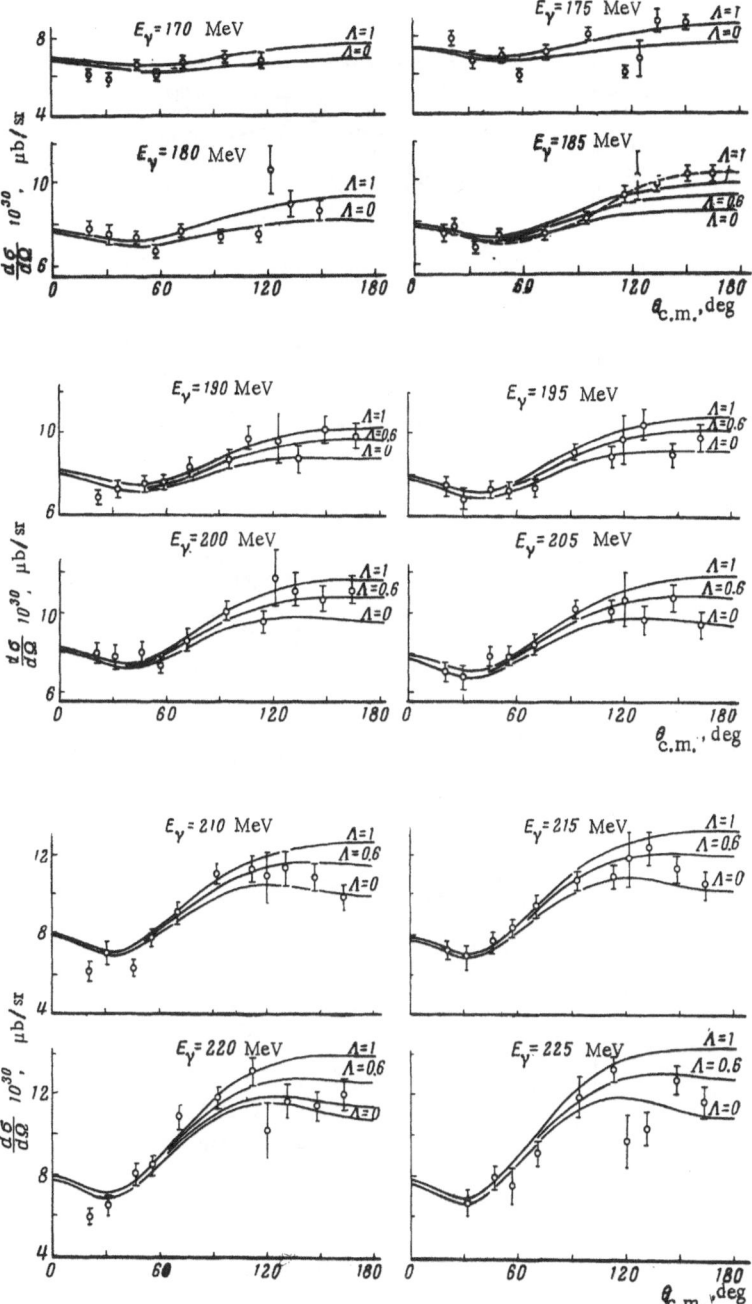

Fig. 9. Angular distribution of π^+-mesons for various photon energies.

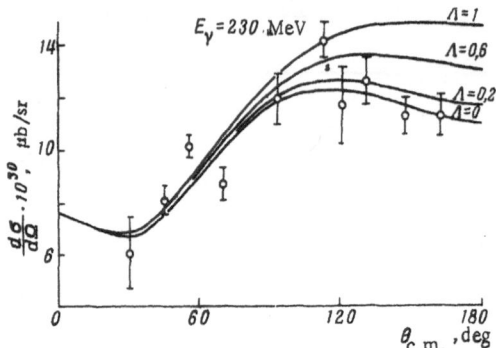

Fig. 9a. Angular distribution of π^+-mesons for photon energy
E_γ = 230 MeV.

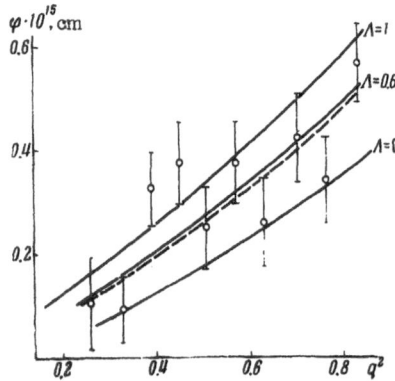

Fig. 10. φ as a function of q^2.

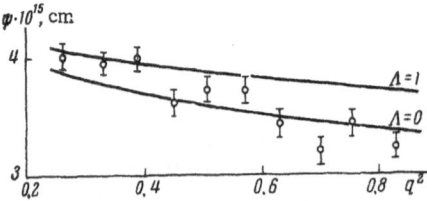

Fig. 11. ψ as a function of q^2.

by the method of least squares and satisfying the relationship $\mu = 0.589 \cdot 10^{-15}$ cm $\cdot$ q^2. It represents the best energy dependence according to the χ^2 criterion.

Figure 11 gives the value of $\psi = \frac{1}{2}\sqrt{k/q}[\sqrt{(d\sigma/d\Omega)(163°)} + \sqrt{(d\sigma/d\Omega)(20°)}]$ as a function of q^2. ψ corresponds to the sum of the photoproduction amplitudes for pions in the S and D states.

Figure 12 shows $(1/\chi)(d\sigma/d\Omega)$ as a function of q^2 for $E_\gamma = 152\text{-}162$ MeV. Here, $\chi = (q/k)(1 + \omega/M)^{-2}$. The solid line represents the same relationship calculated from unidimensional dispersion relations. The dashed line represents $(1/\chi)(d\sigma/d\Omega)$ as a function of q^2 obtained by the method of least squares according to theoretical law. The threshold value of $(1/\chi)(d\sigma/d\Omega)$ (for $q^2 = 0$) found in this way was $22.3 \pm 3.8\ 10^{-30}$ cm^2/sr. The same graph shows the bubble-chamber data obtained by McPherson and Kenney [50].

Figure 13 shows $(k/q)(d\sigma/d\Omega)(90°_{c.m.s.})$ as a function of q^2. The solid line is the curve obtained by the method of least squares

$$\frac{k}{q}\frac{d\sigma}{d\Omega}(90°) = 15.09 - 1.08\,q^2 + 3.92q^4.$$

The error matrix for this curve is

$$A^{-1} = \begin{pmatrix} 2.24 & -9.77 & 9.22 \\ & 47.98 & -48.21 \\ & & 50.70 \end{pmatrix} \sigma_0^2.$$

$\sigma_0 = \sqrt{s^2/(m-n)} = 0.667$; $\chi^2 = 12.40$, the expected $\chi^2_{exp} = m - n = 10$. The probability that χ^2 will exceed the obtained value 12.40 is $P(\chi^2 > 12.40) = 0.260$.

The threshold value of the square of the matrix element is

$$E_{0+}^2 = \frac{k}{q}\frac{d\sigma}{d\Omega}(90°) = 15.09 \pm 0.99.$$

Figure 13 also shows data obtained in other laboratories.

Figure 14 gives $(k/q)(d\sigma/d\Omega)(56°_{lab})$ at angle 56° in the laboratory system as a function of the square of the pion momentum in the center-of-mass system. The same figure shows the data of Lewis et al. [59], obtained in Glasgow with a telescope of scintillation counters. As the figure shows, the experimental data are in good agreement with one another.

To determine the energy dependence by the method of least squares, we used our experimental points in the range 165-210 MeV. We found that if the energy dependence is sought in the form $(k/q)(d\sigma/d\Omega) = a_0 + a_1 q + a_2 q^2 + a_3 q^3 + a_4 q^4 + \ldots$, series containing odd powers of q do not describe the experimental data well. The best fit to the experimental data is obtained when $(k/q)(d\sigma/d\Omega)$ is expanded in a series of even powers of the pion momentum q. Expansion to q^2 gives the following relationship:

$$\frac{k}{q}\frac{d\sigma}{d\Omega}(56°_{lab}) = (15.38 - 3.31\,q^2)\cdot 10^{-30},$$

and $\chi^2 = 6.063$ with the expected $\chi^2_{exp} = m - n = 8$, and

$$P(\chi^2 > 6.063) = 0.648.$$

The experimental points are best described by a polynomial

$$\frac{k}{q}\frac{d\sigma}{d\Omega}(56°_{lab}) = (16.99 - 12.36\,q^2 + 10.73\,q^4)\cdot 10^{-30},$$

and $\chi^2 = 4.678$ with the expected $\chi^2_{exp} = 7$. The probability that $\chi^2 > 4.678$ is 0.6986. This relationship is

M. I. ADAMOVICH ET AL.

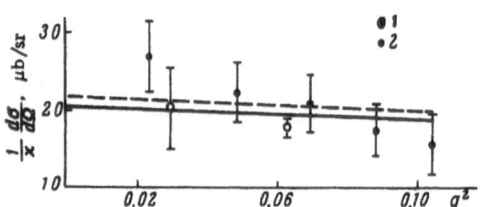

Fig. 12. $(1/\chi)(d\sigma/d\Omega)$ as a function of q^2. Data of: (1) McPherson and Kenney [50]; (2) the present authors.

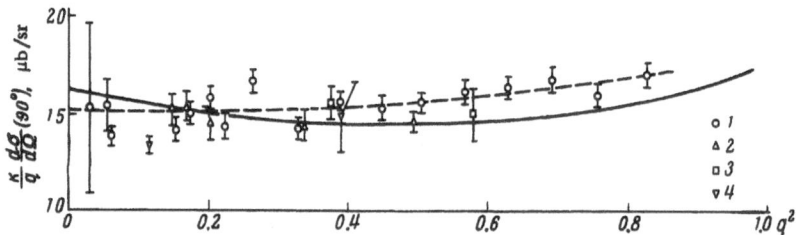

Fig. 13. $(k/q)(d\sigma/d\Omega)$ $(90°)$ as a function of q^2. Data of (1) [50]; (2) [54]; (3) [62]; (4) [53].

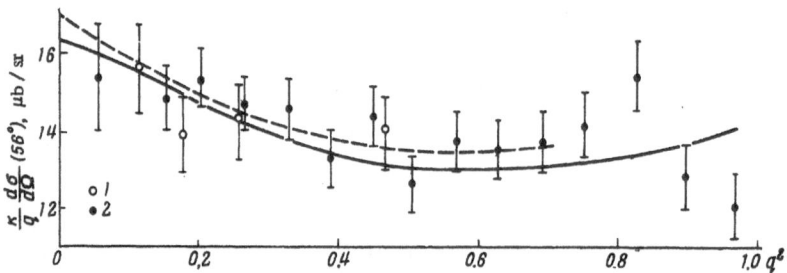

Fig. 14. $(k/q)(d\sigma/d\Omega)$ $(56°_{lab})$ as a function of q^2. Data of: (1) Lewis et al. [59]; (2) the present authors.

shown by the dotted line in Fig. 14. The error matrix for this case is

$$A^{-1} = \begin{pmatrix} 4.51 & -22.48 & 24.62 \\ & 121.36 & -138.99 \\ & & 164.68 \end{pmatrix} \sigma_0^2 \text{ with } \sigma_0 = \sqrt{\frac{s^2}{m-n}} = 0.5886.$$

Thus, the threshold value of the square of the matrix element

$$E_{0+}^2 = (16.99 \pm 1.25) \cdot 10^{-30} \text{ cm}^2.$$

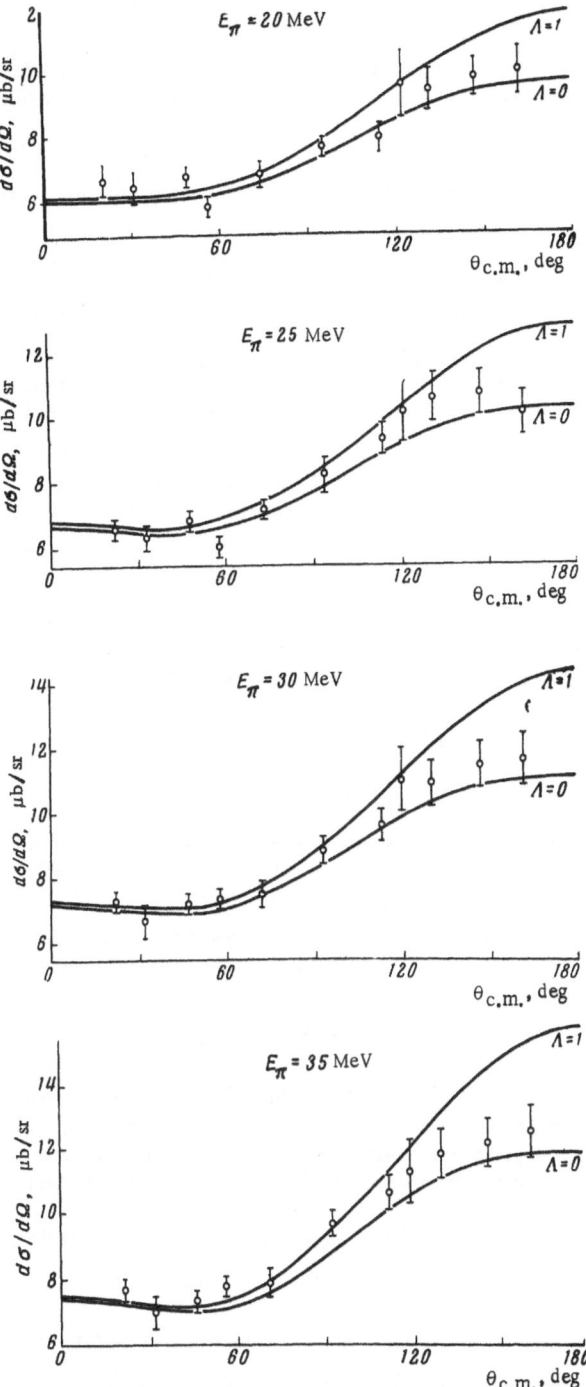

Fig. 15. $d\sigma/d\Omega$ as a function of $\theta_{c.m.}$ for various fixed π^+-meson energies.

Table 10. Values Obtained for Electric Dipole
Amplitude E_{0+} of Photoproduction of Positive Pions
at Threshold from Analysis of Data
for Different Angles

θ_{lab}, deg	$E_{0+} \cdot 10^{15}$, cm	$\theta_{c.m.}$, deg	$E_{0+} \cdot 10^{15}$, cm
76	3.89±0.13	90	4.15±0.18 *
56	4.11±0.15		4.07±0.13 †
36	4.10±0.15		3.92±0.10 ‡
24	4.00±0.07		
16	4.31±0.11		
30	4.10±0.35		

*Results of [54].

†The same at angle $\theta_{c.m.}$ satisfying relationship (26).

‡Born approximation calculated from $f^2 = 0.080 \pm 0.002$.

A similar treatment for angle 24° in the laboratory system gives the following energy dependence:

$$\frac{k}{q} \frac{d\sigma}{d\Omega} (24^\circ_{lab}) = (16.02 - 7.82q^2 + 0.06q^4) \cdot 10^{-30} \ \text{cm}^2/\text{sr}.$$

The error matrix for this angle is

$$A^{-1} = \begin{pmatrix} 0.557 & -1.095 & 0.314 \\ & 2.982 & -1.446 \\ & & 1.337 \end{pmatrix} \sigma_0^2,$$

and $\sigma_0 = 0.790$ and $\chi^2 = 9.75$ with the expected $\chi^2_{exp} = 11$, which corresponds to a probability $P(\chi^2 > 9.75) = 0.553$.

The threshold value of the square of the matrix element is:

$$E^2_{0+} = (16.02 \pm 0.59) \cdot 10^{-30} \ \text{cm}^2.$$

A summary of the threshold values of the electric dipole amplitude of pion photoproduction E_{0+} is given in Table 10.

Using the method of least squares, we determined the energy dependence of

$$\varphi = \frac{1}{2} \sqrt{\frac{k}{q}} \left[\sqrt{\frac{d\sigma}{d\Omega} (163^\circ)} - \sqrt{\frac{d\sigma}{d\Omega} (20^\circ)} \right],$$

which corresponds to the combination of amplitudes $F_{11} - F_{20}$.

The experimental data are best described by a relationship of the form $\varphi = 0.589q^2 \cdot 10^{-15}$ cm, and $\chi^2 = 7.056$ with the expected $\chi^2_{exp} = 7$. In Fig. 10 the empirical dependence of φ on q^2 is shown by a broken line. Thus, the threshold value $(F_{11} - F_{20})/q^2 = -(0.589 \pm 0.03) \cdot 10^{-15}$ cm.

Using the method of least squares, we also analyzed the angular distribution of π^+-mesons produced by 185-MeV photons. The angular dependence was determined in the form of an expansion of $(1 - \beta \cos\theta)^2 \cdot (d\sigma/d\Omega)(\theta)$ in powers of $\cos\theta$:

$$(1 - \beta \cos\theta)^2 \frac{d\sigma}{d\Omega}(\theta) = \Sigma a_j \cos^j \theta.$$

The best fit to the experimental data is obtained by the use of polynomials containing the cosine to the third or fifth power. It should be noted that the experimental points show a tendency to be described better by polynomials stopped at an odd power of the cosine than by polynomials with a last term containing a cosine to an even power. The best approximation of the experimental data has the following form:

$$10^{30} \cdot (1 - \beta \cos\theta)^2 \frac{d\sigma}{d\Omega}(\theta) = 8.09 - 11.05 \cos\theta + 6.30 \cos^2\theta - 1.00 \cos^3\theta - 1.37 \cos^4\theta + 0.79 \cos^5\theta.$$

From this we obtain the following values of $F_{11} - F_{20}$ and $F_{10} + F_{12}$ for photon energy 185 MeV:

$$F_{11} - F_{20} = -0.295 \cdot 10^{-15} \ \text{cm}, \quad F_{10} + F_{12} = 4.03 \cdot 10^{-15} \ \text{cm}.$$

In Figs. 2 and 9 the broken line represents the obtained angular dependence.

Figure 15 gives the relationship $d\sigma/d\Omega(\theta)$ for fixed pion energies 20, 25, 30, and 35 MeV. This allows a comparison of the experimental data with the theoretical calculations involving the same corrections for nuclear interaction of the photons for each energy.

6. Calculation of Differential Cross Sections of π^+-Meson Photoproduction

The calculation of the amplitudes and cross sections of the process $\gamma + p \rightarrow n + \pi^+$ was based on unidimensional dispersion relations. In the calculations of the real parts of the amplitudes, we took into account the contributions of the imaginary parts of the resonance and nonresonance S- and P-wave amplitudes to the dispersion integrals. To describe the polar parts of the amplitudes and kinematic functions in the dispersion integrals, we used their exact relativistic expressions. It is obvious that the value of the results of the analysis of the experimental data on meson photoproduction on the basis of the theory of dispersion relations depends significantly on how well we know the imaginary part of the photoproduction amplitudes in a wide energy range.

Imaginary Part of the Photoproduction Amplitude. In γ-ray region up to 1 BeV, the transitions $M_{1+}^{3/2}$, $E_{2-}^{1/2}$, and $E_{0+}^{(-)}$ make the main contributions to the photoproduction amplitude.

The transition to the state $P_{3/2, 3/2}$ is usually described by the relationship

$$M_{1+}^{3/2} = \frac{\mu_p - \mu_n}{2Mf} e \frac{e^{i\alpha_{33}}}{q} \sin \alpha_{33}, \tag{28}$$

where the resonance phase α_{33} is given by the effective length formula. Then the imaginary part of the amplitude $M_{1+}^{3/2}$ will have the form

$$\text{Im } M_{1+}^{3/2} = \frac{\mu_p - \mu_n}{2f} \frac{k}{q} \frac{f^2 q^2}{\omega} \frac{\frac{4}{3} f^2 \frac{q^3}{\omega}}{\left(1 - \frac{\omega}{\omega_2}\right)^2 + \left(\frac{4}{3} f^2 \frac{q^3}{\omega}\right)^2}. \tag{29}$$

In several works, the expression (29) is replaced by the approximation

$$\frac{\text{Im } M_{1+}^{3/2}}{kq} = a\delta(\omega - \omega_r), \quad a = 1.16. \tag{30}$$

However, as can easily be seen by direct integration of expressions (29) and (30), such an approximation leads to uncertainties of up to 20-30% in the values of the dispersion integrals and can be regarded only as a rough estimate.

Expressions (28) and (29) are suitable in the energy range up to the first resonance, but at high energies they give too high a value. In this energy range it is more correct to use the results of relativistic generalization of the effective length formula [21]. These, however, give too high values for the total cross sections of photoproduction processes in the region of the first resonance.

Hence, we will attempt, avoiding the uncertainties of theoretical approximations, to determine Im $M_{1+}^{3/2}$ from the experimental data on the photoproduction and scattering of π-mesons on nucleons. Since the total cross section of the process $\gamma + p \rightarrow p + \pi^0$, on one hand, is known in the region of the first resonance with good accuracy [64, 65] and, on the other hand, is expressed in terms of the squares of the multipole amplitudes, and estimates indicate a small (about 2%) contribution of nonresonance amplitudes in the region of the $P_{3/2, 3/2}$ resonance, $M_{1+}^{\pi^0}$ can be obtained from $\sigma_n(\gamma + p \rightarrow p + \pi^0)$ to an accuracy of a few percent,

$$M_{1+}^{\pi^0} = \sqrt{\frac{k}{q} \frac{\sigma_n(\gamma + p \rightarrow p + \pi^0)}{8\pi}}. \tag{31}$$

If we assume that the amplitude $M_{1+}^{\pi^0}$ is wholly due to the transition to the isotopic state with $T = {}^3/_2$, then, in accordance with formulas (10)-(12)

$$\text{Im } M_{1+}^{3/2} = \frac{3}{2} \sqrt{\frac{k}{q} \frac{\sigma(\pi^0)}{8\pi}} \sin \alpha_{33}. \tag{32}$$

The resonance $\pi - N$ scattering phase α_{33} is now known in the resonance region to an accuracy of 2-3% [65].

In the energy region $E_{\pi \, kin} > 250$ MeV, we will use the experimental evidence on the exponential decrease of $\sin \alpha_{33}$ [66].

The value of $\text{Im } M_{1+}^{3/2}$ obtained from relationship (32) is illustrated in Fig. 16 by the broken curve. If it is not assumed that $M_{1+}^{\pi^0} = \frac{2}{3} M_{1+}^{3/2}$, then in the determination of $M_{1+}^{3/2}$ from $\sigma_A^{\pi^0}$ the transitions to states with isotopic spin $T = \frac{1}{2}$ must be taken into account:

$$M_{1+}^{\pi^0} = \frac{2}{3} M_{1+}^{3/2} + \frac{1}{3} M_{1+}^{1/2} + \delta M_{1+}^{1/2}. \tag{33}$$

The nonresonance additions $M_{1+}^{1/2}$ and $\delta M_{1+}^{1/2}$ vary smoothly with energy and are involved in different ways in the π^+- and π^0-meson photoproduction cross sections. However, the insufficient accuracy of the experimental data does not enable us to find these amplitudes from the cross sections of the given processes. Hence, we calculate $M_{1+}^{1/2}$ and $\delta M_{1+}^{1/2}$ on the basis of static dispersion relations with the introduction of corrections $1/M$, and we take phase α_{31} from [67]. The inclusion of the nonresonance M_{1+} amplitudes insignificantly alters the maximum value of $\text{Im } M_{1+}^{3/2}$ and slightly shifts the maximum into the lower energy region. The results of calculations of $\text{Im } M_{1+}^{3/2}$ with due regard to the amplitudes $M_{1+}^{1/2}$ and $\delta M_{1+}^{1/2}$ are shown in Fig. 16 by the solid curve. The corrections are significant in the energy region where the amplitude $M_{1+}^{3/2}$ ceases to be dominant. Looking ahead, we mention that the results of calculations of the dispersion integrals which contain only the resonance amplitude are not greatly affected by the consideration of nonresonance amplitudes. In the calculations we considered the contribution of the second resonance of the $\pi - N$ system to the dispersion integrals. For the estimates we express $E_{2-}^{1/2}$ in terms of the total cross section of the process $\gamma + p \rightarrow p + \pi^0$

$$\frac{1}{3} E_{2-}^{1/2} = \sqrt{\frac{k}{q} \frac{1}{8\pi} \left(\sigma_n^{\pi^0} - \sigma_{P_{3/2, \, 3/2}} \right)}. \tag{34}$$

It is assumed here that E_{2-} is dominant in the region of the second resonance. In addition, from the total cross section we deduct the contribution due to the "tail" of the first resonance $P_{3/2, \, 3/2}$.

For the estimates we can assume that in the region of the second resonance there is the usual relationship between photoproduction and scattering: $E_{2-}^{1/2} = |E|_{2-}^{1/2} e^{i\delta_{3/2, 1/2}}$, where the phase of $\pi - N$ scattering in the D state $\delta_{j, T}$ in the region of the second resonance ($E_{\pi \, kin} \sim 600$ MeV) passes through 90° The interaction of the π-meson with the nucleon in the final state of the reaction enables us to estimate this phase from the relationship $\sin \delta_{3/2, 1/2} \simeq |E_{2-}^{1/2}| \, / \, |E_{2-}^{1/2}|_{max}$, whence we obtain the following expression:

$$\frac{1}{3} \text{Im } E_{2-}^{1/2} = \frac{\frac{k}{q} \frac{1}{8\pi} \left(\sigma_n^{\pi^0} - \sigma_{P_{3/2, \, 3/2}} \right)}{\left[\sqrt{\frac{k}{q} \frac{1}{8\pi} \left(\sigma_n^{\pi^0} - \sigma_{P_{3/2, \, 3/2}} \right)} \right]_{max}}. \tag{35}$$

The imaginary part of the amplitude $\text{Im } E_{2-}^{1/2}$, determined from the experimental data, is shown in Fig. 16 by a dot-dash curve.

Since the inaccuracy of the experimental data for the total cross sections of the process $\gamma + p \rightarrow p + \pi^0$, used in the calculations of the imaginary parts of the resonance amplitudes, does not exceed 2-3%, we can expect that the uncertainty in the values of the dispersion integrals of the imaginary resonance amplitudes does not exceed 5%. The main source of uncertainty in the dispersion integrals is the contribution of nonresonance S- and P-wave amplitudes. Unfortunately, the experimental data indicate only the smallness of these amplitudes

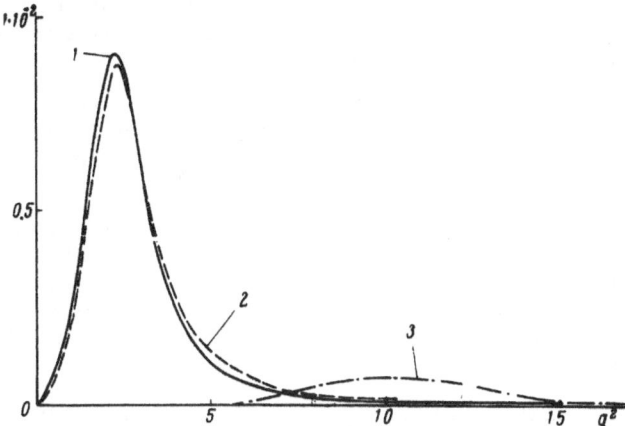

Fig. 16. Values of Im $M_{1+}^{3/2}$/kq and Im $E_{2-}^{1/2}$/kq obtained from total cross sections for π^0-meson photoproduction. (1) Amplitude obtained with all isotopic components included in $M_{1+}^{\pi^0}$; (2) on the assumption that $M_{1+}^{\pi^0}$ = $^2/_3 M_{1+}^{3/2}$; (3) amplitude $E_{2-}^{1/2}$.

and do not allow any definite conclusions regarding their size. In this connection, since dispersion theory gives a qualitative description of photoproduction processes, we calculate these amplitudes on the basis of static dispersion relations with corrections 1/M introduced [68].

We will calculate the dispersion integrals in the polar approximation for Im $M_{1+}^{3/2}$. We will assume that in the high-energy region the usual relationship between photoproduction processes and π-meson scattering is valid and we find the imaginary parts of the nonresonance amplitudes by using relationships (14).

To describe the small S- and P-wave $\pi-N$ scattering phases, we will use the results of Hamilton's analysis and will assume that at high energies all the phases tend to zero.

It should be noted that the real part of the amplitude Re $E_{1+}^{3/2}$ calculated in this way does not pass through zero in the region of the first resonance, which is a result of the approximations used. Hence, the imaginary part of the amplitude of the electric quadrupole transition was determined from the relationship

$$\text{Im } E_{1+}^{3/2} \simeq \text{Re } E_{1+}^{3/2} \sin \alpha_{33}.$$

Calculation of Amplitudes and Cross Sections. The contributions of the first and second resonances in the $\pi-N$ system and nonresonance S- and P-wave amplitudes to the dispersion integrals were calculated separately. In the calculation of the integrals we used the following formula for the principal value:

$$F(x_0) = P \int_a^b \frac{f(x)}{x_0 - x} \, dx = \int_a^b \frac{f(x) - f(x_0)}{x_0 - x} \, dx + f(x_0) \ln \frac{x_0 - a}{b - x_0}. \tag{36}$$

The nonsingular expression was integrated numerically by Simpson's formula with the region of integration divided into 80 intervals. The upper limit of integration was chosen as W' = 12.5, which corresponds to a γ-ray energy $E_\gamma \sim 1$ BeV. The inaccuracy due to the approximate integration and cutoff of the integral did not exceed 1%.

For the calculation of the polar (Born) parts of the amplitudes we used for the constant of $\pi-N$ interaction the value $f^2 = 0.08$, which agrees well with the data for $\pi-N$ and N-N scattering [69].

The cross sections were calculated from formula (10). The imaginary parts of the amplitudes were determined by the procedure described above. The differential cross sections and amplitudes were calculated for all the investigated γ-ray energies for meson emission angles θ_π from 0 to 180° at 15° intervals.

To assess the effect of the employed form of the imaginary part of the amplitude $M_{1+}^{3/2}$ on the calculated cross sections we conducted calculations for the two Im $M_{1+}^{3/2}$ curves given in Fig. 16 and for the amplitude used in [21]. We found that for γ-ray energies up to 230 MeV all three calculations gave results agreeing to within 2-3%.

Consideration of the second resonance in the dispersion integrals leads to a reduction of the π^+-meson photoproduction cross sections at large angles and energies. For instance, at 230 MeV, this reduction reaches 10-12% for angle 180°. We should also mention the high sensitivity of the S-wave amplitude of π^+-meson photoproduction to separate contributions of the imaginary parts of the multipole amplitudes $E_{0+}^{(-)}$ and $E_{1+}^{3/2}$. Each of them can introduce changes of up to 10% in the differential cross sections, but their combined contribution to the cross section is about 3%.

Effect of Resonance 2π Interaction of the Differential Cross Section of the Process $\gamma + p \rightarrow n + \pi^+$. To investigate the effect of resonance $\pi - \pi$ interaction (ρ-meson) on the differential cross section for charged meson photoproduction, and also to derive information about the constant Λ of $\gamma - \pi - \rho$ interaction, which characterizes the contribution of this interaction to the photoproduction amplitudes, we included the addition ΔT (24) in the calculations of the real parts of the amplitudes.

Expression (24), which characterizes the contribution of $\gamma - \pi - \rho$ interaction to the photoproduction amplitudes, was obtained in the so-called bipion approximation. However, the amplitude associated with the invariant form M_A can contain an additional deduction constant in the t channel. We will discuss the effect of $\pi - \pi$ interaction on the differential cross sections for photoproduction in the bipion approximation. As Figs. 2, 9-11, and 15 show, this effect is greatest for emission of mesons at large angles. With $\Lambda = 1$ ef, the differential cross sections at energy $E_\gamma = 170$-230 MeV for meson emission angles 0, 90, and 180° differed from the case of $\Lambda = 0$ by about 2, 7, and 15-30%, respectively.

Thus, $\gamma - \pi - \rho$ interaction has little effect on the differential cross section $d\sigma/d\Omega$ for $\theta_\pi = 90°$ and on the amplitude for meson production in the S state. It is obvious that in view of its peripheral character, the contribution of the ρ-meson will appear in the amplitudes for meson production in the higher states of orbital momentum l and will be responsible for the shape of the differential cross-section curve. On the other hand, the above-discussed deduction constant in the t channel can make a contribution mainly to the isotropic component of the cross section and has little effect on the shape of the curve.

The calculations were made for several values of Λ, varying in the range from -0.2 ef to $+1.0$ ef at steps of 0.2 ef. The results of these calculations were used in the next section for an analysis of the experimental data.

7. Discussion

The values of the electric dipole amplitude of π^+-meson photoproduction at threshold E_{0+}, obtained by extrapolation of the experimental values of $(k/q)(d\sigma/d\Omega)(\theta)$ to threshold, are given in units of 10^{-15} cm in Table 10. The weighted mean value is $E_{0+} = (4.07 \pm 0.05) \cdot 10^{-15}$ cm. The same table gives our results, obtained with a polyethylene target at 30° in the range 152-162 MeV, and the results obtained from an analysis of the data of Bazin and Pine [54]. In the comparison of the data of different investigations, a further inaccuracy of 2.5% due to uncertainty of the absolute calibration must be taken into account. For comparison with the experimental results, the table also gives the threshold value calculated in the Born approximation (polar term) on the basis of the known constant of pion−nucleon interaction $f^2 = 0.08 \pm 0.002$. As a comparison of the tabulated data shows, the results are in good agreement with the value of the polar term of the amplitude at threshold. Since the S-wave photoproduction amplitude for the process $\gamma + p \rightarrow \pi^+ + n$ has the form (neglecting the imaginary part)

$$E_{0+}^{\pi^+} = \sqrt{2}(E_{0+}^{(0)} + E_{0+}^{(-)}),$$

and for the process $\gamma + n \to \pi^- + p$

$$E_{0+}^{\pi^-} = \sqrt{2}(E_{0+}^{(0)} - E_{0+}^{(-)}),$$

it would be of interest to compare these two amplitudes to determine the isotopic structure of the scattering matrix. According to the results of [70], the electric dipole amplitude of π^--meson photoproduction on neutrons is $E_{0+}^{\pi^-} = (4.51 \pm 0.15) \cdot 10^{-15}$ cm. [*] Using this value and the weighted mean $E_{0+}^{\pi^+} = (4.07 \pm 0.049) \cdot 10^{-15}$ cm for π^+-mesons, we obtain for the isovector part of the pion photoproduction amplitude

$$E_{0+}^{(-)} = (3.03 \pm 0.064) \cdot 10^{-15} \text{ cm}$$

and for the isoscalar part

$$E_{0+}^{(0)} = -(0.156 \pm 0.064) \cdot 10^{-15} \text{ cm}.$$

The ratio of the cross sections for π^-- and π^+-meson photoproduction at threshold is $r = \sigma^-/\sigma^+ = 1.23 \pm 0.06$.

It is also of interest to compare the obtained results with recent data for the Panofsky ratio and the difference in scattering lengths $(\alpha_3 - \alpha_1)/q$. An analysis, carried out by Hamilton and Woolcock [13, 67], of the data for the scattering of negative and positive pions in hydrogen gave the following value for the difference in S phases of pion–nucleon scattering

$$\alpha_3 - \alpha_1 = (0.245 \pm 0.007)q.$$

The use of this value and the results of a recent determination of the Panofsky ratio $P = 1.53 \pm 0.02$ [71] gives the following value for the electric dipole amplitude for negative pion photoproduction:

$$E_{0+}^{\pi^-} = \frac{\alpha_3 - \alpha_1}{3} \sqrt{\frac{\beta_0}{P k_{\text{thr}}}} = (4.40 \pm 0.41) \cdot 10^{-15} \text{ cm},$$

where β_0 is the velocity of the π^0-meson produced as a result of π^- charge exchange on the proton; $k_{\text{thr}} = 0.927$ is the pion photoproduction threshold.

As can be seen, the main parameters of low-energy pion physics (E_{0+}, σ^-/σ^+, P, $\alpha_3 - \alpha_1$), which are connected by single-valued relationships, agree satisfactorily with one another. The effect of $\gamma - \pi - \rho$ interaction on the threshold photoproduction amplitude is very slight. The error in the isoscalar part of the electric dipole amplitude does not allow an estimate of the constant of $\gamma - \pi - \rho$ interaction on the basis of the threshold parameters alone. Moreover, there is considerable uncertainty in the theoretical estimates of the correction to the polar term for small nonresonance scattering phases.

The effect of $\gamma - \pi - \rho$ interaction can greatly affect the shape of the angular distributions of π^+-mesons. The shape of the angular distribution for $E_\gamma = 185$ MeV (see Figs. 2 and 9 for $E_\gamma = 185$ MeV), analyzed by the method of least squares, indicates that the shape of the angular distribution of π^+-mesons is described satisfactorily by the theoretical curve with $\Lambda \simeq 1$ ef. As Fig. 9 shows, the effect of $\gamma - \pi - \rho$ interaction is greater at large angles and is practically zero at angles less than 90°.

Figure 17 shows theoretical curves with $\Lambda = 0$, but with different assumptions regarding the contributions of different multipoles to the dispersion integrals. As this figure shows, the different contributions of multipoles do not alter the shape of the angular distribution but merely shift it parallel to itself. An exception is the contribution of the second resonance and $\gamma - \pi - \rho$ interaction to the dispersion integrals. Hence, it seems reason-

---

[*] The experimental value $(E_{0+}^{\pi^-})^2 = 20.9 \pm 2.1$, obtained by extrapolation to threshold, must be increased by 7.5% owing to the change in calibration.

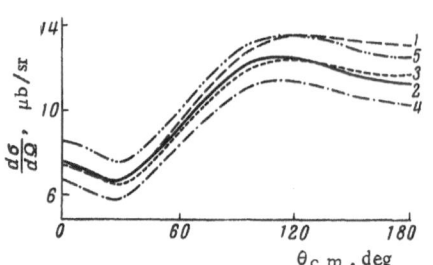

Fig. 17. Calculated angular distribution of π^+ mesons for energy E_γ = 230 MeV with due regard to the imaginary parts of the multipole amplitudes. (1) Resonance amplitude $M_{1+}^{3/2}$; (2) S- and P-wave amplitudes and quadrupole $E^{1/2}$; (3) $M_{1+}^{3/2}$ and $E_{2-}^{1/2}$; (4) the same as 2, except for E_{1+}; (5) all P-wave amplitudes and $E_{2-}^{1/2}$.

able to compare the theory and experiment in such a way that the various uncertainties are reduced to a minimum. The relationship between φ and the square of the pion momentum is most suitable for this purpose.

Figure 10 compares the experimental values of φ with the theoretical values. As this figure shows, the curve with Λ = 0 lies well below the set of experimental points. The curve with Λ = 1 ef lies slightly above. The best agreement with the experimental points, according to the χ^2 criterion, is obtained with Λ = 0.6 ef. The 95% confidence limits for Λ are 0.4 and 0.8 ef. It should be noted that the obtained result is practically independent of the contribution of the various small phases of pion−nucleon scattering. In addition, the deduction constant of the process $\gamma + \pi \rightarrow \pi + \pi \rightarrow N + \overline{N}$ has very little effect on the value of φ. However, the result depends greatly on the contribution of the second resonance to the dispersion integrals. For instance, if the second resonance is ignored, the experimental points are best described by a theoretical curve with Λ = 0.

The behavior of the relationship between ψ and q^2 in Fig. 11 is such that at low photon energies the experimental points lie closer to the curve with a high value of Λ, whereas at energies near 220 MeV they lie closer to the theoretical curve with Λ = 0. In the comparison of the experimental and theoretical values we might have expected a constant difference between the theoretical and experimental results at all energies, since ψ strongly depends on the calibration. In fact, however, we find that the behavior of the experimental curve with energy is different from that of the theoretical curve. This is probably due to the inadequacy of our knowledge of pion−nucleon scattering at high energies. This also emerges from a comparison with the theoretical angular distribution of pions. For instance, the angular distribution of pions at photon energy 185 MeV indicates that the constant Λ = 1.2 ef, whereas the data for φ gives Λ = (0.6 ± 0.2)ef in the energy range 185-220 MeV. This can hardly be due to experimental errors. We have already discussed the question of calibration. As regards other possible errors, the graphs in Fig. 15, which show the behavior of the differential cross sections with photon energy for a fixed pion energy, rule out this possibility. It appears that theory still does not give due regard to certain multipoles and pion−nucleon scattering phases. This may lead to variation of Λ with energy.

Hence, there is sense in comparing all the experimental data with the theoretical calculations. Table 9 sums up the results of this comparison. As the table shows, the best agreement between theory and experiment is observed when Λ = 0.4 ef and k = 1. It should be noted that the value χ^2 = 150 with expected χ^2 = 102, indicates that the agreement obtained is not the best. However, when the uncertainty in the absolute calibration and theoretical calculations is considered, the situation is much better. Further comparison of experiment with theory and the extraction of useful information regarding the role of various effects, will necessitate an increase in the accuracy of the experimental data and of the dispersion integrals in the theoretical calculations. It should be noted that the experimental accuracy now attainable is comparable with, and often even better than, the accuracy of theoretical estimates. Hence, as a whole, the photoproduction data can be regarded as valuable material providing a criterion for the selection of the best combinations of sets of pion−nucleon scattering phases and contributions of different multipoles.

In conclusion, the authors express their thanks to Corresponding Member of the Academy of Sciences of the USSR P. A. Cherenkov and Professor A. M. Baldin for their interest and valuable discussions, to the team operating the synchrotron for invaluable assistance, to R. Uvarova for carrying out the numerical calculations, and to

A. P. Belokopytova, R. G. Voropaeva, L. I. Ivanova, L. G. Kasparova, M. V. Kalugina, T. A. Lebedeva, S. V. Minina, and V. D. Orekhova for the processing and scanning of the emulsions.

LITERATURE CITED

1. K. M. Watson, Phys. Rev., 95: 228 (1954).
2. E. Fermi, Lectures on Pions and Nucleons [Russian translation], IL (1956).
3. M. Gell-Mann and K. M. Watson, Annual Rev. Nucl. Sci., 4: 219 (1954).
4. A. A. Logunov, A. N. Takhvelidze, and L. D. Solov'ev, Nucl. Phys., 4: 427 (1957).
5. G. F. Chew, M. L. Goldberger, F. E. Low, and Y. Nambu, Phys. Rev., 106: 1345 (1957).
6. R. E. Marshak, Phys. Rev., 82: 313 (1951).
7. H. L. Anderson and E. Fermi, Phys. Rev., 86: 794 (1952).
8. A. Baldin, Nuovo Cim., 8: 569 (1958).
9. M. I. Adamovich, É. G. Gorzhevskaya, V. G. Larionova, V. M. Popova, S. P. Kharlamov, and F. R. Yagudina, Zh. Eks. i Teor. Fiz., 38: 580 (1960).
10. G. Puppi, Proc. 1958 Internat. Conf. on High-Energy Physics, CERN. Geneva (1958), pp. 46-51.
11. M. Cini, R. Gatto, E. L. Goldwasser, and M. Ruderman, Nuovo Cim., 10: 234 (1958).
12. J. M. Cassels, Nuovo Cim. Suppl., 14: 259 (1959).
13. J. Hamilton and W. S. Woolcock, Phys. Rev., 118: 291 (1960).
14. M. I. Adamovich, É. G. Gorzhevskaya, V. G. Larionova, N. M. Panova, V. M. Popova, S. P. Kharlamov, and F. R. Yagudina, Zh. Eks. i Teor. Fiz., 41: 1811 (1961).
15. R. G. Vasil'kov, B. B. Govorkov, and V. I. Gol'danskii, Zh. Eks. i Teor. Fiz., 37: 11 (1959).
16. P. D. Luckey, L. S. Osborne, and J. J. Russell, Phys. Rev., 117: 1364 (1959).
17. A. M. Baldin and B. B. Govorkov, Dokl. Akad. Nauk SSSR, 127: 993 (1959).
18. A. M. Baldin, Zh. Eks. i Teor. Fiz., 38: 579 (1960).
19. L. Castillejo, R. H. Dalitz, and F. J. Dyson, Phys. Rev., 101: 453 (1956).
20. K. Dietz, G. Höhler, and A. Müllensiefen, Z. Phys., 159: 77 (1960).
21. J. S. Ball, Phys. Rev., 124: 2014 (1961).
22. L. D. Solov'ev, G. Belyavski, and A. Yurevich, Zh. Eks. i Teor. Fiz., 40: 839 (1961).
23. N. V. Demina, V. L. Evteev, V. A. Kovalenko, L. D. Solov'ev, R. A. Khrenova, and Ch'en Ts'ung-mo, Zh. Eks. i Teor. Fiz., 44: 272 (1963).
24. M. Gourdin, D. Lurie, and A. Martin, Nuovo Cim., 18: 933 (1960).
25. M. Cini and S. Fubini, Ann. Phys., 10: 352 (1960).
26. M. Kowaguchi and H. Yokomi, Progr. Theoret. Phys. Suppl., 21: 71 (1962).
27. L. D. Solov'ev and Ch'en Ts'ung-mo, Zh. Eks. i Teor. Fiz., 42: 527 (1962).
28. J. Steinberger and A. S. Bishop, Phys. Rev., 78: 494 (1950).
29. J. Steinberger and A. S. Bishop, Phys. Rev., 86: 171 (1952).
30. R. S. White, M. J. Jacobson, and A. G. Schulz, Phys. Rev., 88: 836 (1952).
31. B. T. Feld, D. H. Frish, I. L. Lebow, L. S. Osborne, and J. S. Clark, Phys. Rev., 85: 680 (1952).
32. G. S. Janes and W. L. Kraushaar, Phys. Rev., 93: 900 (1954).
33. T. L. Jenkins, D. Luckey, T. R. Palfrey, and R. R. Wilson, Phys. Rev., 95: 179 (1954).
34. G. Bernardini and E. L. Goldwasser, Phys. Rev., 94: 729 (1954).
35. J. E. Liess, C. S. Robinson, and S. Penner, Phys. Rev., 98: 201 (1955).
36. J. E. Liess and C. S. Robinson, Phys. Rev., 95: 638 (1954).
37. A. V. Tollestrup, J. C. Keck, and R. M. Worlock, Phys. Rev., 99: 220 (1955).
38. R. L. Walker, J. G. Teasdale, V. Z. Peterson, and J. I. Vette, Phys. Rev., 99: 210 (1955).
39. M. Beneventano, G. Bernardini, D. Carlson-Lee, G. Stoppini, and L. Tau, Nuovo Cim., 4: 323 (1956).
40. E. L. Goldwasser, Proc. 1960 Internat. Conf. on High-Energy Physics, Rochester, New York (1960), p. 26.
41. G. Bernardini and E. L. Goldwasser, Phys. Rev., 95: 857 (1954).
42. K. M. Watson, J. C. Keck, A. V. Tollestrup, and R. L. Walker, Phys. Rev., 101: 1159 (1956).
43. J. H. Malmberg and C. S. Robinson, Phys. Rev., 109: 158 (1958).
44. A. J. Lazarus, W. K. H. Panofsky, and F. R. Tangherlini, Phys. Rev., 113: 1330 (1958).

45. A. Barbaro, E. L. Goldwasser, and D. Carlson-Lee, Bull. Amer. Phys. Soc., 4: 23 (1959).
46. É. G. Gorzhevskaya, V. M. Popova, and F. R. Yagudina, Zh. Eks. i Teor. Fiz., 38: 276 (1960).
47. M. I. Adamovich, É. G. Gorzhevskaya, V. G. Larionova, V. M. Popova, S. P. Kharlamov, and F. R. Yagudina, Zh. Eks. i Teor. Fiz., 38: 1978 (1960).
48. G. M. Lewis, R. E. Azuma, E. Gabathuler, G. W. G. S. Leith, and W. R. Hogg, Phys. Rev., 125: 378 (1962).
49. M. I. Adamovich, É. G. Gorzhevskaya, and F. R. Yagudina, Zh. Eks. i Teor. Fiz., 43: 1113 (1962).
50. D. McPherson and R. Kenney, Bull. Amer. Phys. Soc., 16: 523 (1961).
51. E. A. Knapp, R. W. Kenney, and V. Perez-Mendez, Phys. Rev., 114: 605 (1959).
52. M. I. Adamovich, É. G. Gorzhevskaya, V. G. Larionova, N. M. Panova, V. M. Popova, S. P. Kharlamov, and F. R. Yagudina, Zh. Eks. i Teor. Fiz., 41: 1811 (1961).
53. I. K. Walker and J. P Burq, Phys. Rev. Letters, 8: 37 (1962).
54. M. J. Bazin and J. Pine, Phys. Rev., 132: 830 (1963).
55. C. S. Robinson, Tables of Cross Sections for Photoproduction from Hydrogen. Phys. Res. Lab. Univ. of Illinois (1959).
56. K. Althoff, Z. Phys., 175: 34 (1963).
57. B. H. Patrick, J. M. Paterson, J. G. Rutherglen, and I. Garvey, Phys. Rev. Letters, 10: 157 (1964).
58. M. I. Adamovich, É. G. Gorzhevskaya, V. G. Larionova, N. M. Panova, S. P. Kharlamov, and F. R. Yagudina, Proc. 1962 Internat. Conf. on High-Energy Physics. Geneva (1962).
59. G. M. Lewis, G. W. G. S. Leith, D. L. Thomas, R. Little, and E. M. Lawson, Nuovo Cim., 27: 384 (1963).
60. C. S. Robinson, P. M. Baum, L. Criegee, and J. M. McKinley, Phys. Rev. Letters, 9: 349 (1962).
61. R. J. Walker, T. R. Palfrey, Jr., R. O. Haxby, and B. M. K. Nefkens, Phys. Rev., 132: 2656 (1963).
62. G. W. G. S. Leith, R. Little, and E. M. Lawson, Phys. Rev. Letters, 8: 355 (1964).
63. I. Kirk and I. McElroy, Nuovo Cim., 31: 705 (1964).
64. K. Berkelman and J. A. Waggoner, Phys. Rev., 117: 1364 (1959).
65. B. B. Govorkov, S. P. Denisov, A. M. Lebedev, and E. V. Minarik, Zh. Eks. i Teor. Fiz., 44: 1464 (1963).
66. G. Höhler, Nuovo Cim., 16: 585 (1960).
67. I. Hamilton and W. S. Woolcock, Rev. Mod. Phys., 35: 737 (1963).
68. L. D. Solov'ev, Zh. Eks. i Teor. Phys., 33: 801 (1957).
69. B. M. Pontecorvo, in: Proceedings of the International Conference on High-Energy Physics, Kiev, 1959. Izd. Akad. Nauk SSSR (1960), p. 83.
70. M. I. Adamovich, V. G. Larionova, and S. P. Kharlamov, Tr. Fiz. Inst. Akad. Nauk SSSR, 19: 37-65 (1963).
71. V. T. Cocconi, T. Tazzini, G. Fidecaro, M. Legros, N. H. Lipman, and A. W. Merrison, Nuovo Cim., 22: 494 (1961).

NUCLEAR PHOTOEFFECT IN THREE-PARTICLE NUCLEI

A. T. Varfolomeev, A. N. Gorbunov, and V. N. Fetisov

INTRODUCTION

One of the most important questions in the physics of the atomic nucleus is this: Is it possible to describe both the static and dynamic properties (nuclear reactions) of heavier systems in simple terms of a pair-type nucleon−nucleon interaction, such as that used to describe the ground state and scattering of an isolated pair of nucleons? If such an approach proves unsuccessful even for H^3 and He^3 nuclei, then we must refine our views on the nature of nuclear forces. In particular, we shall require a more detailed study of problems relating to the existence of many-particle forces. Published estimates of the part played by three-particle forces [1, 2] indicate that these have very little influence on the binding energy of tritium.

Serious mathematical difficulties arise when analyzing systems containing only a few nucleons, such as H^3, He^3, and He^4 (difficulties typical of the many-body problem), since, on dispensing with the concept of the shell structure we are forced to return to a direct solution of Schrödinger's equation for the nucleon system.

In problems relating to the scattering of nucleons by deuterons, these methods have been worked out for simple forms of central interaction: Skornyakov and Ter-Martirosyan [3] used δ-forces, Danilov [4] carried out an expansion with respect to the radius of action of the forces, while Kharchenko [5] calculated the scattering lengths and obtained the wave function of tritium with a divided central potential.

Recently, a number of authors [6-8] have proposed complex potentials successfully describing experiments on the scattering of nucleons over a wide range of energies. Only variational methods are effective for calculating with these potentials, which allow for the repulsion of nucleons at small distances and tensor and spin-orbital forces. Werntz [9], Dawson [10], and Blatt et al. [11] have studied the static properties of nuclei heavier than the deuteron by means of the new potentials.

Owing to a certain arbitrariness in selecting the form of the interaction (central forces, divided potential, etc.) and in choosing the form of test functions [12], and also owing to the admission of various kinds of special approximations (the validity of which is not always easily verified) for solving the equations, a study of the static characteristics of H^3 and He^3 alone (binding energy, Coulomb energy) is quite insufficient to establish the complex nature of the wave functions. Additional information on the system of three bodies is given by a study of nuclear reactions involving H^3 and He^3.

Experiments on the scattering of electrons at H^3 and He^3 and experiments on the photo-disintegration of these nuclei may prove useful in solving these complicated problems. Reactions involving π- and K-mesons and nucleons are more difficult to interpret, since these introduce an additional complication, associated with the strong interaction of the incident particle with the target nucleus, whereas an electromagnetic interaction, which is known to a greater accuracy, allows perturbation theory to be used.

Hofstadter [13] studied the scattering of electrons by H^3 and He^3 nuclei. In order to explain the observed difference in the electric and magnetic form factors of He^3, Schiff [14] proposed a phenomenological approach

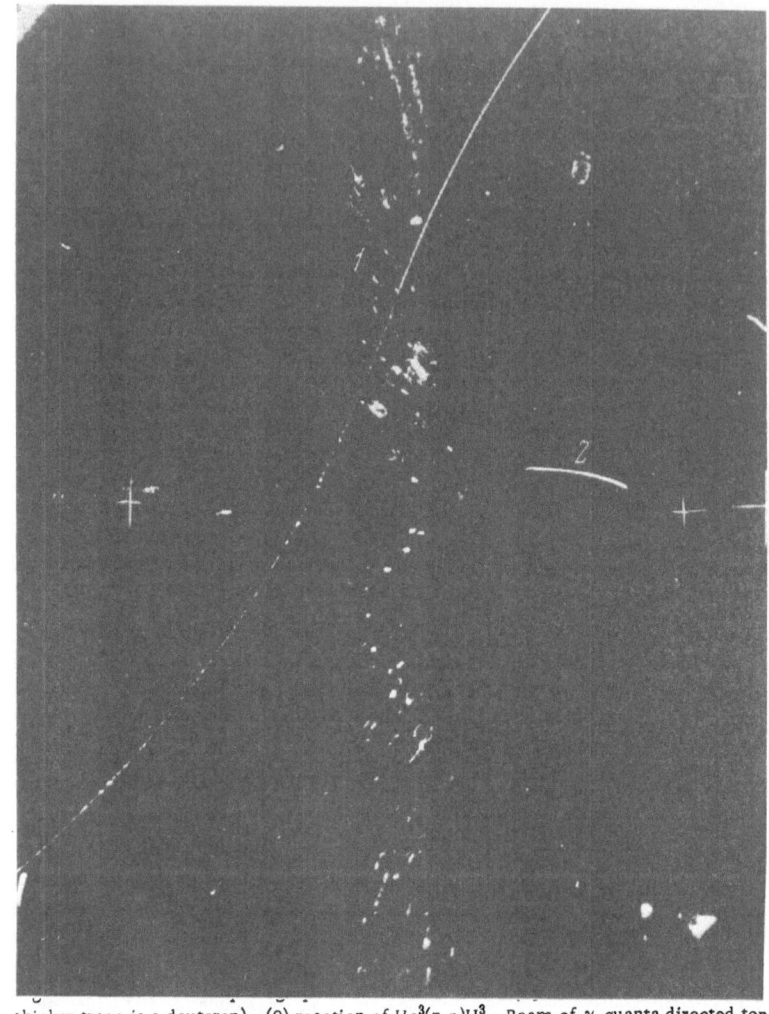

thicker trace is a deuteron); (2) reaction of $He^3(n,p)H^3$. Beam of γ-quanta directed top to bottom.

to the determination of the parameters of the ψ-function. The proportion of the mixed-symmetry S state found by Schiff in addition to the completely symmetric S state exceeded the estimates obtained by the variational method.

A systematic study of the photo-disintegration of He^3 is being carried out in the P. N. Lebedev Physical Institute of the Academy of Sciences of the USSR and in laboratories in other countries (see Sec. 5). The experimental data set out in Sec. 2 extend considerably beyond existing theoretical calculations. We hope that the new experimental information will facilitate both an explanation of the mechanism of γ-quanta absorption by the He^3 nucleus and the establishment of the structure of wave functions of different parity. We would note that the theoretical description of angular distributions in the photo-disintegration of the deuteron only became

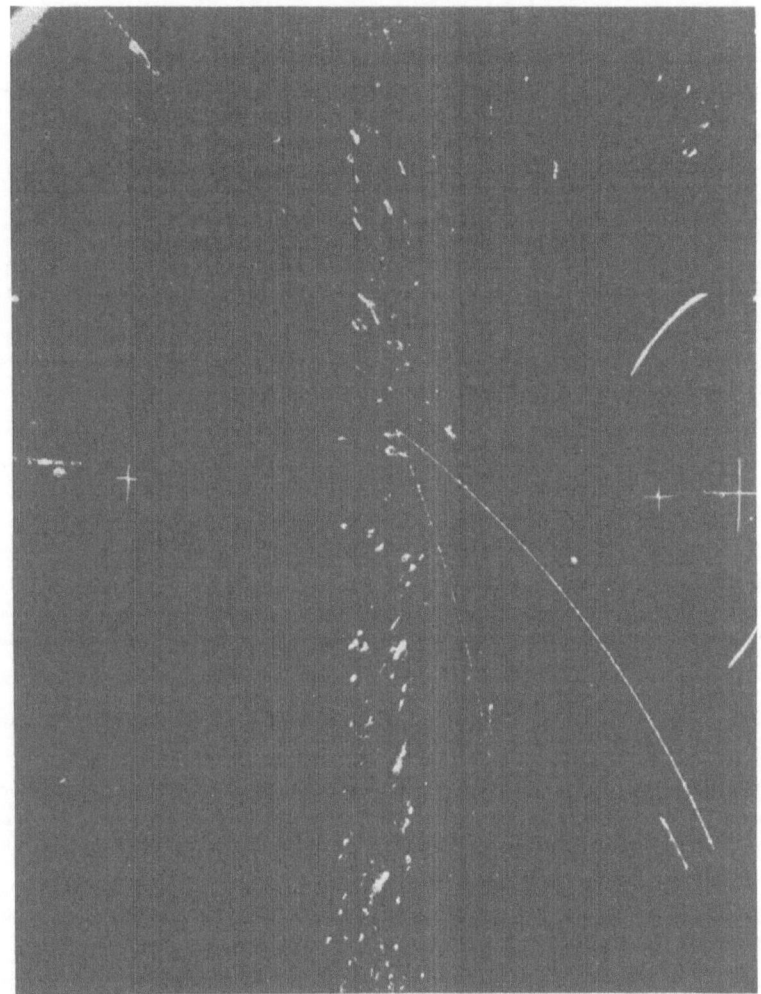

Fig. 2. Wilson-chamber photograph. Reaction of He$^3(\gamma,n)$2p.

possible as a result of considering the D-state contribution, the weight of the D state being much greater than would be expected from previous estimates, namely, 7% instead of 4% [15]. The wave function of the ground state of H^3 and He^3 is, generally speaking, a superposition of S, P, and D components [16]. In carrying out calculations it is usual to consider only the completely symmetric S state, and the effect of the other minor contributions on the characteristics of photo-disintegration has been little studied, as may be gathered from a review of the theoretical literature.

In contrast to experiments on the photo-disintegration of He^4 [17], experiments with He^3 may give considerably fuller information regarding the system of nucleons since, in this case, recording the particles with a Wilson chamber in a magnetic field enables us, in principle, to measure all the differential characteristics and to take strict account of the kinematics of the disintegration of the particles.

SEC. 1. EXPERIMENTAL METHOD

A Wilson chamber 30 cm in diameter and 8 cm high, filled with He^3 to an excess pressure of 0.9 atm and placed in a magnetic field of 10,400 Oe, operated in a collimated beam of bremsstrahlung with a maximum energy of 170 MeV (cross section of beam in chamber = 4 × 9 mm). The chemical purity of the He^3 used in the experiment was better than 99.99%, and the concentration of tritium was no greater than 10^{-11}%. The arrangement of the apparatus and the circuit synchronizing the operation of the Wilson chamber with the synchrotron were analogous to those described earlier [18].

The intensity of the radiation pulses recorded by the Wilson chamber was measured by means of a pulse-type ionization chamber calibrated in two different ways: with a quantum meter [19] and with respect to the yield of the $C^{12}(\gamma,n)C^{11}$ reaction [20]. The difference between the results of these calibrations was no greater than 3%.

The magnetic field was measured to an accuracy of 0.5% with a fluxmeter calibrated by the nuclear-resonance method. The current in the magnet was stabilized to an accuracy of 0.5%.

Only two photonuclear reactions are possible for the He^3 nucleus at energies below the meson threshold:

$$He^3 \ (\gamma, p) \ d \ \text{(reaction threshold } Q = 5.49 \text{ MeV)}, \tag{1}$$

$$He^3 \ (\gamma, \ n) \ 2p \ \text{(reaction threshold } Q = 7.72 \text{ MeV)}. \tag{2}$$

Photographs of typical cases of these reactions appear in Figs. 1 and 2. In identifying reactions (1) and (2) from Wilson-chamber photographs, we started from considerations of conservation. Since two particles occur in the final state (proton and deuteron) in the $He^3(\gamma,p)d$ reaction, this reaction is characterized by the following features:

1. In the center-of-mass system the two particles are emitted in opposite directions with different momenta;

2. In the laboratory system the two particles acquire a slightly forward motion, so that the vertex of the angle between the traces is oriented in opposition to the direction of the beam, the magnitude of the angle lying between 170 and 180° (approximately), depending on the direction of flight of the particles.

3. The directions of the traces of the two particles are coplanar with the axis of the γ-quantum beam.

4. Since the momenta of the proton and deuteron are also approximately equal in the laboratory system of coordinates, the density of ionization in the deuteron trace is on average four times that in the proton trace.

In $He^3(\gamma,n)2p$ reactions, there are three particles in the final state, only the two protons being recorded in the chamber. Hence, in this reaction the traces of the protons may be oriented arbitrarily, both relative to each other and relative to the direction of the γ-quantum beam.

The characteristic features of the kinematics of reactions (1) and (2) enabled these processes to be clearly identified in the overwhelming majority of cases, even by visual inspection of the photographs. In the comparatively small number of cases in which visual identification of reaction (1) seemed doubtful, the disintegration kinetics were analyzed quantitatively, and identification then became obvious. A certain indeterminacy in the identification of reactions of type (2) arose when one of the protons stopped in the gas of the Wilson chamber and had a path less than 10 mm long. In these cases the short trace of the proton might be taken for the trace of a recoil nucleus in reactions of the type

$$\gamma + x \rightarrow x' + p + n, \tag{3}$$

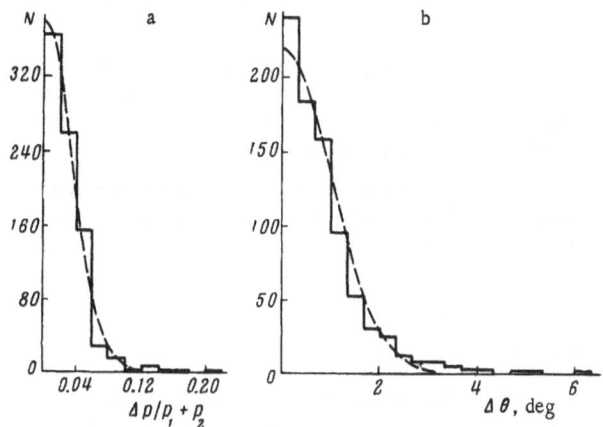

Fig. 3. Distribution of errors in measuring the momenta of the particles (a) and the recoil angles of the particles with respect to the axis of the γ-quantum beam (b).

$$\gamma + x \rightarrow x'' + p \qquad (4)$$

at carbon and oxygen nuclei forming part of the composition of water and alcohol vapor. Conversely, some reactions of types (3) and (4) might sometimes be ascribed to reaction (2). The number of such doubtful cases was fairly small. In addition to this, the expected number of reactions of types (3) and (4) at impurities could easily be estimated from the observed yield of "stars" with the emission of more than two charged particles, the identification of which caused no difficulty, and from data relating to the relative yields of different photonuclear reactions at carbon [21] and oxygen [22]. The correction thus found to the number of reactions (2) was 2.5%.

In addition to reactions (1)-(4), each Wilson photograph showed on average one $He^3(n, p) H^3$ reaction, this having a great probability of occurring under the influence of slow neutrons. In 99% of cases, however, this reaction occurred outside the region of the γ-ray beam, and furthermore its external features were so characteristic that they could be identified with practical certainty.

The energies of the photons producing the disintegrations were determined from the conservation laws, using the directions and momenta of the particles emitted in the reaction. The directions of the particles relative to the axis of the γ-ray beam were measured with a protractor, enabling the spatial picture of the event to be established from stereoscopic photographs. The radii of curvature of the traces in the magnetic field were determined by comparing with standard circles. Since the Wilson-chamber photographs were taken through an aperture 16 cm in diameter in the upper pole of the magnet (interpole distance 25 cm), the magnetic field in the working space of the chamber was not entirely homogeneous. The inhomogeneity of the magnetic field was approximately ±2%. In view of this, the value of the magnetic field averaged over the trace was used in calculating the momentum of the particle. Small corrections were also introduced in order to allow for the change in curvature along the traces of slow protons with energies up to 4.5 MeV and deuterons up to 6 MeV.

Errors in measuring the recoil angles and momenta of the particles were estimated as follows. In reaction (1), the momenta of the proton p_1 and deuteron p_2 in the center-of-mass system should be the same, and the sum of the angles θ_1 and θ_2 between the directions of p_1 and p_2, respectively, and the direction of the γ-quantum beam should be 180°. In order to characterize the measuring errors we took the ratio $\Delta p/(p_1 + p_2)$, where $\Delta p = |p_1 - p_2|$, and the value of $\Delta \theta = |\theta_1 + \theta_2 - 180°|$. The experimental distributions of these characteristics for 830 reactions of type (1) appear in Fig. 3. Both distributions have a Gaussian character with values of σ equal to 0.035 and 1°, respectively. The corresponding error in the determination of the energy of the

γ quanta causing reaction (1) varies from 3% for E_γ = 8 MeV to 9% for E_γ = 100 MeV. The errors in determining the energies of the photons producing reaction (2) are of the same order.

In order to secure a fair accuracy, we only selected disintegrations occurring in the central part of the chamber (18 cm long), for which the angle between the axis of the chamber and the trace of each of the charged particles was less than 30°. The total number of disintegrations with corresponding kinematic characteristics was then determined by introducing geometrical corrections.

For a two-particle reaction, the geometrical correction was easily introduced from the condition of the azimuthal symmetry of the process relative to the axis of the beam. For each measured case of reaction (1) with proton escape angle θ_p relative to the axis of the beam, the total number of cases of this type was

$$N_1 = \frac{\pi}{2 \arcsin\left(^1/_2 \sin \theta_p\right)} \cdot$$

The value of N_1 varies from 1 to 3 with an average of 2.35.

The geometrical factor for the three-particle reaction was calculated on the assumption of equal probability of all values of the two parameters φ and φ'. The parameter φ is the azimuthal angle of rotation of the plane containing the proton-momentum vectors p_1 and p_2 around the vector $P = p_1 + p_2$; the parameter φ' is the azimuthal angle of rotation of the plane containing vectors P and $\varkappa$ (photon momentum) around vector $\varkappa$. By an appropriate choice of origin for angles φ and φ', all possible inclinations β_1 and β_2 of vectors p_1 and p_2 to the plane of the chamber are exhausted by varying φ and φ', respectively, between 0° and 180° and between 0° and 90°. The geometrical factor was calculated on an electronic computer. The ranges of variation of φ and φ' were split into equal intervals $\Delta\varphi = \Delta\varphi' = 2°$, and for each of the possible 4050 combinations of these intervals the angles β_1 and β_2 were calculated. Since the kinematic characteristics of the reactions in our experiment were only measured for cases in which β_1 and β_2 were less than 30°, the geometric factor N_2 for each case measured was given by

$$N_2 = \frac{\pi^3}{2 \sum\limits_{i=1}^{45} \sum\limits_{j=1}^{90} \Delta\varphi_j \Delta\varphi'_i \delta_{ij}},$$

where

$$\delta_{ij} = \begin{cases} 1, \text{ if } \beta_1; \text{ and } \beta_2 \leqslant 30°, \\ 0, \text{ if } \beta_1 \text{ or } \beta_2 > 30°. \end{cases}$$

The value of N_2 varied from 1 to 8.6, the average being $\overline{N}_2$ = 3.82. A criterion of the accuracy of the geometrical factors so determined was obtained by comparing the total number of reactions (1) and (2) (M_1 and M_2) observed in an 18-cm column of gas in the chamber with the sums of $(N_1)_i$ and $(N_2)_i$ calculated for each of the measured cases of these reactions. It was found that M_1 = 1964 ± 37 coincided with $\Sigma(N_1)_i$ = 1963 ± 70, while M_2 = 1970 ± 37 was a little smaller than $\Sigma(N_2)_i$ = 2125 ± 102, although even in this case agreement was very good. Although the difference between M_2 and $\Sigma(N_2)_i$ was hardly outside the limits of statistical error, in calculating the effective cross sections of reaction (2) we introduced an additional factor b = $M_2 / \Sigma(N_2)_i$ = 0.93, in order to eliminate the effects of this discrepancy.

After measuring the escape angles and momenta of the protons emitted in the three-particle disintegration, we used the laws of conservation to determine the escape angle and momentum of the neutron as well. All the numerical calculations were carried out on the electronic computer.

SEC. 2. RESULTS OF MEASUREMENTS

1. Yields of He$^3(\gamma,p)$d and He$^3(\gamma,n)$2p Reactions

In 22,300 Wilson photographs we observed 2771 cases of an He$^3(\gamma,p)$d reaction and 2780 cases of an He$^3(\gamma,n)$2p reaction. We thus see that the yields of the two- and three-particle photo-disintegrations of He^3

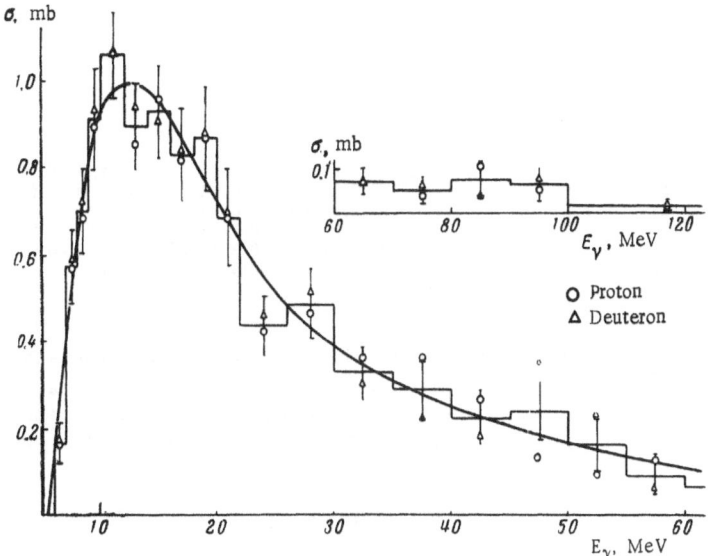

Fig. 4. Effective cross section of the He³(γ, p) d reaction.

were identical within the limits of statistical error. The absolute value of the yield of each of the reactions (for E_γ^{max} = 170 MeV) equalled

$$Y_1 = Y_2 = \int_0^{170\ MeV} \sigma(E_\gamma)\,\eta(E_\gamma)\,dE_\gamma = (1.47 \pm 0.03)\ mb,$$

where $\sigma(E_\gamma)$ is the effective cross section of reaction (1) or (2), and $\eta(E_\gamma)$ is the bremsstrahlung spectrum (we used the spectrum of Davies et al. [23], calculated with due allowance for absorption and twofold emission in a synchrotron target [24] of thickness t = 0.12 rad. units). Only the statistical error is given. The error associated with intensity measurement is approximately 6%.

2. Two-Particle Disintegration [He³(γ,p)d Reaction]

(a) Effective Cross Section. The effective cross section of reaction (1) was determined from measurements of 830 cases observed in the central part of the chamber, 18 cm long, for which the angle between the traces and the chamber axis was no greater than 30°. Since only two particles are formed in reaction (1), it is sufficient to measure the escape angle and momentum of one of the particles (proton or deuteron) in order to determine the energy of the photon producing the reaction. For checking purposes, however, the escape angles and momenta of both particles were measured, so that the energy of the photon could be determined twice. The effective cross sections obtained by analyzing the traces of the protons and deuterons are given separately in Fig. 4. The mean value of the effective cross section in each of the energy intervals selected is shown by a histogram in Fig. 4. For convenience in comparing with theoretical calculations, the histogram is based approximately on the area of the smooth curve. As seen from Fig. 4, the effective cross section of the He³(γ,p)d reaction has a wide momentum at a photon energy of about 12-13 MeV. The value of the effective cross section at the maximum is approximately 1.0 mb.

Table 1

Photon energy range, MeV	A, mb/sr	β	γ	δ	$\varepsilon = \dfrac{\sigma}{\frac{4}{3}\pi A}$
6—12	0.072±0.065	0.69±0.14	0.1±0.3	0.03±0.04	1.06±0.09
12—16	0.086±0.013	0.95±0.25	0.9±0.6	0.02±0.08	1.21±0.17
16—22	0.091±0.011	0.57±0.18	0.1±0.4	0.02±0.06	1.05±0.12
22—170	0.028±0.004	0.57±0.24	1.1±0.6	0.09±0.11	1.35±0.20
6—22	0.123±0.008	0.7±0.1	0.4±0.2	0.01±0.02	1.09±0.05
6—170	0.153±0.006	0.66±0.10	0.5±0.2	0.03±0.02	1.14±0.05

(b) Angular Distribution of Protons. The angular distributions of the protons emitted in the $He^3(\gamma, p)$ d reaction in the center-of-mass system are shown for several ranges of γ-quantum energy in Fig. 5. The x axis gives the escape angle of the protons in the center-of-mass system, and the y axis gives the cross section averaged over the energy range $d\sigma/d\Omega$ in mb/sr. The smooth curve in the graph is calculated by the method of least squares from the formula

$$f(\theta) = A(\sin^2\theta + \beta \sin^2\theta \cos\theta + \gamma \sin^2\theta \cos^2\theta + \delta).$$

The values of coefficients A, β, γ, and δ are given in Table 1. We see from the table and from Fig. 5 that at photon energies of less than 22 MeV the isotropic component of the angular proton distributions is very small. For large energies the isotropic part becomes considerably larger. Asymmetry of the angular distributions, leading to a forward displacement of the maximum, appears quite near the threshold of the reaction and reaches its greatest value at photon energies of 12-16 MeV, afterwards falling off gradually. The coefficient γ has a tendency to rise with increasing photon energy.

3. Three-Particle Disintegration [$He^3(\gamma, n)$ 2p Reaction]

(a) Effective Cross Section. The effective cross section of the $He^3(\gamma, n)$ 2p reaction determined from measurements of 515 cases is shown in Fig. 6. The histogram shows the mean value of effective cross section in each photon-energy interval. For convenience in comparing with theoretical calculations, the histogram is based approximately on the area of the smooth curve. We see from Fig. 6 that the effective cross section has a very wide maximum for photon energies between 16 and 19 MeV; as in the case of the two-particle channel, the cross section at the maximum equals $\sigma_{max} \simeq 1.0$ mb. The curves giving the effective cross sections of the two- and three-particle processes differ little in form at low energies. At high energies, however, the difference becomes considerable, the effective cross section of three-particle photo-disintegration falling much more slowly with increasing photon energy than that of the two-particle type. The half-width of the curve giving the effective cross section of the (γ,n) reaction is about 30 MeV, while that of the curve representing the (γ,p) reaction equals 18 MeV. For excitation energies in the range 25 to 100 MeV, three-particle disintegration is roughly twice as probable as two-particle. For photon energies above 100 MeV, the ratio of the cross sections of the three- and two-particle channels rises to 5 or 10.

In order to compare the effective cross sections of the γ,p and γ,n reactions, Table 2 shows the integral cross sections σ_0 together with the cross sections weighted in accordance with the bremsstrahlung spectrum σ_{-1}, for various ranges of photon energy.

(b) Angular Distributions. The angular distributions of the neutrons and protons emitted in the $He^3(\gamma, n)$ 2p reaction are shown in Figs. 7 and 8 for two ranges of γ-quantum energy: from the reaction threshold to 30 MeV and from 30 to 170 MeV. The x axis shows the escape angles of the particles in the center-of-mass system relative to the axis of the beam, and the y axis gives the yield in mb/sr. The curves in the form of $A \sin^2\theta + B$ in Fig. 7 are calculated by the method of least squares.

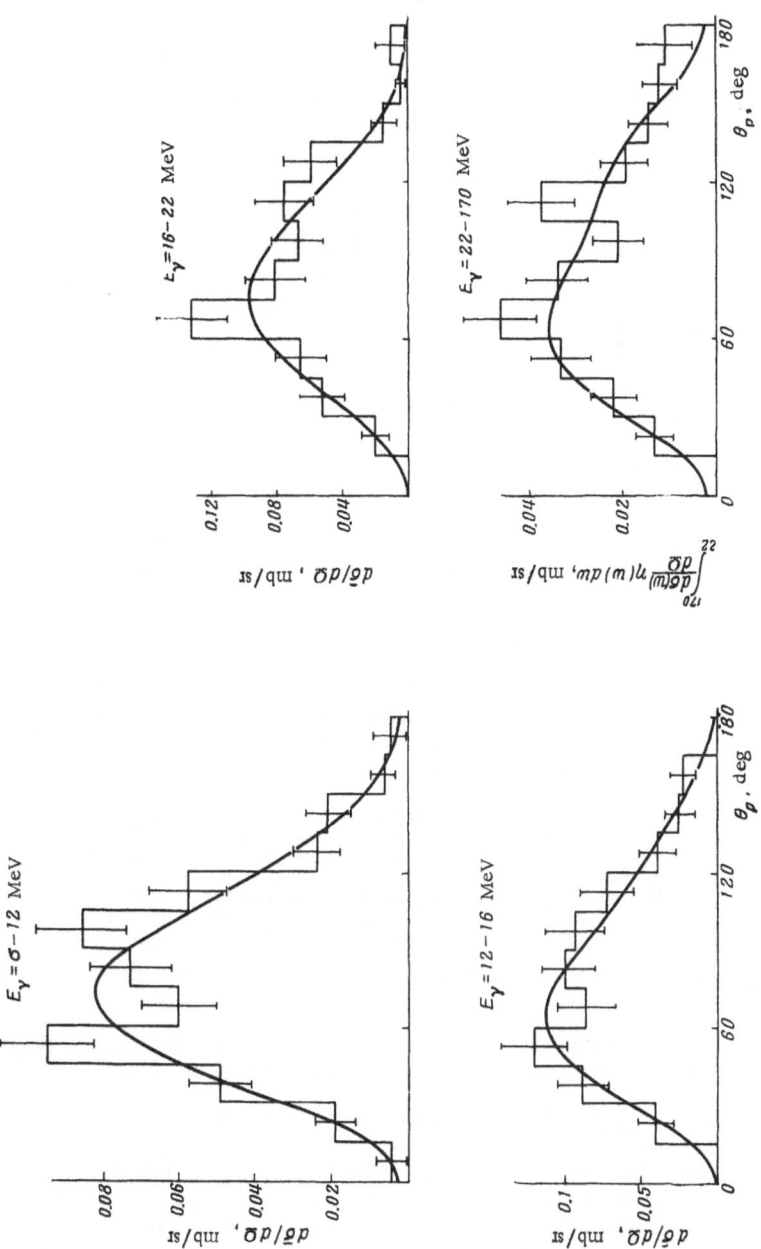

Fig. 5. Angular distribution of protons emitted in the γ-reaction (center-of-mass system).

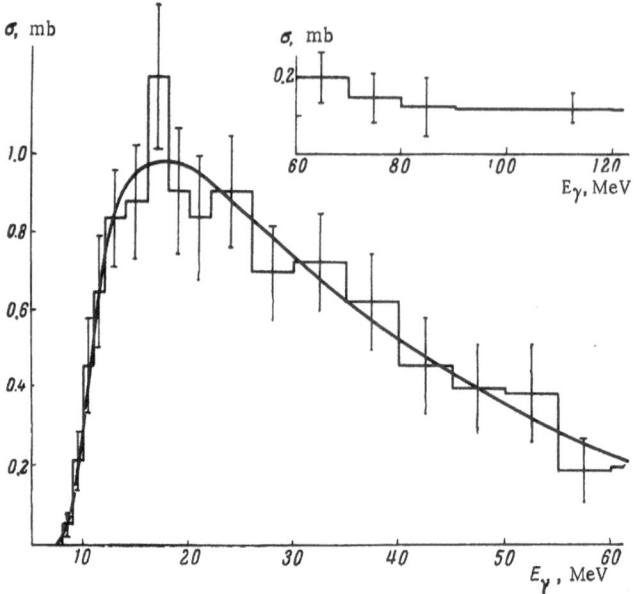

Fig. 6. Effective cross section of the $He^3(\gamma,n)$ 2p reaction.

Table 2*

Range of photon energy ΔE_γ, MeV	$\sigma_0 = \int \sigma(E_\gamma)\, dE_\gamma$, MeV · mb		$\sigma_{-1} = \int \sigma(E_\gamma)\, \dfrac{dE_\gamma}{E_\gamma}$, mb	
	γ, p	γ, n	γ, p	γ, n
0—22	12.9±0.6	10.7±0.7	1.00±0.04	0.69±0.05
0—30	16.6±0.7	17.1±1.0	1.14±0.04	0.94±0.05
0—40	19.7±0.8	23.8±1.4	1.23±0.05	1.14±0.06
0—100	25.9±1.2	38.0±2.3	1.34±0.05	1.37±0.07
0—170	26.5±1.3	43.6±2.7	1.34±0.05	1.42±0.07
22—100	13.0±1.1	27.3±2.2	0.34±0.03	0.68±0.05
100—170	0.6±0,4	5.6±2.2	0.005±0.003	0.04±0.02

*Only statistical errors shown.

We see from Figs. 7 and 8 that the angular distribution of the neutrons in the two energy ranges is quite close to a $\sin^2 \theta$ distribution, while that of the protons has a considerable isotropic component.

(c) E n e r g y   S p e c t r u m   o f   t h e   N e u t r o n s .   Figure 9 shows the energy spectrum of the neutrons for two proton-energy ranges: 8-12 and 12-170 MeV. The x axis gives the values of E_n/E_{max}, where E_n is the observed neutron energy in the center-of-mass system and E_{max} is the maximum possible neutron energy. The y axis gives the number of observed cases for the range $\Delta(E_n/E_{max}) = 0.1$. (Since E_n/E_{max} is directly determined by the relative energy of the protons in a system of coordinates associated with their center of mass, the graphs in Fig. 9 also constitute spectra of the relative energies of the two protons.) The γ-quantum energy ranges 8-12 and 12-170 MeV were chosen in view of the fact that the form of the neutron energy spectrum in the narrower ranges ΔE_γ = 12-16, 16-22, 22-30, 30-170 MeV was almost identical. We see from Fig. 9 (lower part) that the neutron energy spectrum has a sharp maximum at $E_n = E_{max}$ (0.85-0.9). (We note that $E_n/E_{max} = 1$ in the case

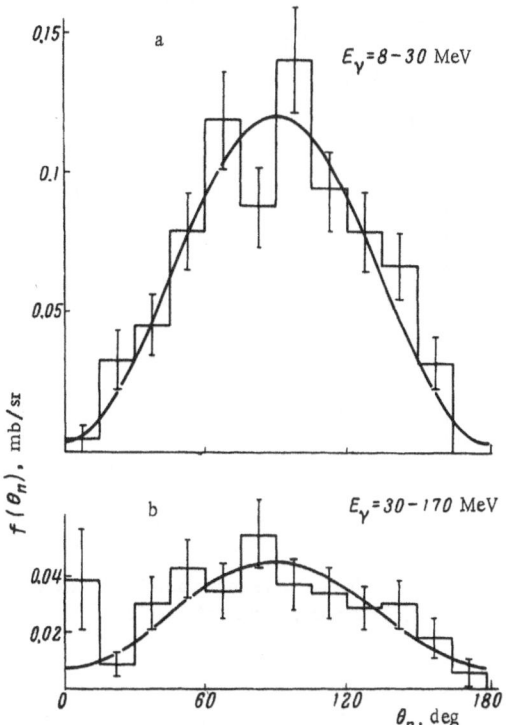

Fig. 7. Angular distributions of neutrons emitted in the $He^3(\gamma,n)\,2p$ reaction (center-of-mass system). Smooth curves calculated by the method of least squares: (a) $f(\theta_n) = (0.116 \pm 0.009)\,[\sin^2\theta_n + (0.03 \pm 0.03)]$ mb/sr; (b) $f(\theta_n) = (0.038 \pm 0.006)\,[\sin^2\theta_n + (0.18 \pm 0.12)]$ mb/sr.

Table 3

Photon energy range ΔE_γ, MeV	$\overline{E_n/E_{max}}$	γ_{np}	γ_{pp}	$\overline{P_n/P_{max}}$	$\overline{P_p/P_{max}}$
8—12	0.53	121°45′	116°30′	0.73	0.70
12—16	0.64	129 30	101	0.80	0.66
16—22	0.65	130 30	99	0.81	0.66
22—30	0.68	132 50	94 20	0.82	0.65
30—170	0.65	130 20	99 20	0.81	0.66

in which both protons are emitted in the same direction with equal momenta, as a "biproton.") In addition to this, in the energy range $E_n = E_{max}$ (0.2-0.3) there is still another maximum, as we might have expected on considering that the three-particle process proceeds as a "quasi-γ, p" reaction in which the neutron and one of the protons are emitted with equal momenta in the same direction. The neutron energy in this case is easily seen to be $E_n = \frac{1}{4}E_{max}$. The second maximum is especially clear if we subtract the theoretically expected neutron spectrum (continuous lines in Fig. 17), calculated on the assumption that the protons are emitted in the S state, from the over-all spectrum in Fig. 9 (lower part).

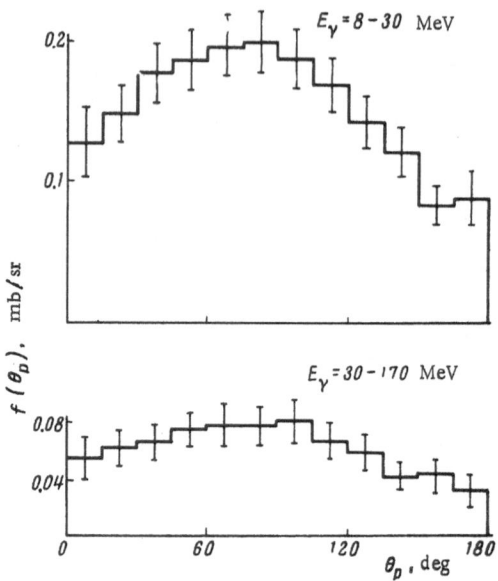

Fig. 8. Angular distributions of protons emitted in the $He^3(\gamma,n)2p$
reaction (laboratory system of coordinates).

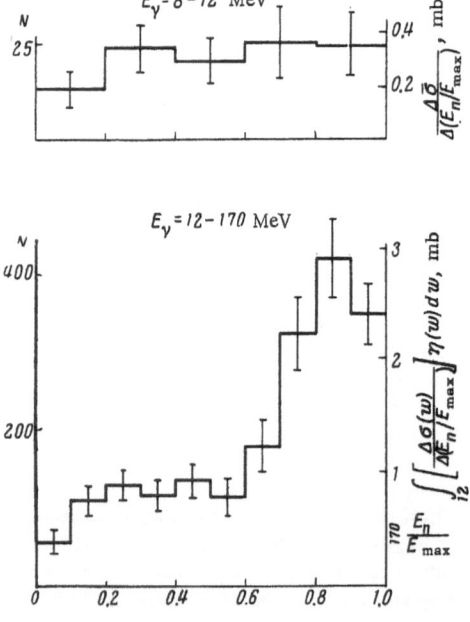

Fig. 9. Energy spectra of neutrons emitted in the $He^3(\gamma, n)2p$
reaction.

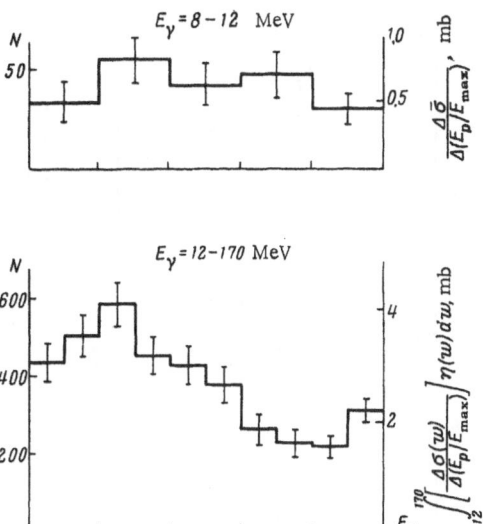

Fig. 10.   Energy spectra of protons emitted in the He³(γ,n) 2p
reaction.

Table 3 shows the average values of neutron energy for several photon energy ranges. We see from the table that the value of $\overline{(E_n/E_{max})}$ remains constant over a wide energy range from 12 to 170 MeV.

(d) Energy Spectrum of the Protons. The energy spectrum of the protons is shown in Fig. 10 for photon energy ranges 8-12 and 12-170 MeV. The x axis gives the values of E_p/E_{max}. We see from the figure that for photon energies near the threshold of the reaction the proton spectrum coincides with the spectrum calculated on the basis of the distribution of phase volume, without considering the interaction in the final state. For photon energies above 12 MeV, the proton spectrum acquires a wide maximum at an energy of $E_p = E_{max} \cdot 0.25$, as would be expected on the basis of the characteristics of the neutron spectrum. We also see from the graph that in a small number of cases the proton is emitted with the maximum possible energy. The mean proton energy for the photon-energy ranges shown in Table 3 may be found from the average neutron energies, remembering that

$$2\,\overline{(E_p/E_{max})} + \overline{(E_n/E_{max})} = 1.5.$$

(e) Distribution of Relative Angles between the Escape Directions of the Particles. The distribution of the angles between the neutron and proton escape directions (for the photon energy range 8-170 MeV) is shown in Fig. 11a. The x axis shows the relative angles in the center-of-mass system of the three particles, and the y axis gives the number of cases with relative inter-trace angles between γ_{ij} and $\gamma_{ij} + 15°$. {In order to obtain the distribution referred to unit solid angle, the ordinates in Fig. 11a must be divided by $2\pi[\cos \gamma_{ij} - \cos (\gamma_{ij} + 15°)]$} We see from the graphs that the distribution of the relative angles between the neutron and proton escape directions has a sharp maximum for large angles γ_{np}, while the distribution of angles between the directions of proton escape is practically independent of angle.

Table 3 shows the mean values of the relative angles for several photon-energy ranges. We see from the table that near the reaction threshold all the escape directions are on average equally probable. For photon energies above 12 MeV, the relative angle between the neutron and proton rises to 130° and then remains constant over a wide range of photon energies.

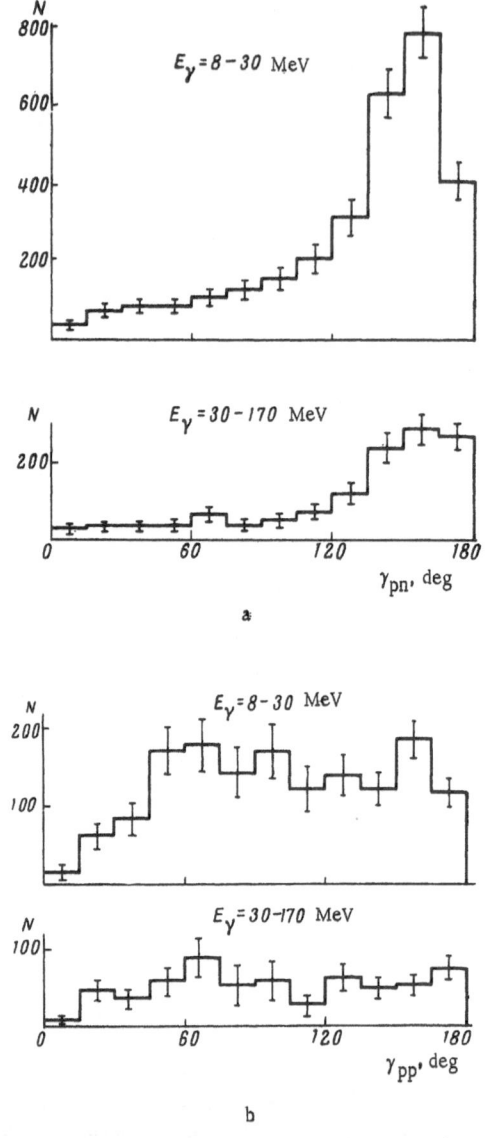

Fig. 11. Distribution of angles between the directions of the momenta in the center-of-mass system. (a) Proton and neutron; (b) two protons.

(f) Mean Values of Momenta. Table 3 shows the mean square values of momenta for particles emitted in the $He^3(\gamma,n)\,2p$ reaction, calculated in units of the maximum possible momentum of a particle.

Figure 12 shows a scheme for the three-particle photo-disintegration in two γ-quantum energy ranges, 8-12 and 12-170 MeV. The relative angles and momenta of the escaping particles are shown in the figure.

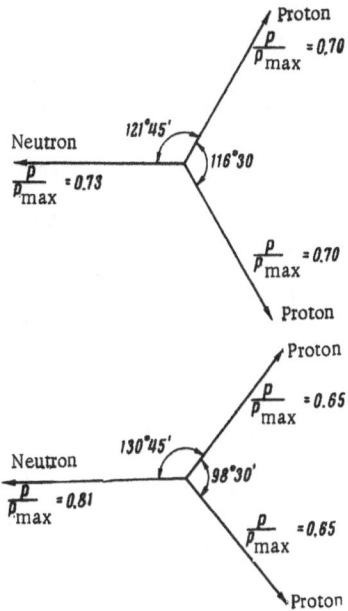

Fig. 12. Kinematic schemes for the three-particle photo-disintegration of He^3 for γ-quantum energies between 8 and 12 MeV (top) and between 12 and 170 MeV (bottom).

SEC. 3. RULE OF SUMS AND DIMENSIONS OF THREE-PARTICLE NUCLEI

In order to calculate the differential and total cross sections of various photonuclear processes (and, in particular, photonuclear processes with three-particle nuclei), we must know the wave functions of the initial and final states of the system; unfortunately, these are not too well known. This difficulty may be partly avoided if we sum the cross sections over all possible final states of the system. The rule of sums for E1, E2, and M1 transitions were considered by Levinger and Bethe [25, 26], Migdal [27], Khokhlov [28], and others.

1. Rule of Sums for the Integral Cross Section

If we use the formula of Levinger and Bethe [25] for the integral cross section of the electric-dipole absorption of photons by a nucleus with charge Z and mass A

$$\sigma_0 = \int_0^\infty \sigma_{E1}(E_\gamma)\, dE_\gamma = \frac{2\pi^2 e^2 \hbar}{Mc}\, \frac{NZ}{A}\, (1 + 0.8\, x), \tag{5}$$

(the formula being obtained on the independent-particle approximation) in order to secure a coarse estimate of σ_0 for nuclei with A = 3 (H^3 and He^3), and allow for a proportion of exchange forces x = 0.5, we shall obtain $\sigma_0 = $ 56 MeV · mb. In the present investigation, the experimental value of the integral cross section for the absorption of photons by an He^3 nucleus was found by summing the integral cross sections of the two- and three-particle processes for He^3 given in Table 2:

$$(\sigma_0)_{\gamma p} = (26.5 \pm 1.3)\ \text{MeV} \cdot \text{mb}, \qquad (\sigma_0)_{\gamma n} = (43.6 \pm 2.7)\ \text{MeV} \cdot \text{mb}.$$

Table 4

Authors	$\sigma_0 = \int \sigma(E_\gamma)\, dE_\gamma$, MeV · mb	$\sigma_{-1} = \int \sigma(E_\gamma)\, \dfrac{dE_\gamma}{E_\gamma}$, MeV · mb
	Experiment	
Varfolomeev and Gorbunov (present work)	for σ_{tot}	
	70 ± 5	2.76 ± 0.18
	for σ_{E_1}	
	62 ± 6	2.53 ± 0.19
	Theory	
Levinger and Bethe [25]	56	—
Rustgi [31]	51.0 (Serber mixture)	1.32
	57.6 (Rosenfeld mixture)	
Mathur et al. [35]	54.4 (Serber mixture)	2.85
	63.0 (Rosenfeld mixture)	
Fetisov [39]	—	2.32 (Kikuta potential)
Davey and Valk [37]	65.0	2.36
	(core + tensor forces, Serber mixture)	
	64.0	2.40
	(Hu and Massey potential, Serber mixture)	
Davey and Valk [42]	—	1.99
	(Schiff [14] wave functions containing 3.5% S' state)	

Thus, $(\sigma_0)_{\text{total}} = (70 \pm 5)\,\text{MeV} \cdot \text{mb}$ (the total error is given, including the statistical errors and the error associated with intensity measurement). In order to compare the experimental value of σ_0 with the results of calculations based on the rule of sums, we must subtract the contribution due to E" absorption from the latter. The contribution of E2 absorption may be estimated by comparing the angular distributions of protons emitted in reaction (1) with the theoretical calculations of Eichmann [29] (see Sec. 4). It was found that the contribution of E2 absorption to $(\sigma_0)_{\gamma p}$ was 11 ± 4%. If the contribution of E2 absorption to $(\sigma_0)_{\gamma n}$ is taken as being the same, then for $(\sigma_0)_{E_1}$ we obtain the value $(\sigma_0)_{E_1} = (62 \pm 6)\,\text{MeV} \cdot \text{mb}$ (total error given).

Rustgi [30, 31] obtained σ_0 for a proportion of Majorana exchange forces x and Heisenberg exchange forces y in the form

$$\sigma_0 = 40\left[1 + 0.55\left(x + \tfrac{1}{2}\,y\right)\right]\,\text{MeV} \cdot \text{mb},$$

taking as ground state of tritium the Irving wave function $\Psi \sim e^{-\sqrt{2\alpha}(\rho^2 + {}^3/{}_4\,r^2)^{1/2}}$, where $\rho = \mathbf{r}_3 - (\mathbf{r}_1 + \mathbf{r}_2)/2$, $\mathbf{r} = \mathbf{r}_1 - \mathbf{r}_2$, where $\mathbf{r}_1, \mathbf{r}_2$, and $\mathbf{r}_3$ are the radii vectors of the nucleons ($\alpha = 0.92\,f^{-1}$ is the variation parameter corresponding to the minimum H^3 energy).

For an exchange system of central Rosenfeld [32] or Inglis [33] forces, $x + \tfrac{1}{2}y = 0.8$, and hence $\sigma_0 = 57.6$ MeV · mb. In the case of Serber [34] exchange mixture, $x + \tfrac{1}{2}y = 0.5$ and $\sigma_0 = 51$ MeV · mb.

Mathur et al. [35] calculated σ_0 using a central spin-dependent exponential potential allowing for the hard core of the nucleon and a tritium wave function in the form proposed by Kikuta et al. [36]:

$$\Psi = \sqrt{N}\,(e^{-\mu(r_{12}-d)} - e^{-\nu(r_{12}-d)})\,(e^{-\mu(r_{13}-d)} - e^{-\nu(r_{13}-d)})\,(e^{-\mu(r_{23}-d)} - e^{-\nu(r_{23}-d)}), \tag{6}$$

where

$$r_{12} = |\mathbf{r}_1 - \mathbf{r}_2|, \quad r_{13} = |\mathbf{r}_1 - \mathbf{r}_3|, \quad r_{23} = |\mathbf{r}_2 - \mathbf{r}_3|.$$

The variation parameters $\mu = 0.4\,F^{-1}$ and $\nu = 4.5\,F^{-1}$, with a hard-core radius of d = 0.4 F, give a tritium binding energy and He^3 Coulomb energy close to the experimental values. It was found that on the basis of this model of the nucleus, σ_0 exceeded the previous results by 8% [31].

Davey and Valk [37] calculated $\sigma_0 = 65.0$ MeV · mb with a potential containing a core and having a tensor component. The potential parameters were taken so as to give a satisfactory description of the static properties of the H^2, H^3, He^3, and He^4 nuclei.

The values of σ_0 obtained for the three-particle nuclei (H^3 or He^3) are shown in Table 4. We see from the table that the experimental value of $(\sigma_0)_{E1}$ agrees closely with calculations based on the rule of sums. Unfortunately, the indeterminacy of the theoretical calculations and the experimental errors prevent us from choosing one form of calculation as being superior to the rest, although it would seem that experiment agrees quantitatively better with the last two forms [35, 37].

2. Rule of Sums for the Integral Cross Section Weighted with Respect to the Bremsstrahlung Spectrum

In Levinger and Bethe's paper [25], another rule of sums was obtained for the cross section of electric-dipole absorption:

$$(\sigma_{-1})_{E1} = \int_{\text{thresh}}^{\infty} \frac{\sigma_{E1}(E_\gamma)}{E_\gamma}\,dE_\gamma = \frac{4\pi^2}{3}\left(\frac{1}{\hbar c}\right)\langle \overline{D}^2\rangle_{00}, \tag{7}$$

where $\mathbf{D} = (ZN/A)\,e\mathbf{R}_{pn}$ is the electric-dipole moment of the nucleus, $\mathbf{R}_{pn} = \mathbf{R}_p - \mathbf{R}_n$, $\mathbf{R}_p$ is the radius vector of the center of mass of the protons in the nucleus, and $\mathbf{R}_n$ is the radius vector of the center of mass of the neutrons. The brackets $<\ >_{00}$ indicate averaging over the ground state of the nucleus.

The experimentally measured quantity $(\sigma_{-1})_{E1}$ is more suitable for comparing with theory than σ_0. This is partly because the experimental value of $(\sigma_{-1})_{E1}$ is much less sensitive than σ_0 to errors in measuring the effective cross section at high energies (owing to the factor $1/E_\gamma$). In addition to this, for every specific photonuclear reaction the value of σ_{-1} is approximately equal to the integral $\int \sigma \eta dE_\gamma$ representing the yield of the reaction, since the bremsstrahlung spectrum is not very different from $1/E_\gamma$. Hence it is sufficient to measure, not the cross section, but only the yields of all the photonuclear reactions at the given nucleus, in order to determine the value of σ_{-1} to fair accuracy. On the other hand, as seen from (7), the quantity $(\sigma_{-1})_{E1}$, in contrast to $(\sigma_0)_{E1}$, is not explicitly dependent on the character of the forces between the nucleons, and is determined solely by the properties of the wave function of the ground state of the system; in particular, it depends on the mean square value of $\mathbf{R}_{pn}$.

Foldy [38] showed that, if the wave function of the ground state of the nucleus were completely symmetric relative to the spatial coordinates of all the nucleons, then

$$\langle R_{pn}^2\rangle = \frac{A^2}{ZN(A-1)}\langle R^2\rangle, \tag{8}$$

where $<R^2>^{\frac{1}{2}}$ is the mean-square radius of the charge distribution in the nucleus (for point nucleons). Hence

$$(\sigma_{-1})_{E1} = \frac{4\pi^2}{3}\left(\frac{e^2}{\hbar c}\right)\frac{NZ}{A-1}\langle R^2\rangle. \tag{9}$$

It is an important fact that this expression is obtained without any special assumptions regarding the nuclear forces and wave functions (apart from their spatial symmetry).

In order to determine the experimental value of σ_{-1} for the He^3 nucleus from the effective cross sections of the reactions (γ,p), (γ,n), the integrals

$$(\sigma_{-1})_{\gamma p} = \int_0^{170\,\text{MeV}} \sigma_{\gamma p}\frac{dE_\gamma}{E_\gamma} = (1.34 \pm 0.05)\ \text{mb}$$

and

$$(\sigma_{-1})_{\gamma n} = \int_0^{170\ \text{MeV}} \sigma_{\gamma n}\,\frac{dE_\gamma}{E_\gamma} = (1.42 \pm 0.07)\ \text{mb}$$

were calculated (only the statistical errors are given) and the value of $(\sigma_{-1})_{\text{total}} = (\sigma_{-1})_{\gamma p} + (\sigma_{-1})_{\gamma n} = (2.76 \pm 0.18)$ mb (total error given) was found. In order to find $(\sigma_{-1})_{E1}$, we then subtracted the term due to E2 absorption from $(\sigma_{-1})_{\text{total}}$. For the quantity $(\sigma_{-1})_{\gamma p}$ this contribution equalled $(8.5 \pm 2)\%$. Assuming the same E2-absorption contribution to $(\sigma_{-1})_{\gamma n}$, we find $(\sigma_{-1})_{E1} = (2.53 \pm 0.19)$ mb (total error given).

The theoretical values of $(\sigma_{-1})_{E1}$ calculated with different variational wave functions of the H^3 nucleus are given in Table 4. We see from the table that the experimental value of $(\sigma_{-1})_{E1}$ is almost twice the theoretical value calculated by Rustgi [30] with the Irving wave function, but it agrees much better with the value obtained by Mathur et al. [35]. It should be noted, however, that the calculation is [35] carried out for quite coarse values of the parameters μ and ν of the Kikuta wave function, $\mu = 0.4\ F^{-1}$ and $\nu = 4.5\ F^{-1}$, with d = 0.4 F. For this reason, the author of [39] calculated σ_{-1} for different values of the hard-core radius d, with the optimum parameters μ and ν given in Kikuta's paper.

The results of the calculations for two variants of the effective singlet radius show that the introduction of the hard core leads to a rise in σ_{-1}, while on varying d from zero to 0.4 F, σ_{-1} rises from 1.19 to 2.32 mb. Closer values of σ_{-1} were obtained by Davey and Valk by considering a core together with tensor forces [37] or by using the Hu−Massey potential [40].

We see from Table 4 that the experimental values of $(\sigma_{-1})_{E1}$ are in satisfactory agreement with theoretical calculations based on potentials with a hard core.

3. Radii of Charge Distribution in Three-Particle Nuclei

Using the experimental value of $(\sigma_{-1})_{E1}$ and formula (9), the mean-square radius of the charge distribution of the He^3 nucleus (for point nucleons) was calculated as $<R^2>^{\frac{1}{2}} = (1.62 \pm 0.06)$F; considering the proton charge-distribution radius $R_p = (0.805 \pm 0.011)$ F [41] and the relation $<R_C^2> = <R^2> + <R_p^2>$, we have $<R_C^2>^{\frac{1}{2}} = (1.81 \pm 0.06)$F. The latter value is in satisfactory agreement with the value of $<R_C^2>^{\frac{1}{2}}_{He^3} = (1.97 \pm 0.1)$F obtained in Hofstadter's experiments [13] on the scattering of electrons by He^3. We note that the value of $<R_C^2>^{\frac{1}{2}}$ obtained in experiments on the nuclear photoeffect in He^3 is rather smaller than the value of $<R_C^2>^{\frac{1}{2}}$ found by experiments on e−He^3 scattering. This difference may constitute an additional indication of the incomplete symmetry of the He^3 wave function. Schiff [14] showed that the difference in the form factors of the H^3 and He^3 nuclei (or the mean-square radii of these nuclei) observed in Hofstadter's experiments may be due to a small amount of a state with mixed symmetry (S' state) added to the completely symmetric S state.

Davey and Valk [42] calculated the value of $(\sigma_{-1})_{E1}$ with the Gaussian wave functions (containing 3.5% of the S' state) used by Schiff [14] in order to describe the results on e−H^3 (or He^3) scattering; they found that the 3.5% admixture of the S' state reduced $(\sigma_{-1})_{E1}$ by 8.5%. It was also shown that in this case the $(\sigma_{-1})_{E1}$ formula for the H^3 and He^3 nuclei could be expressed in the form

$$(\sigma_{-1})_{He^3} = (\sigma_{-1})_{H^3} = \frac{4\pi^2}{3}\left(\frac{e^2}{\hbar c}\right)\frac{NZ}{A-1}\,<R^2>_{H^3}. \qquad (10)$$

If the S' state is absent, this coincides with Foldy's formula, since, in this case, $<R^2>_{H^3} = <R^2>_{He^3}$. If, however, the S' state is present, then, by using formula (10) and the measured value of $(\sigma_{-1})_{E1}$ for the He^3 nucleus, we actually determine the charge-distribution radius for the H^3 and not the He^3 nucleus. It thus follows that our value of $<R_C^2>^{\frac{1}{2}} = (1.81 \pm 0.06)$F should be compared with the value $<R_C^2>^{\frac{1}{2}}_{H^3} = (1.68 \pm 0.17)$F measured by the Hofstadter group for e−H^3 scattering. The satisfactory agreement between these two quantities shows that the assumption of an added component of S' states in the ground state of the three-particle nuclei is in no way contrary to existing experimental data.

SEC. 4. COMPARISON BETWEEN EXPERIMENTAL DATA AND THEORETICAL CALCULATIONS
OF EFFECTIVE CROSS SECTIONS

The cross section for the photo-disintegration of H^3 and He^3 nuclei is calculated by the standard method

$$d\sigma = \frac{2\pi}{\hbar c}|M_{if}|^2 \rho_f,$$

where ρ_f is the density of the final states, and $M_{if} = \int \psi_f^* \hat{H}_{int} \psi_i dV$ is the matrix element of the transition of the system from the $\psi_i(H^3)$ state to the final state (n, d, or p, 2n), in the first order of perturbation theory with respect to the electromagnetic interaction $\hat{H}_{int}$ of the radiation with the nucleus. Comparison of $d\sigma$ with the experimental cross section enables us in principle to draw specific conclusions as to the correctness of the wave functions ψ_i, ψ_f chosen.

In comparing the experimental cross sections with the theoretical curves, we must remember that the cross-section calculations discussed below are made for H^3. An estimate of the part played by Coulomb repulsion in the two-particle channel [43] shows that the cross section is most strongly deformed near the reaction threshold; in the neighborhood of the maximum, the He^3 cross section is 10-15% below that corresponding to H^3.

1. Two-Particle Channel for Absorption of γ-Quanta by the Nucleus

The first work in this direction was carried out by Verde [43], who produced (in the isotopic-spin formalism) a classification of H^3 and He^3 wave functions in accordance with the group of permutations of three particles. Verde showed that the ground-state function of H^3 and He^3 with spin $S = \frac{1}{2}$ and isotopic spin $T = \frac{1}{2}$ was a combination of a completely symmetric state φ^s, a completely antisymmetric state φ^a, and two other states φ' and φ'' of mixed symmetry

$$\psi_i(S = \frac{1}{2}, \ T = \frac{1}{2}) = \varphi_i^s u^a + \varphi_i^a u^s + \varphi_i' u'' - \varphi_i'' u',$$

where u^s, u^a, u', and u'' are the spin-charge functions of corresponding symmetry.

An important result of [43] is the fact that in the case of a magnetic-dipole transition the spin-charge function u^a is an eigenfunction of the magnetic-moment operator,* and the matrix element of the M1 transition from the ground state $\varphi_i^s u^a$ is strictly equal to zero (if we neglect the φ^a, φ', φ'' components) on account of the orthogonality of the functions of the discrete and continuous spectra. Thus, the magnetic transition should be sharply suppressed if the amount of φ^a, φ', φ'' states mixed with the completely symmetric state is small. This is evidently one reason for the smallness of the cross section representing the capture of nucleons by deuterons.

In contrast to Verde's paper [43], a great part of which is of a methodical character, more detailed calculations of the cross sections of the electric-dipole disintegration of H^3 were carried out by Gunn and Irving [44] (without considering interaction in the final state) for two variants of ground-state wave functions

$$\varphi_1^s \sim e^{-\mu_{T_1}^2 \left(2\rho^2 + \frac{3}{2}r^2\right)}, \quad \varphi_2^s \sim \frac{e^{-\mu_{T_1}\sqrt{2\rho^2 + \frac{3}{2}r^2}}}{\sqrt{2\rho^2 + \frac{3}{2}r^2}}, \qquad (11)$$

where

$$\rho = -r_1 + \frac{r_2 + r_3}{2}, \quad r = r_2 - r_3,$$

r_1, r_2, and r_3 are the radii vectors of the nucleons. In calculating the cross sections, Gunn and Irving chose a deuteron wave function for the φ_1^s variant in the Gaussian form $e^{-\mu_D^2 r^2}$; for the φ_2^s variant they took a function in the form $e^{-\gamma r}/r$ corresponding to zero radius of action of the forces. The ratio μ_D/μ_{T_1} was taken as

---

*We assume that the magnetic-moment operator of the H^3 (He^3) nucleus equals the sum of the operators of the magnetic moments of the free nucleons.

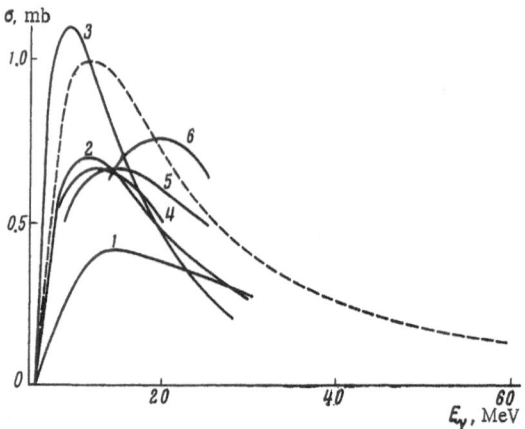

Fig. 13. Effective cross section of the reaction He³(γ,p)d, calculated with the wave functions of Gunn and Irving (curves 1, 2, 3, 4,
and 6) and that of Kikuta et al. (curve 5). Broken line gives the
experimental cross section.

1.12. The parameter μ_{T_1} was determined by a variation method with central forces of Gaussian form. The values of $1/\mu_{T_1}$ = 4.2 and 3.65 F were obtained for the forces of Margenau and Warren [45] and Rosenfeld [32]. The parameter μ_{T_2} was determined by a variation method with central forces in the Yukawa form. The best value was $1/\mu_{T_2}$ = 2 F [47]. The binding energy E_B,* the Coulomb energy E_C, and the mean-square radius $\langle R^2 \rangle^{\frac{1}{2}}$ are shown as functions of $1/\mu_{T_2}$ in Table 5.

Curves 1-3 of Fig. 13 [44] illustrate the sensitivity of the cross sections to $\langle R^2 \rangle^{\frac{1}{2}}$. †

In accordance with the results of Sec. 3, in which it was shown that $\langle R^2 \rangle^{\frac{1}{2}}_{\exp}$ = (1.62 ± 0.06) F, curves 4 and 6 of Fig. 13 show the cross section calculated with the Hulthen deuteron function [48] and Irving's φ_2^S, φ_1^S functions, which give $\langle R^2 \rangle^{\frac{1}{2}}$ = 1.75 F, quite close to $\langle R^2 \rangle^{\frac{1}{2}}_{\exp}$. The φ_2^S function obtained in this way, for example, differs from the best variational function with which curve 1 was plotted; the method of determining the parameter of the function from $\langle R^2 \rangle^{\frac{1}{2}}$ (or from the Coulomb energy) should be considered as phenomeno- logical.

The variational calculations of Pease and Feshbach with central and tensor forces [49] and those of Irving with central forces [47] failed to give the correct Coulomb energy of He³; three-particle nuclei with reasonable values of binding energy E_B proved to be too constricted. A possible way out of this situation was indicated by Kikuta, Morita, and Yamada [36], who used central potentials with a hard core and a variational wave function of form (6) (see Sec. 3).

---

* By binding energy E_B for $1/\mu_{T_2}$ = 2.5 and 3 F we mean the integral

$$E_B = -\int \bar{\psi}_{H^3}^* \left(-\sum_i \frac{\hbar^2}{2\mu_i} \Delta_i + \frac{1}{2} \sum_{ij} V_{ij} \right) \psi_{H^3} dV.$$

† The sensitivity of the cross section to the radius of the H³ nucleus was discussed by Rossetti [46], who used one of the Irving-function variants as ground-state function [47]. We omit discussion of this article, as it contains nothing new as compared with the calculations of Gunn and Irving. In addition to this, the author made a mistake in calculating the cross section, obtaining a value of 1.8 mb at the maximum, whereas the correct value is about 0.65 mb.

Table 5

$1/\mu T_s$, F	E_B, MeV	E_c, MeV	$\langle R^2 \rangle^{1/2}$, F
2	7.1	0.995	1.49
2.5	6.3	0.796	1.86
3	5.0	0.664	2.23

Table 6

$E_\gamma - 8.48$, MeV	σ_{M1}, mb	σ_{E1}, mb
10	$4.45 \cdot 10^{-4}$	$8.59 \cdot 10^{-1}$
0.31	$1.9 \cdot 10^{-3}$	$1.57 \cdot 10^{-2}$
0.0031	$2.34 \cdot 10^{-4}$	$1.57 \cdot 10^{-5}$

One of the authors of [39], using the function of Kikuta, Morita, and Yamada, with three sets of variational parameters

$$d = 0.4 \text{ F}, \qquad \mu = 0.4 \text{ (F)}^{-1}, \qquad \nu = 4.5 \text{ (F)}^{-1}, \qquad (12)$$

$$d = 0.4, \qquad \mu = 0.5, \qquad \nu = 4.5, \qquad (13)$$

$$d = 0.6, \qquad \mu = 0.5, \qquad \nu = 4.5, \qquad (14)$$

calculated the cross section of the $H^3(\gamma,n)d$ reaction with the Hulthen deuteron functions without considering the interaction of the reaction products in the final state. The mean-square radius $\langle R^2 \rangle^{1/2}_{H^3}$ was close to the value of $\langle R^2 \rangle^{1/2} = 1.68 \text{ F}$ for variant (12). Curve 5 in Fig. 13 corresponds to this variant. The cross sections for variants (13) and (14) are close to curve 1. Comparing the calculations obtained with different variants of the ground state, we note that functions giving similar values of $\langle R^2 \rangle^{1/2}$ also give similar cross sections.

Comparing the cross sections obtained with the Hulthen deuteron function and the H^3 wave functions of Gunn and Irving (curves 4 and 6) and the H^3 function of Kikuta et al. (variants 12, 13, 14), we see that the calculated cross sections lie 20-30% below the experimental values indicated by the broken curve. Considering the sensitivity of the cross section to the dimensions of H^3 (curves 1-3 in Fig. 13) and also remembering the Coulomb factor (see note on page 99), we can scarcely expect to raise the cross section into the region of experimental values by modifying the form of the ground-state wave function without giving too large dimensions for He^3. This fact clearly indicates the necessity of refining the wave function of the nucleon and deuteron in the final state.

The results of Kikuta, Morita, and Yamada [36] hardened the view that the existence of a core to the nucleon was necessary in order to obtain correct values of binding energy and Coulomb energy for three-particle nuclei. * Some authors have nevertheless recently disputed this view. Davey and Valk [37] used the Hu and Massey potential [40] without a hard core and obtained a function satisfactorily describing the dimensions of the nucleus. On the other hand, Völkel [51] used a central interaction, and by means of a variational procedure for a combination of Gaussian functions

$$\varphi^s = \sqrt{N} \sum_{\nu = 1, 2} A_\nu e^{-C_\nu (\rho^2 + 3/4 \, r^2)} \qquad (15)$$

with three parameters $\eta = (A_1/A_2) = 0.15$, $C_1 = 0.3 f^{-2}$, $C_2 = 0.07 f^{-2}$, obtained $E_B \sim 6.6$ MeV and $E_c \sim E_c^{exp}$.

Eichmann [29] used Völkel's wave functions of the ground S state, allowing for the component of mixed symmetry φ', φ''

$$\psi_i \sim \varphi^s u^a + \delta \frac{1}{\sqrt{2}} (\varphi' u'' - \varphi'' u') \qquad (16)$$

---

*This question was recently discussed by Pappademos [50], who studied the effect of the hard core of the nucleon on the mean-square radius of tritium.

in order to calculate the cross section of the (γ, n) reactions at the H^3 and He^3 nuclei. In order to simplify the calculations, the deuteron function was taken in the form proposed by Ernst and Flügge [52]: $g(r) = \sqrt{N} \sum_{\mu} q_{\mu} e^{-\alpha_{\mu} r}$. The parameter δ was taken as approximately 0.1 in accordance with the cross section for the capture of thermal neutrons by deuterons. Magnetic disintegration of H^3 was possible owing to the admixture of φ', φ'' states.

We see from Table 6 (see [29]) that the E1 transition predominates over the M1 transition over a wide energy range, nearly up to the reaction threshold. The cross sections calculated by Eichmann are shown in Fig. 14. By comparing the theoretical cross sections with the experimental curve we see that allowing for the states of mixed symmetry considerably improves agreement with experiment in the region of the maximum, but slightly reduces the cross section in the region of $E_\gamma \gtrsim 28$ MeV. We notice that calculations based on the improved wave function of the final state, allowing for the phase of the scattering of a nucleon at a deuteron in the $P_{\frac{1}{2}}$ state, increase the cross section by 20-25% in the neighborhood of the maximum (curve 4 in Fig. 14) and lift the cross section into the region of the experimental values. The effect of interaction in the final state also clearly brings curves 4 and 5 of Fig. 13 toward the experimental curve.

The angular distribution of the reaction products in the E1 approximation in the calculations under consideration has the simple form $\sim \sin^2 \theta$, where θ is the escape angle of the particle with respect to the beam of γ-quanta. According to the interpretation of Eichmann, the displacement of the angular-distribution maximum in the low-angle direction is due to interference of the E1 and E2 transitions.

Figure 15 shows the angular distribution calculated by Eichmann in the (E1 + E2) approximation for energy $E_\gamma \sim 20$ MeV. For comparison, the same graph shows the angular distribution obtained experimentally for the γ-ray energy range $E_\gamma = 16$-22 MeV (see Table 1). We see from Fig. 15 that the curves are of very similar form.

2. Three-Particle Disintegration of H^3

The cross section of the three-particle disintegration of H^3 with plane waves in the final state was considered by Verde [43]* and Gunn and Irving [44]. The cross section obtained by Gunn and Irving with the wave function φ_2^S ($1/\mu T_2 = 2.5$ F, $\langle R^2 \rangle^{\frac{1}{2}} = 1.86$ F) is shown by the broken (dashed) line in Fig. 16. In contrast to the case of two-particle disintegration, in which the deviation between theory and experiment was no greater than 20 to 50%, the experimental curve of the three-particle channel was 2.5 to 3 times lower than the calculated curve for reasonable wave-function parameters. This probably indicates the urgent necessity of allowing for internucleon interaction in the final state.

Delves [53] first considered the effect of interaction in the final state on the total-disintegration cross section of H^3. A transformation from the ordinary Jacobi variables

$$\mathbf{r}_n = \mathbf{r}_1 - \mathbf{r}_2, \quad \mathbf{r}_p = -\mathbf{r}_3 + \frac{\mathbf{r}_1 + \mathbf{r}_2}{2}, \tag{17}$$

where $\mathbf{r}_1$, $\mathbf{r}_2$, and $\mathbf{r}_3$ are the radii vectors of the neutrons and the proton, to the new variables

$$\mathbf{r}_n = (^4/_3)^{1/2} \mathbf{r} \sin \alpha, \quad \mathbf{r}_p = (^3/_4)^{1/2} \mathbf{r} \sin \alpha, \quad \theta_n, \ \varphi_n, \theta_p, \ \varphi_p \tag{18}$$

enabled Delves, after making some assumptions regarding the quantum numbers of the ground state, to obtain two one-dimensional differential equations of the second order, both for the bound state and for the state of the continuous spectrum. It was found that for a central nucleon-nucleon interaction of Gaussian form the calculated cross section in the neighborhood of the maximum was roughly 10 times greater than experimental.

---

*Verde obtained general formulas for the complete-disintegration cross section assuming a Gaussian ground-state wave function, but made no analysis of the cross-section curves.

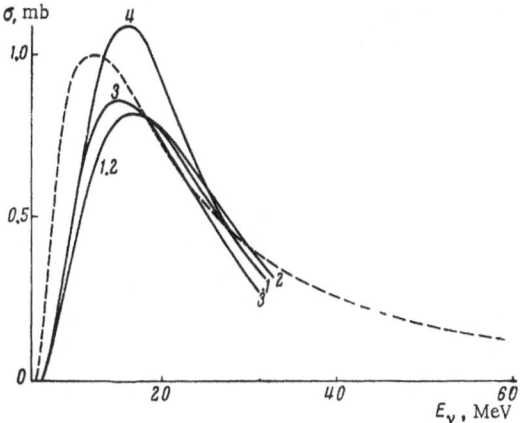

Fig. 14. Effective cross section of the He³(γ,p)d reaction calculated by Eichmann (curves 1-4). Broken line gives experimental cross section.

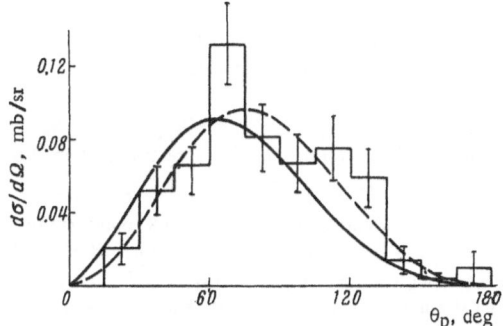

Fig. 15. Angular distribution of protons emitted in the He³(γ,p)d reaction as calculated by Eichmann, allowing for the interference of electric-dipole and quadrupole transitions (continuous line). Broken line gives experimental angular distribution.

The author obtained the wave functions and cross sections by numerical methods. This fact, together with the author's large number of special assumptions, makes it difficult to give a direct indication of the defects in the Delves theory.

The theory developed in [53] considers an average inter-nucleon interaction distorting the wave function of the three particles. It is well known, however, that in reactions involving three particles (for example, in the disintegration of a deuteron by a nucleon [54]), an important part is played by the interactions of separate pairs of nucleons, leading to considerable changes in the energy spectra of the products of the reaction p + d → p + p' + n. Let us consider this question in relation to the present case. If the spatial part of the integrand in M_{if} is written in Jacobi variables [20], then in the case of dipole absorption the operator for the interaction of radiation with the nucleus, $\hat{H}_{int}$, has the simple form

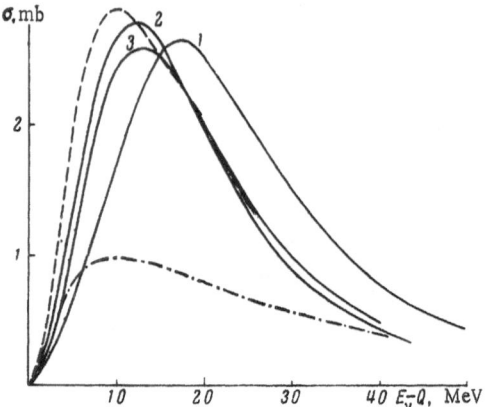

Fig. 16. Effective cross section of three-particle disintegration calculated in this paper (continuous lines). The broken (dashed) line shows
the effective cross section calculated by Gunn and Irving, and the
other (dot-dashed) line gives the experimental effective cross section.

$$\hat{H}_{int} \sim \begin{cases} ur_p & \text{for} \quad H^3, \\ ur_n & \text{for} \quad He^3, \end{cases} \tag{19}$$

where u is the radiation polarization vector. Since the operator $\hat{H}_{int}$ is independent of the radius vector $r = r_1 - r_2$ (1, 2 being identical nucleons), the orbital moment of the relative motion of the identical particles remains constant during the transition $\psi_i \to \psi_f$.

It is usually assumed that the ground state ψ_i contains only an S component in the variables $r_{n(p)}$ and r. As a result of this, the final state ψ_f also contains only an S wave in the variable of the relative motion of the neutrons, $r = r_1 - r_2$, and a P wave in the variable $r_p = -r_3 + (r_1 + r_2)/2$. In the ensuing calculation we consider the scattering of a neutron at a neutron in the singlet S state but not the interaction of the proton with the neutrons.

The H^3 photo-disintegration cross section in the $H^3(\gamma, n)d$ channel was described very satisfactorily by Eichmann, as shown earlier, by means of a wave function (16) including not only the completely symmetric S state φ_i^S, but also the S states $\varphi_i^!$, $\varphi_i^"$ of mixed symmetry. As ground-state function we shall take function (16). The state of the continuous spectrum will be taken in the form

$$\psi_f = \frac{1}{\sqrt{3}} \hat{a} \psi_f(12, 3) a_3 b_1 b_2 \chi_{1/2\,\sigma}(3) \chi_{00}(12), \tag{20}$$

where $\hat{a}$ is the operator of antisymmetrization of particle 3 with particles 1 and 2, a and b are the isotopic-spin wave functions, and $\chi_{s\sigma}$ are the spin functions [55]. The spatial part of function (20), $\psi_f(12, 3)$, is the product of a plane wave e^{iKr_p} (K is the momentum of the proton relative to the neutrons) and the wave function of the relative motion of the neutrons, satisfying Schrödinger's equation for e = 0, with potential $V(r)$ taken, for simplicity of calculation, in the form [56]

$$V(r) = \begin{cases} -V_0 & \text{for} \quad r \leqslant R, \\ \\ 0 & \text{for} \quad r > R. \end{cases} \tag{21}$$

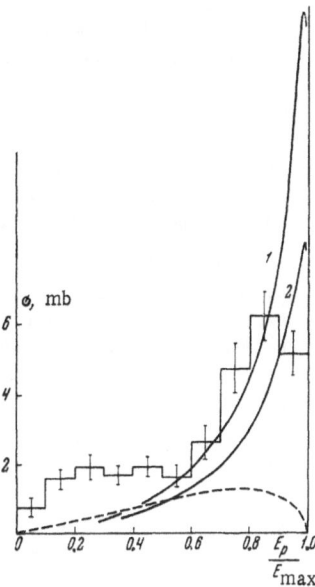

Fig. 17. Energy spectra of protons in the $H^3(\gamma,p)$ 2n reaction. Curves 1 and 2 are calculated for values of $E_\gamma - Q$ equal to 10 and 6 MeV, respectively. The histogram shows the experimental neutron spectrum in the reaction $He^3(\gamma,n)$ 2p.

Potential (21) with parameters $V_0 = 16.8$ MeV, $R = 2.37f$, correctly describes the scattering of low-energy nucleons in the singlet state. Omitting the simple derivations, let us give the final formula (in the linear approximation with respect to δ) for the cross section, integrated over the particle escape angles

$$\sigma(\beta) = 87.5 \cdot 10^{-6} E_\gamma E^2 e^{-0.0536 \, E \sin^2 \beta} \sin^4\beta \, \frac{1}{1 + \operatorname{ctg}^2 \delta_0} \{[A_1(\beta) +$$

$$+ 5.7 \, e^{-0.088 E \sin^2 \beta} A_2(\beta)]^2 - \delta \, [A_1(\beta) + 5.7 \, e^{-0.088 \, E \sin^2 \beta} A_2(\beta)] \times$$

$$\times \, [(1 - 0.0107 \, E \sin^2 \beta) A_1(\beta) + 24.4 \, (1 - 0.046 \, E \sin^2 \beta) e^{-0.088 \, E \sin^2 \beta} A_2(\beta) -$$

$$- 0.11 \, (B_1(\beta) + 5.7 \, e^{-0.088 \, E \sin^2 \beta} B_2(\beta))]\} \text{ mb},$$

(22)

$$A_1(\beta) = \frac{[\cos kR + \operatorname{ctg} \delta_0 \sin kR]}{\sin Q'R} \int_0^R e^{-0.75 \, c_1 \xi^2} \sin(Q'\xi)\xi \, d\xi +$$

$$+ \int_R^\infty e^{-0.75 \, c_1 \xi^2} [\cos k\xi + \operatorname{ctg} \delta_0 \sin k\xi] \, \xi \, d\xi.$$

Here δ_0 is the neutron–neutron scattering phase, and $Q' = \sqrt{(MV_0/\hbar^2) + k_2}$, $B_1(\beta)$ differs from $A_1(\beta)$ by an additional factor ξ under the integrals. Terms $A_2(\beta)$, $B_2(\beta)$ are obtained from $A_1(\beta)$, $B_1(\beta)$ by the substitution $c_1 \to c_2$. Momenta k and K are related to the total energy $E = E_\gamma - Q$ by relations $k = \sqrt{(ME/\hbar^2)} \cos\beta$, K = $\sqrt{(4ME/3\hbar^2)} \sin\beta$. All the integrals required were calculated on an electronic computer.

Figure 17 shows the energy spectra of the protons for a certain set of E values. For comparison purposes, the broken line gives the proton spectrum (E = 6 MeV) obtained for V(r) = 0. We notice the severe distortion

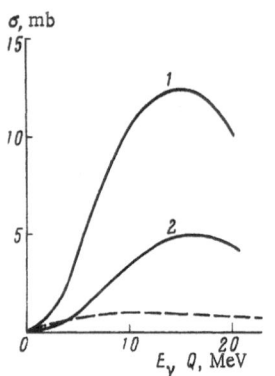

Fig. 18. Effective cross section of the $H^3(\gamma,p)2n$ reaction obtained in [55]. Curves 1 and 2 calculated for $V(r) \neq 0$ and $V(r) = 0$, respectively. Broken line shows the effective cross section of the $He^3(\gamma,n)2p$ reaction.

of the spectrum for large values of E_p. In the present investigation, the neutron spectrum was measured for the disintegration of He^3. Comparing the experimental and theoretical spectra for $E_p/E_{max} = 0.5-0.9$, we see that a considerable discrepancy occurs for very small relative momenta of the protons. This evidently reflects the important part played by Coulomb interaction. In the region $E_p/E_{max} \sim 0.25$ there is an additional maximum, probably due to interaction between proton and neutron. *

The angular distribution of protons (in the case of the H^3 nucleus) or neutrons (in the case of the He^3 nucleus) calculated on the El approximation is proportional to $\sin^2\theta$ (owing to the fact that the nonidentical nucleon is formed in the P state relative to the identical nucleons) and agrees closely with the experimental values (see Fig. 7).

The total cross sections for $V(r) \neq 0$ ($\delta = 0$, $\delta = 0.1$) and $V(r) = 0$ ($\delta = 0$) are shown in curves 2, 3, and 1 in Fig. 16. The photo-disintegration maximum is displaced in the direction of lower γ-quantum energies on taking account of $V(r)$. Curves 2 and 3 give the position of the maximum better, but the calculated cross section remains approximately 2.7 times greater than the experimental, which is shown by the broken line. Calculation of the total cross section with other forms of the (nn) potential ($e^{-\alpha r}$, $e^{-\alpha r}/r$) gives results similar to those given above for a rectangular potential well. This indicates that the effects of the (np) interaction are also important in understanding the mechanism of the complete photo-disintegration of three-particle nuclei.

The effects of neutron—neutron interaction in the total photodisintegration of H^3 were considered by Györgyi and Hraskó [57], who used the Gaussian form of the wave function

$$\psi(H^3) = \left(\frac{4\alpha^2}{3\pi^2}\right)^{3/4} e^{-\frac{\alpha}{3}\left(2\rho^2 + 3/2\,r^2\right)}.$$

The proton spectra calculated for $\alpha = 0.22$, 0.17, and 0.11 F^{-2} (mean-square radius of nucleus $<R^2>^{\frac{1}{2}} = 1/\sqrt{2\alpha}$) were similar in form to those of Fig. 17, but the total cross sections were considerably different, as may be seen by comparing Figs. 16 and 18. The maximum value of the cross section changed very little in our calculation on including the interaction $V(r)$. Moreover, for high energies ($E_\gamma \gtrsim 23$ MeV) the presence of V_r reduced the cross section in comparison with the case of $V(r) = 0$. The large cross sections in Fig. 18 are probably due to the use of an oversized H^3 ($<R^2>^{\frac{1}{2}} = 2.13$ F). The use of numerical integration in obtaining the cross sections prevents us from understanding the reasons for the fact (illustrated in Figs. 1 and 2 of [57]) that the cross section obtained with due allowance for (nn) interaction is considerably larger than that obtained with plane waves for $E_\gamma = 43.48$ and 36.48 MeV, if we use the real dimensions of H^3 ($<R^2>^{\frac{1}{2}} = 1.5$ F and $<R^2>^{\frac{1}{2}} = 1.72$F, respectively).

SEC. 5. EXPERIMENTAL RESULTS OF OTHER AUTHORS

1. Two-Particle Disintegration

The first experiment on the photo-disintegration of He^3 was carried out by Cranberg in 1958 [58], using a 22-MeV betatron. Cranberg recorded the protons and deuterons arising in the $He^3(\gamma,p)d$ and $He^3(\gamma,n)2p$ reactions by means of nuclear photo-emulsions placed inside a gas target containing He^3. Since the traces of

_____

*A similar effect also arises in the $p + d \rightarrow p + p' + n$ reaction [54].

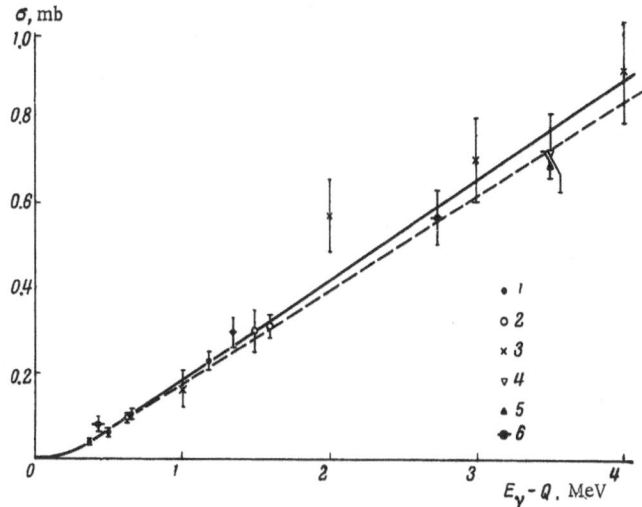

Fig. 19. Effective cross sections of the two-particle photo-disintegration of H^3 and He^3 measured near the threshold. (1) [63]; (2) [64]; (3) present investigation; (4) [67]; (5) [66]; (6) [65].

protons and deuterons were not distinguished, Cranberg assumed, in order to separate out the cases associated with the two-particle reaction, that all protons with energies greater than 5 MeV arose from a two-particle reaction. We see from our own proton-energy spectrum for photon energies below 22 MeV that protons with energies right up to the maximum possible for the reaction in question arise in the three-particle process: $(E_p)_{max} = {}^2/_3 (E_\gamma - 7.72)$ MeV [for $E_\gamma = 22$ MeV, $(E_p)_{max} = 9.5$ MeV]. Thus, Cranberg's method of distinguishing the two-particle process is not entirely reliable. The angular distribution of protons with $E_p \geq 4$ MeV found by Cranberg has a distinct asymmetry, the coefficient 0.79 of the $\sin^2 \theta \cos \theta$ term in the angular distribution being close to our own value (Table 1) for $E_\gamma = 6$-22 MeV. Cranberg's large isotropic component of the angular distribution of protons with $E_p \geq 4$ MeV is probably associated with the protons emitted in the three-particle reaction. We see from Fig. 8 that the angular distribution of the protons in this reaction has a considerable isotropic component.

The first information on the two-particle photo-disintegration of He^3 was obtained still earlier in [59-63] when studying the reverse process, the radiative capture of protons by deuterons, $p + d \rightarrow He^3 + \gamma$ for low-energy protons. It was shown in these papers that the angular distribution of the γ-rays emitted in this reaction obeyed a $\sin^2\theta$ law with a very small isotropic component added. This indicated that electric-dipole radiation was emitted as a result of the radiative-capture reaction. The same conclusion was reached by Wilkinson [61] when studying the polarization of this radiation. By using the detailed-balance rule and the effective cross section of the $p + d \rightarrow He^3 + \gamma$ reaction, we can find the effective cross section of the forward process $\gamma + He^3 \rightarrow p + d$. Unfortunately the experiments on radiative capture were carried out with very low proton energies, so that they only enabled the cross section of the $He^3(\gamma,p)$ d reaction to be obtained up to an energy of $E_\gamma \simeq 1.2$ MeV. The cross section of the $He^3(\gamma,p)d$ reaction calculated from the results of Griffits et al. [63] is shown in Fig. 19. The same figure gives the total effective cross sections of the two-particle photo-disintegration near the reaction threshold as obtained later by other authors.

Warren et al. [64] measured the total effective cross section for the two-particle photo-disintegration of He^3 at energies $E_\gamma = 6.14$, 6.97, and 7.08 MeV by means of a grid-containing ionization chamber, filled with a mixture of He^3, methane, and argon. For control purposes, an experiment was made with an identical chamber filled with He^4, methane, and argon. The intensity of the γ-ray flux was measured with an NaI (Tl) scintillation counter.

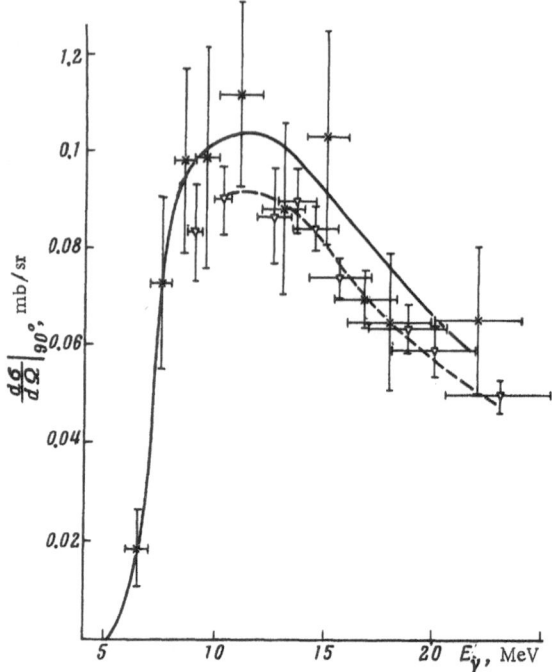

Fig. 20. Differential effective cross section of two-particle photo-disintegration measured at an angle of 90° to the direction of the γ-ray beam (in the laboratory system).
Continuous line obtained from points determined in the present investigation (crosses);
broken line from points determined in [67] (triangles).

Bösch et al. [65] measured the total effective cross section of the two-particle photo-disintegration of
H^3 [$H^3(\gamma,n)$d reaction], using monochromatic γ-rays obtained in (n,γ) reactions with Ni (E_γ = 9 MeV), Fe
(E_γ = 7.6 MeV), and Ti (E_γ = 6.7 MeV). The photoneutrons were recorded with a BF_3 counter. The effective
cross section was determined from comparative measurements with a deuterium target, since the photodisintegration cross section of the deuteron was well known, and the angular distribution of photoneutrons from H^2 had
the form $\sin^2 \theta$, in the same way as the angular distribution of protons from the $He^3(\gamma,p)$d reaction near the
threshold.

Owing to the charge independence of the nuclear forces, the effective cross sections of two-particle reactions at H^3 and He^3 nuclei should be identical for electric-dipole absorption, except in a range of about 2 MeV
near the threshold, where the Coulomb barrier reduces the photo-disintegration cross section of helium. We see
from Fig. 19 that, except for points with $E_\gamma - Q$ = 0.4 MeV, the effective cross sections of the $H^3(\gamma,n)$d reaction agree with the results for He^3. So far there have been no other direct measurements of the total effective
cross section of the two-particle reaction.

Figure 19 shows the effective cross sections of the $He^3(\gamma,p)$ d reaction calculated from differential cross
sections $(d\sigma/d\Omega)_{90°}$ measured by Berman et al. [66] and Stewart et al. [67]. In [66], the value of $(d\sigma/d\Omega)_{90°}$
for the two-particle photo-disintegration of He^3 was measured for photon energies of E_γ = 8.5-21.5 MeV by recording the $(p-d)$ coincidences in scintillation counters placed inside the He^3-containing gas target in a plane
perpendicular to the axis of the γ-ray beam and at an angle of 180° to each other. Pulses from the two
counters were photographed on the screen of an oscillograph. The energies of the particles were determined
from the amplitudes of the pulses. Identification of the disintegration was based on the observed ratio of the

Table 7

Literature source	σ_0, MeV · mb	σ_{-1}, mb
[66]	13	0.87
[67]	14 ± 1.4	1.0 ± 0.1
Present work	19.7 ± 1.4	1.23 ± 0.09

Note. Upper limit of integration $E_\gamma = 40$ MeV.

pulse amplitudes, which should be equal to 2 if a proton falls into the first counter and a deuteron into the other. The intensity of the radiation was measured with a Dural ionization chamber. The authors indicate that the maximum error in the absolute value of the cross section is 6% or under.

In [67] crystal counters were also used to measure the differential effective cross section of the two-particle reaction at an angle of 90° in the laboratory system for the energy range 8.5 to 46 MeV.

We calculated the total effective cross section from the values of $(d\sigma/d\Omega)_{90°}$ given in [66, 67] by using the formula $\sigma_{total} = (8\pi/3)(d\sigma/d\Omega)_{90°}\varepsilon'$, the value of $\varepsilon' = \varepsilon/(1 + \delta) = 1.03$ being found from the angular distribution of the protons (see Fig. 5, Table 1) for the range $E_\gamma = 6$-22 MeV. The continuous straight line was calculated by the method of least squares from the results of Griffits et al. [63] and Warren et al. [64], and the broken line from all the points given in the graph except the point $E_\gamma = 6.7$ MeV for H^3. We see from Fig. 19 that all the results agree within 10% except for the obviously erratic point at $E_\gamma - Q = 2$ MeV.

The total cross sections were not estimated from the $(d\sigma/d\Omega)_{90°}$ for higher γ-ray energies, since (as shown in Table 1), the angular distributions have a considerable asymmetry for energies above 12 MeV. Hence, in order to estimate the cross section we must know how the form of the angular distributions changes with the energy of the γ-rays. Instead of this we estimated the differential cross sections $(d\sigma/d\Omega)_{90°}$ in the center-of-mass system from the results of our own work and compared them in Fig. 20 with the measurements of the Yale group [67] (the results of the Illinois [66] and German [68] groups almost coincided with the Yale results). We see from the graph that the two curves are very similar in form, while in absolute magnitude they differ by some 10%. The discrepancy diminishes by several percent if we consider that our own data were obtained for an angle of 90° in the center-of-mass system, whereas Stewart's data were for 90° in the laboratory system.

There are considerably greater differences between our own integral cross sections obtained by direct integration of the total effective cross section and those of Berman et al. and Stewart et al. determined from

$$\sigma_0 = \frac{8\pi}{3} \int_0^E \left(\frac{d\sigma}{d\Omega}\right)_{90°} dE.$$

their differential cross section at an angle of 90° from the formula   Table 7 shows

our own data together with the values of σ_0 and σ_{-1} obtained in the other two papers, the results of Berman et al. being extrapolated to $E_\gamma = 40$ MeV.

The differential cross section $(d\sigma/d\Omega)_{90°}$ for He^3 in the range $E_\gamma = 11$-30 MeV was also recently measured by Becchi et al. [69], who used crystal counters to measure either the coincidence of protons and deuterons emitted in the $He^3(\gamma,p)d$ reaction in opposite directions, or else the particles emitted at an angle of 90° to the beam, the first, thin counter measuring dE/dx and the second, thick counter, the energy of the particle. In the first experiment the $He^3(\gamma,p)d$ reaction was identified by the ratio of the pulse amplitudes in the two counters. Becchi's article gives the total cross section of the two-particle photo-disintegration of He^3 calculated from the value of $(d\sigma/d\Omega)_{90°}$. (No indication of the method used for calculating σ appears in the article.) The relation between the effective cross section and the photon energy obtained in this paper by the second method (in the range $E_\gamma = 16$-30 MeV) is of the same form as that given in Fig. 4, although Becchi's cross section is 40-50% larger in this range of energies. For $E_\gamma = 12$ MeV, Becchi's cross section is twice ours.*

---

*Note added in proof. In view of the disagreement with other authors, we repeated the calibration of our apparatus for measuring the intensity of the bremsstrahlung, using the $\gamma + d \rightarrow n + p$ reaction as standard. The results of this calibration agreed to within 3% with the earlier data.

2. Three-Particle Disintegration

Published information on the three-particle reaction is extremely limited. Berman's paper [66] gives the $d^2\sigma/(d\Omega_1 d\Omega_2)$ cross section for the case in which the protons are emitted at an angle of 180°. The cross section has a maximum at an energy of 17 MeV in the same way as our own total cross section (Fig. 6), but falls more rapidly than the latter with increasing energy. Berman et al. also found that the $(d\sigma/d\Omega)_{90°}$ for the three-particle reaction was 3.3 times larger than the $(d\sigma/d\Omega)_{90°}$ for the two-particle reaction; they concluded from this that the integral cross section of the three-particle reactions should be at least three times that of the two-particle reaction, in agreement with the calculations of Gunn and Irving [44]. Thus, the cross section of the three-particle reaction, integrated up to 40 MeV, would, according to [66], be expected to be at least 39 MeV · mb. According to our own data, the value is (23.8 ± 2) MeV · mb.

Stewart et al. [67] subtracted from the total number of protons recorded for a given energy E_γ the number of deuterons recorded for the same energy, and obtained the proton spectrum for $E_\gamma^{max} = 40$ MeV. By comparing this with the spectrum calculated from the theoretical effective cross section of the three-particle disintegration given by Gunn and Irving (without considering the interaction of the protons in the final state), Stewart found (in contrast to Berman et al.) that the experimental proton spectrum differed by a factor of 2 in absolute magnitude from the theoretical, although the two spectra had the same shape. Stewart attributed this discrepancy to the fact that the Gunn and Irving calculations failed to allow for the interaction of protons in the S state.

CONCLUSIONS

1. We have measured the effective cross sections, yields, angular distributions, and energy spectra of the nucleons for the two possible channels involved in the photo-disintegration of He³, the (γ,p), and the (γ,n). The integral cross section $(\sigma_0)_{E1} = (62 \pm 6)$ MeV · mb agrees closely with the theoretical rule of sums, being little dependent on the model of the nucleus.

2. The mean-square charge-distribution radius of He³, calculated from the integral cross section σ_{-1} on the assumption that the spatial part of the ground-state function is a completely symmetric S state, equals $\langle R_C^2 \rangle^{\frac{1}{2}} = (1.81 \pm 0.06)$F. This is slightly less than the value obtained by the Hofstadter group from experiments on the scattering of electrons in He³: $\langle R_C^2 \rangle_H^{\frac{1}{2}} = (1.97 \pm 0.1)$F; it agrees better with the value of $\langle R_C^2 \rangle_H^{\frac{1}{2}} = (1.68 \pm 0.17)$F obtained from experiments on e–H³ scattering. In other words, our value of $\langle R_C^2 \rangle^{\frac{1}{2}}$ indirectly measured by photoeffect experiments is in no way contradictory to the assumption that a certain admixture of an S component of mixed symmetry exists in the ground state of He³.

3. The experimental effective cross section of two-particle disintegration in the neighborhood of the maximum $\sigma_{max} \approx 1.0$ mb is 20-30% higher than the theoretical value obtained without allowing for the interaction of the nucleon with a deuteron for a wide class of ground-state wave functions correctly describing the dimensions of the nucleus. If we take the interaction in the final state into account, then, according to estimates given in [29], this discrepancy vanishes, although the position of the maximum (Fig. 14) is still a little too high (by 5-6 MeV). The calculations of [29] are closest to the experimental results in the range $E_{thresh} \le E_\gamma \le 25$ MeV, but give too low a value of cross section at higher energies of the γ-quanta.

The angular distribution of protons calculated with due allowance for dipole and quadrupole absorption agrees with experiment to within the limits of measuring error.

4. The kinematics of the escape of particles in the three-particle disintegration channel, according to the experimentally determined average angles γ_{np}, γ_{pp}, and the neutron energy spectra (Sec. 2), are such that there is a high probability of two protons escaping with a low relative momentum. It is true that allowance for the singlet S scattering of identical particles in the mirror reaction (H³ + $\gamma \rightarrow$ n + n + p) in general gives the correct form of the spectrum. The drop in the neutron spectrum at very small relative momenta of the protons is probably due to the Coulomb interaction of the protons.

5. In the energy spectrum of the neutrons (Sec. 2, Part 3c, Fig. 9) there is a second maximum at $(E_n/E_{max}) \sim 0.25$; this is evidently associated with the formation of a singlet (np) pair in the S state with a low relative momentum. On this assumption the ratio E_n/E_{max} is easily seen to be of the order of 0.25.

6. The absolute value of the three-particle disintegration cross section at the maximum is almost identical with the two-particle disintegration cross section (about 1.0 mb). However, the half-width of the cross section in the three-particle channel is 1.6 times that in the two-particle channel. Hence, the $\sigma_0 = 43.6$ MeV $\cdot$ mb and $\sigma_{-1} = 1.42$ mb in this channel slightly exceed the $\sigma_0 = 26.5$ MeV $\cdot$ mb and $\sigma_{-1} = 1.34$ mb in the two-particle channel (Table 2). The calculated effective cross section (allowing for singlet interaction of identical particles) gives the correct position for the cross-section maximum but, as in the case of plane waves in the final state, gives much too high a value of the cross section (at least a factor of 2.7). This is a direct indication of the need to improve the wave function of the system of three particles in the final state, since the ground-state function of H^3 (He^3) gives a fair prediction of the integral cross sections σ_0 and σ_{-1} and satisfactorily describes the cross section of the two-particle channel.

The form of the angular distribution of neutrons, which is proportional to $\sin^2\theta$ in the E1 approximation, is not sensitive to the detailed structure of the wave functions and agrees (within the limits of experimental error) with measured values.

In conclusion, the authors wish to thank V. E. Yakushkin for a great deal of work in planning and making the apparatus, A. M. Ivanov and K. G. Kuvatov for help in making and preparing the system, V. A. Dubrovina, A. I. Orlova, V. A. Osipova, and G. G. Taran for assistance in the experiments, V. S. Silaeva, and M. S. Starichenko for help in analyzing the results, and T. D. Kruglova and V. P. Fomina for the computer calculations.

The authors are grateful to Professor H. S. Valk for a number of valuable comments.

LITERATURE CITED

1. B. Gruber, Phys. Letters, 5:278 (1963).
2. E. V. Teodorovich and N. N. Kolesnikov, Zh. Eks. i Teor. Fiz., 32:392 (1957).
3. G. V. Skornyakov and K. A. Ter-Martirosyan, Zh. Eks. i Teor. Fiz., 31:775 (1956).
4. G. S. Danilov, Zh. Eks. i Teor. Fiz., 43:1436 (1962)..
5. V. F. Kharchenko, Ukr. Fiz. Zh., 7:573 (1962); 7:582 (1962); A. G. Sitenko and V. F. Kharchenko, Izv. Akad. Nauk SSSR, Ser. Fiz., 28:41 (1964).
6. J. L. Gammel and R. M. Thaler, Phys. Rev., 107:291 (1957); 107:1359 (1957).
7. P. S. Signell and R. E. Marshak, Phys. Rev., 106:832 (1957); 109:1229 (1958); P. S. Signell, R. Zinn, and R. E. Marshak, Phys. Rev. Letters, 1:416 (1958).
8. G. Breit, M. H. Hull, K. Lassila, and K. D. Pyatt, Phys. Rev., 120:2227 (1960).
9. C. Werntz, Phys. Rev., 121:849 (1961).
10. J. F. Dawson and J. D. Walecka, Ann. Phys., 22:133 (1963).
11. J. M. Blatt, G. H. Derrick, and J. N. Lyness, Phys. Rev. Letters, 8:323 (1962).
12. R. C. Herndon, E. W. Schmid, and Y. C. Tang, Nucl. Phys., 42:113 (1963).
13. H. Collard and R. Hofstadter, Phys. Rev., 131:416(1963); H. Collard, R. Hofstadter, A. Johansson, R. Parks, M. Ryneveld, A. Walker, M. R. Yearian, R. B. Day, and R. T. Wagner, Phys. Rev. Letters, 11:132 (1963); L. I. Schiff, H. Collard, R. Hofstadter, A. Johansson, and M. R. Yearian, Phys. Rev. Letters, 11:387 (1963).
14. L. I. Schiff, Phys. Rev., 133:B802 (1964).
15. J. J. deSwart and R. E. Marshak, Phys. Rev., 111:272 (1958); M. L. Rustgi, W. Zernik, G. Breit, and D. I. Andrews, Phys. Rev., 120:1881 (1960).
16. G. H. Derrick and J. M. Blatt, Nucl. Phys., 8:310 (1958).
17. A. N. Gorbunov and V. M. Spiridonov, Zh. Eks. i Teor. Fiz., 34:866 (1958).
18. A. N. Gorbunov and V. M. Spiridonov, Zh. Eks. i Teor. Fiz., 33:21 (1957).
19. R. Wilson, Nucl. Instrum., 1:101 (1957).

20. W. C. Barber, W. D. George, and D. D. Reagan, Phys. Rev., 98:73 (1955).
21. G. G. Taran and A. N. Gorbunov, Zh. Eks. i Teor. Fiz., 46:1492 (1964).
22. A. N. Gorbunov and V. A. Osipova, Zh. Eks. i Teor. Fiz., 43:40 (1962).
23. H. Davies, H. A. Bethe, and L. C. Maximon, Phys. Rev., 93:788 (1954).
24. R. Wilson, Proc. Phys. Soc., A66:638 (1953).
25. J. S. Levinger and H. A. Bethe, Phys. Rev., 78:115 (1950).
26. J. S. Levinger, Nuclear Photodisintegration. Oxford Univ. Press (1960).
27. A. B. Migdal, Zh. Eks. i Teor. Fiz., 15:81 (1945).
28. Yu. K. Khokhlov, Dissertation, Physics Institute, Academy of Sciences, USSR (1955).
29. U. Eichmann, Z. Phys., 175:115 (1963).
30. M. L. Rustgi and J. S. Levinger, Phys. Rev., 106:530 (1957).
31. M. L. Rustgi, Phys. Rev., 106:1256 (1957).
32. L. Rosenfeld, Nuclear Forces. Amsterdam (1948).
33. D. R. Inglis, Rev. Mod. Phys., 25:390 (1953).
34. R. Serber, Phys. Rev., 72:114 (1947).
35. V. S. Mathur, S. N. Mukherjee, and M. L. Rustgi, Phys. Rev., 127:1663 (1962).
36. T. Kikuta, M. Morita, and M. Yamada, Progr. Theoret. Phys., 15:222 (1956); 17:326 (1957); 22:34 (1959).
37. P. O. Davey and H. S. Valk, Phys. Letters, 7:155 (1963).
38. L. L. Foldy, Phys. Rev., 107:1303 (1957).
39. V. N. Fetisov, Zh. Eks. i Teor. Fiz., 46:1395 (1964).
40. T. Hu and H. S. W. Massey, Proc. Roy. Soc., A196:135 (1949).
41. L. N. Hand, D. G. Miller, and R Wilson, Rev. Mod. Phys., 35:335 (1963).
42. P. O. Davey and H. S. Valk, Phys. Letters, 7:335 (1963).
43. M. Verde, Helv. Phys. Acta, 23:453 (1950).
44. J. C. Gunn and J. Irving, Philos. Mag., 42:1353 (1951).
45. H. Margenau and D. Warren, Phys. Rev., 52:790 (1937).
46. C. Rossetti, Nuovo Cim., 14:1171 (1959).
47. J. Irving, Philos. Mag., 42:338 (1951).
48. L. Hulthen and M. Sugavara, Handbuch der Physik, Vol. 39. Springer, Berlin (1957), p. 105.
49. R. L. Pease and H. Feshbach, Phys. Rev., 88:945 (1952).
50. J. N. Pappademos, Nucl. Phys., 42:122 (1963).
51. A. Völkel, Dissertation, Marburg (1961).
52. G. Ernst and S. Flügge, Z. Phys., 162:448 (1961).
53. L. M. Delves, Nucl. Phys., 29(2):268 (1962).
54. V. V. Komarov and A. M. Popova, Zh. Eks. i Teor. Fiz., 38:1559 (1960).
55. M. Verde, Handbuch der Physik, Vol. 39. Springer, Berlin (1957), p. 144.
56. L. Hulthen and M. Sugavara, Handbuch der Physik, Vol. 39. Springer, Berlin (1957), p. 55.
57. G. Györgyi and P. Hraskó, Acta Phys. Acad. Scient. Hung., 17:253 (1964).
58. L. Cranberg, Bull. Amer. Phys. Soc., 3:173 (1958); Private communication, December, 1958.
59. S. C. Curren and J. Strothers, Proc. Roy. Soc., 172:72 (1939).
60. W. A. Fowler, C. C. Lauritsen, and A. V. Tollestrup, Phys. Rev., 76:1767 (1949).
61. D. H. Wilkinson, Philos. Mag., 43:659 (1952).
62. G. M. Griffits and J. B. Warren, Proc. Roy. Soc., 68:781 (1955).
63. G. M. Griffits, E. A. Larson, and L. P. Robertson, Canad. J. Phys., 40:402 (1962).
64. J. B. Warren, K. L. Erdman, L. P. Robertson, D. A. Axen, and J. R. Macdonald, Phys. Rev., 132:1691 (1963).
65. R. Bösch, J. Lang, R. Müller, and W. Wölfli, Phys. Rev. Letters, 8:120 (1964).
66. B. L. Berman, L. J. Koester, and J. H. Smith, Phys. Rev. Letters, 10:527 (1963); Phys. Rev., 133:B117 (1964).

67. J. R. Stewart and J. S. O'Connell, Private communication, July, 1963; J. R. Stewart, J. S. O'Connell, and R. C. Morrison, Private communication, April, 1964.

68. E. Finckh, R. Kosiek, K. H. Lindenberger, U. Meyer-Berkhout, N. Nücker, and K. Schlüpmann, Phys. Letters, 7:271 (1963).

69. C. Becchi, G. E. Manuzio, L. Meneghetti, and S. Vitale, Phys. Letters, 8:322 (1964).

STUDY OF NUCLEAR REACTIONS AT HIGH ENERGIES
BY THE RECOIL-NUCLEUS METHOD*

F. P. Denisov

INTRODUCTION

An important means of studying the structure of atomic nuclei is supplied by high-energy particles. The wavelength of the particles falls with increasing energy and at energies of approximately 100 MeV becomes smaller than the mean distance between the nucleons in the nucleus, while the time of interaction with the nucleus is shorter than the characteristic nuclear time. These characteristics of high-energy particles enable us to obtain information on nuclear characteristics not accessible to other methods of study. The study of nuclear reactions at high energies has developed with particular intensity in recent years. Important information has been obtained on the radial distribution of the density of nuclear matter [1], on the shell structure of the nucleus [2], and on the association of nucleons in nuclei [3-5].

An analysis of papers appearing in recent years shows that the study of nuclear reactions at high energies is being carried out in the following main directions:

1. Measurement of the total cross sections of the inelastic interaction of particles with nuclei, and also the angular and energy distributions of fast particles (products of nuclear reactions).

2. Study of reactions of the (p, 2p) type.

3. Study of multicharged particles.

4. Measurement of the cross sections of formation of the individual products of nuclear reactions and the angular and energy distributions of recoil nuclei.

Let us consider the principal results of these investigations.

The total cross sections of inelastic interaction measured experimentally were close to the geometric cross sections of the nuclei, but a little smaller than the latter, owing to the partial transparency of atomic nuclei for high-energy nucleons [6-8]. Measurements also showed that the transparency of the nuclei diminished with increasing atomic number.

The angular distributions of fast particles constituting the products of nuclear reactions showed great anisotropy relative to the direction of the primary particle; the energy spectrum contained a considerable number of particles with high energies [9, 10].

---

*Dissertation in pursuit of the Degree of Candidate in the Physico-Mathematical Sciences. Defended July 19, 1964.

These experimental data are satisfactorily explained by the cascade model, which assumes that a high-energy particle interacts with a nucleus as if it were a gas of nucleons lying in a potential field of a certain configuration [11, 12]. Calculations based on this model [13, 14] show that the nuclear-reaction characteristics depend little on specific assumptions regarding the structure of the nucleus and are determined mainly by the mechanism of paired nucleon–nucleon collisions. Hence, agreement between experimental data and theory should be considered as a confirmation of the fundamental principles of the cascade model regarding the "quasi-free" interaction of the incident particles with the nucleons of the nucleus [15-17].

From the point of view of obtaining information on the structure of the nucleus, experiments with (p, 2p) reactions, investigations into the emission of complex particles, and a study of the ultimate characteristics of nuclear reactions, namely, the effective cross sections and the angular and energy distributions of the recoil nuclei, are of special importance.

Extensive studies of (p, 2p) reactions carried out in recent years [2] have shown that the characteristics of these reactions are due to the shell structure of the nuclei. As a result, it has been possible to obtain a whole series of important data regarding the shell structure of the nuclei, and to determine the energy levels of the shells in nuclei from He^4 to Ca^{40} by the most direct method. It is clear, however, that the shell picture of the nucleus is only very approximate and that the strong interaction between the nucleons should lead to correlation of the nucleons in the nucleus and to the formation of nucleon associations. A study of the characteristics of intranuclear associations is of great importance to nuclear physics. Without such study, we cannot understand the nature of nuclear forces or set up a satisfactory theory of the nucleus.

There have been many papers on the emission of complex particles (from deuterons to fission fragments). We shall not consider the emission of deuterons, in view of the specific mechanism of their formation, nor processes associated with the fission of heavy nuclei by high-energy particles. This will leave us free to give detailed consideration to the emission of particles with $Z \geq 2$.

Experiments involving reactions of the (p, a) [4, 18, 19] and (α, α) [20] types are of considerable interest. In order to understand the results of these experiments, we must assume the existence of α-particle groupings within the nuclei. From the resultant data we may determine, though as yet extremely roughly, such characteristics of these groupings as the probability of their existence in the nucleus and their momentum distribution [5, 21, 22].

Investigations into the emission of particles with $Z \geq 2$ carried out by the photographic-plate or radiochemical methods [23] are of extreme importance. The particles emitted (fragments) have a whole series of specific characteristics which cannot be explained by the evaporation model. There are good grounds for supposing that the emission of fragments takes place in the course of an intranuclear cascade, when considerable energy is imparted to individual groups as a result of elastic [24-27] or inelastic [28] scattering of nucleons at the corresponding groups or as a result of the breaking of the nuclear bonds holding the group to the nucleus [29].

Calculations carried out on the basis of these principles involve the probability of the existence of the corresponding groups in the nucleus, the binding energy of the groups with respect to the nucleus, and the momentum distribution of the groups as parameters. Information relating to these characteristics of nuclear substructure may be obtained by comparing calculations with experiment.

Thus, the interpretation of a whole series of experimental data obtained in high-energy nuclear reactions demands an understanding of the structure of the nucleus, especially as regards the associations of nucleons within it. Experiments involving recoil nuclei are therefore of special interest.

If we neglect the interaction between nucleons escaping in the course of the cascade and the nucleus, then, as required by the law of conservation of momentum, the momentum of a recoil nucleus formed after the cascade is equal and opposite to the sum of the momenta formerly possessed by the escaping nucleons when in the nucleus. To this we should add the momentum received by interaction of the escaping nucleons with the nucleus and by the evaporation of a certain number of nucleons from the excited nucleus. Hence, experiments with recoil nuclei may give information on the momentum structure of the nucleus.

Before proceeding to analyze such investigations, however, we must decide exactly what type of experiments are to be considered as experiments with recoil nuclei.

A number of authors [30, 31] regard the term "recoil studies" extremely widely, including in its meaning not only cases in which the final nucleus differs little in mass from the original, but also cases involving radioactive fragments and fission products. This amalgamation is artificial in the purely methodical sense and impedes the study of each individual phenomenon.

In order to differentiate these processes, we use the following principles. Fragmentation is well known to comprise the emission of a multicharged particle with $A \geq 4$. As a rule, the emission of a fragment is accompanied by the escape of a considerable number of nucleons (say 5 to 10) from the nucleus. Thus, the final nucleus formed by emission of the fragment differs from the original by at least 10 to 15 atomic mass units. Fission products differ from the original nucleus still more. We may therefore suppose that final products differing from the original nucleus by not more than 10 mass units are due to a cascade process, not accompanied by the emission of a fragment or by fission. It is experiments on processes of this kind which we shall be considering as experiments with recoil nuclei.

At the present time there are about 10 papers relating to recoil nuclei. Let us now consider their results.

Hintz [32] determined the yield of recoil nuclei formed in the reactions $Al^{27}(p, 3pn)Na^{24}$ and $C^{12}(p, pn)C^{11}$. The target, irradiated in the inner beam of a phasotron, consisted of a mixture of target and detector foils. The ratio of the activity in the absorbing foil to the total activity in the target and absorbing foils was obtained for proton energies between 20 and 100 MeV. The results of the experiments may be compared with calculations based on the model of the compound nucleus. For determining the ranges of Na^{24} and C^{11} in aluminum and polyethylene, the semi-empirical range-energy relation of Knipp and Teller [33] may be used. The ranges of recoil nuclei in gas mixtures are reduced to those in air by comparing the range of the recoil nucleus with that of fast α particles. The results of calculations up to energies of 45 MeV for the $C^{12}(p, pn)C^{11}$ reaction and up to about 70 MeV for the $Al^{27}(p, 3pn)$ reaction agree with experiment. For large energies the calculated yield is smaller than the experimental; in the author's view this is a result of direct interaction not involving the formation of a composite nucleus.

S.-C. Fung and Perlman [34] made a detailed study of the yield of recoil nuclei from thick Al targets (target thickness exceeded the range of the recoil nuclei). The measurements were made in the inner beam of a phasotron. The bombarding particles included 60- to 340-MeV protons, 40- to 194-MeV deuterons, and 60- to 380-MeV α particles. A pile of target and detector foils (graphite or polystyrene) was placed perpendicular to the beam, and the number of recoil nuclei escaping from the target and held by the detector foil was determined. The authors compare the experimental results with calculations based on the model of the intermediate nucleus. The mean-square momentum received by the nucleus on evaporation was calculated in the same way as that of a Brownian particle, i.e., from the formula $p^2 = \Sigma p_i^2$, where p_i are the momenta of the evaporating particles. The yield of recoil nuclei from the target was calculated numerically. The range of Na^{24} in Al was determined from the Knipp and Teller formula [33, 35], the parameters of which were chosen so that agreement was obtained with experiment at low energies. The results of the calculation only agreed with experiment up to 70 MeV. For large energies, calculation gave overestimated yields. The authors also made a calculation based on a model in which it was assumed that a constant excitation, equal to 70 MeV for protons and 110 MeV for α particles, was imparted to the nucleus, and that the incident particle left the nucleus without changing its own direction. This gave better agreement with experiment than the calculation based on the model of the composite nucleus.

Fung and Turkevich [36] measured the yield of Ni^{57} and Ni^{56} formed on irradiating copper with 100- to 440-MeV protons. The method described in [34] was used. The authors considered that the formation of Ni^{57} took place as a result of the transfer of an excitation energy of about 100 MeV to Cu^{63}.

Sugarman, Campos, and Wielgoz [30] measured the yields of Tl^{198}, Bi^{199}, Bi^{200}, Pb^{200}, Bi^{201}, Pb^{201}, Tl^{202}, and Bi^{203} recoil nuclei formed on disintegrating Bi with 450-MeV protons. The measuring method was analogous to those used in the previous cases. The effective ranges of the recoil nuclei were almost 100 times shorter than those of fission products, and the angular distribution of most of the nuclei was very anisotropic. The authors considered that their recoil nuclei were formed as a result of a transfer of excitation energies between 100 and 150 MeV to the nucleus, the angular anisotropy being due to momentum received from the primary particle.

Ostroumov [37] measured the range of recoil nuclei formed on irradiating Ag and Br nuclei with 130-, 460-, and 660-MeV protons. The photographic-plate method of recording was used. Photographic plates with fine-grain emulsions of the P-9 type and also NIKFI emulsions of the K type were used. The velocity of the recoil nuclei was determined from the range-velocity curve obtained in [38] for light fission fragments. The transfer factor from range in air to range in the emulsion was taken as 1760. The author used the resultant data to estimate the excitation energy. This paper constitutes an important contribution to radiochemical studies.

Borisova et al. [39] studied the angular and energy distributions of Ag^{106}, Ag^{103} + Ag^{104}, Nb^{90}, Zr^{89}, Rb^{81} + Rb^{82}, and Se^{73}, formed on disintegrating silver with 480-MeV protons. A target made of a silver film 0.5 mg/cm^2 thick was irradiated in the inner beam of a synchrocyclotron. The yield of all the recoil nuclei except Se^{73} was obtained in the angular ranges 5 to 58, 63 to 117, and 122 to 175°. The measuring accuracy was about 50%. Special experiments were made to determine the anisotropy of the angular distribution of the recoil nuclei; this agreed within 25% with the value obtained in experiments with thick targets. The range distribution of the recoil nuclei was measured for angular ranges 90 ± 40° to an accuracy of about 30%. In determining the energy of the recoil nuclei, the range-energy relation obtained in [40] was used. The energy distributions were considered from the point of view of a model in which it was assumed that the momentum of the recoil nuclei was due to evaporation. The momentum received in the cascade process was not considered. The value of the momentum acquired on evaporation was calculated (in analogy with multiple scattering) from the formula $\sqrt{\bar{P}^2} = p_{el}\sqrt{n}$, where n is the number of evaporated particles and $\bar{p}_{el}$ is their mean momentum. The range distribution of the recoil nuclei was regarded as Gaussian. The authors came to the conclusion that the calculation agreed satisfactorily with experiment if the evaporation of α particles were taken into account.

Baker and Katcoff [41] used the photographic-emulsion method to determine the angular distribution of recoil nuclei formed on irradiating Ag and Br nuclei with 1-, 2-, and 3-GeV protons. The distribution had a considerable anisotropy in a forward direction. The anisotropy diminished with increasing energy of the incident particles.

Porile [42] calculated the momentum of recoil nuclei formed as a result of the cascade process. The target nuclei chosen were Ru^{100}, Bi^{209}, and U^{238}. The energy of the incident protons was 460, 990, and 1840 MeV. The same calculating system and the same original data were used as in [13]; the nucleus was considered as a homogeneous sphere, the momentum of the nucleons in the nucleus was given a Fermi distribution, and no account was taken of reflection and refraction of the cascade nucleons at the boundary of the nucleus, nor of the emission of composite particles. In calculating the total perpendicular component of the momentum, the sign of one of the perpendicular components was not known (this was taken in a random manner). The calculations provided information on the momentum components of the recoil nuclei parallel and perpendicular to the proton beam and also on the relation between the parallel momentum component and the excitation energy. The author considered that the results of the calculation agreed satisfactorily with the experimental data of [32] relating to the parallel momentum component. The angular distributions of the recoil nuclei also agree with experiment for an energy of 1 GeV, but for 2 GeV there is a serious discrepancy. The relation between the parallel momentum component and the excitation energy differs considerably from the corresponding relations obtained in earlier papers using simplified models.

Crespo [31] measured the yields of Na^{24} on irradiating Al with α particles (320, 660, and 880 MeV) and 660-MeV protons.

Summarizing the results of these papers, we must note the following:

1. Despite the importance of experiments with recoil nuclei, the number of investigations based on this method is quite small.

2. No work has been done on photonuclear reactions.

3. The method used is extremely imperfect. In the majority of cases, the total yield of recoil nuclei from thick targets has been measured instead of ranges and angular distributions. The best from a methodical point of view is [39], but even here the range spectrum was only determined to an accuracy of ± 30%, and the angular resolution was ± 20-25°.

4. In analyzing the experimental data, simplified and inconsistent models have been used. In the majority of the calculations, the momentum transferred in the cascade was not considered. An attempt was made in [42] to calculate this momentum, but the simplifying assumptions made in the calculation (the diffuse boundary of the nucleus, the refraction and reflection of nucleons at the surface of the nucleus, and a number of other factors were not considered), reduce its value considerably.

5. Finally, as we shall see later (Chapter II, Sec. 2), the majority of calculations employed incorrect range-energy relations, making the range of the recoil nucleus too small. In a number of cases this led research workers to false conclusions regarding the validity of their models.

A series of investigations into photonuclear and nuclear reactions was carried out in 1957-1963 in the Photomesonic-Processes Laboratory of the P. N. Lebedev Physical Institute of the Academy of Sciences of the USSR, using the recoil-nucleus method. The aim of this series was as follows:

To obtain new information on recoil nuclei in nuclear reactions.

To extend the field of investigation to photonuclear reactions.

To create a new measuring method so as to be able to measure the angular and energy distributions of the recoil nuclei to a fair statistical accuracy with good angular and energy resolutions.

To carry out a valid and consistent comparison between theory and experiment so as to enable clear conclusions to be drawn regarding the validity of the various models of high-energy nuclear reactions treated in the literature.

This paper represents the results of these investigations.

CHAPTER I. EXPERIMENTAL

1. Method

Various means of disintegrating the target nuclei are possible with incident particles of high energy. For example, a high-energy γ-quantum may produce (γ,n), (γ, p), (γ, pn), $(\gamma, 2n)$, and other types of reaction. The chamber method (Wilson and bubble chambers), the method of photographic plates, and other methods based on counters enable the angular and energy distributions of reaction products to be estimated, but except in special cases the precise reaction responsible for a particular recorded particle cannot be determined.

The radiochemical method, in the form in which it is usually employed, enables a specific reaction to be identified, but gives no information on the energy and angular distributions of the nuclear reactions.

The present research series was based on the recording of radioactive recoil nuclei, which enabled the angular and energy distributions of specific nuclear reactions to be studied, and was thus free from the disadvantages of the methods mentioned earlier. The essence of the method was that the recoil nuclei escaping from the target and passing through an absorber of a given and fairly small thickness were collected in a film, the activity of which was measured by an ordinary counter. Information regarding the angular and energy distribution of the recoil nuclei was derived from the number of disintegrations recorded.

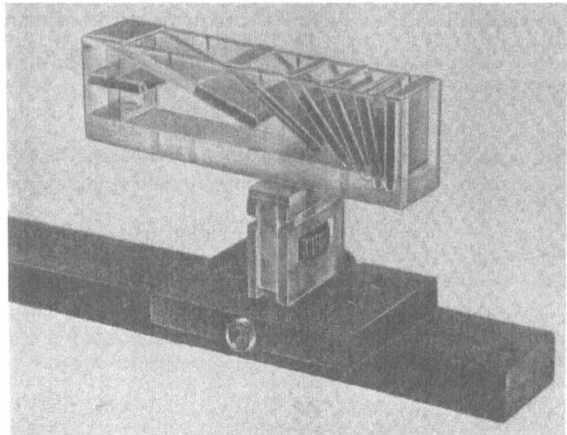

Fig. 1. Apparatus for the irradiation of packets of targets and films placed
at various angles to the beam of radiation. The targets and films in which
the recoil nuclei are collected lie in direct contact.

2. Apparatus

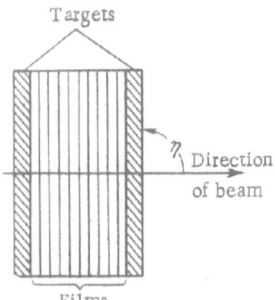

Targets

Direction
of beam

Films

Fig. 2. Arrangement of targets
and films in each of the packets.

(a) Irradiation System. Two types of systems were used for
irradiating targets in a beam of particles. In the first type of system (Figs. 1
and 2), the targets lay in direct contact with the film collecting the recoil nuclei. The target thickness exceeded the maximum range of the
recoil nuclei. Depending on the conditions of the experiment, the number of targets varied from 5 to 50, and the number of films between each
two targets from 3 to 10. The packet of films and targets may be placed
at various angles to the beam of radiation. Most of our measurements
were made with the packets arranged in such a way that the normal to
the targets made angles η of 0, 90, and 180° with the direction of the
beam. In some cases the packet was placed at angles of $\eta = k \cdot 15°$ (k =
0, 1, . . . , 12). The size of the targets and films was 3×3 cm. Apparatus of this type only enabled us to measure the total yield of recoil
nuclei from targets placed at different angles to the beam.

In apparatus of the second type [43], the targets and the films collecting the recoil nuclei were separated. The principal element of this apparatus is a cassette (Fig. 3) containing a target and two collecting
films. The target is fixed in the center of the cassette. The collecting films lie on either side of the target
at distances of 3.5 cm from it. Placed closely behind these are control films made of the same material. The
special collimators of the cassette separate out the recoil nuclei escaping at a particular angle. The angular
resolution of our collimators was about 7°. In cases in which the activity of the targets and the intrinsic activity
of the collecting films differed in half-life, we were able to use cassettes in which collimation was effected by
means of special grids or lattices placed between the targets and the collecting films (Fig. 4).

In order to increase the yield, we simultaneously irradiated 30 to 40 cassettes of the first or second type
set at the same angle to the beam of incident particles and arranged in a special chamber. By varying the
angle of inclination or the gas pressure in the chamber, we were thus able to measure the angular and energy
distributions of the recoil nuclei.

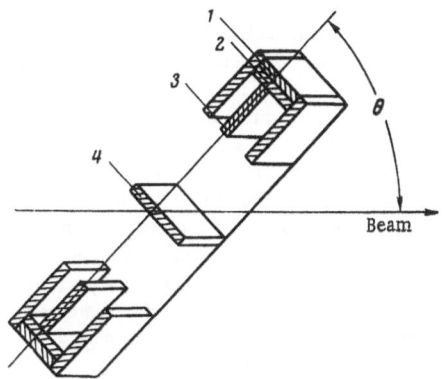

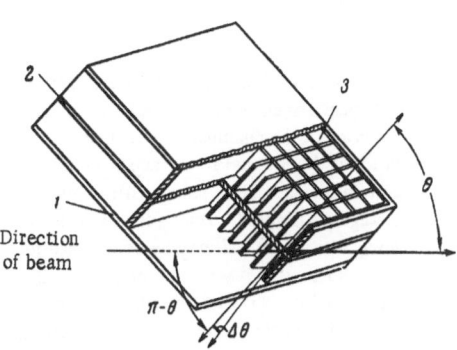

Fig. 3. Cassette with target and collecting films. Collecting films separated from the beam of radiation. (1) Control films; (2) collecting films; (3) collimator; (4) target.

Fig. 4. Cassette with target and collecting films. Collecting films lie in the beam of radiation. Collimation is achieved by means of a grid. (1) Collecting films; (2) target; (3) collimators.

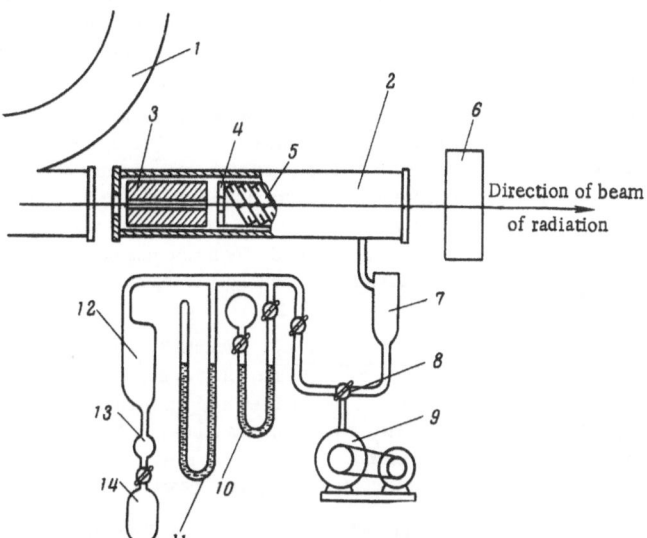

Fig. 5. Arrangement of apparatus for measuring the angular and energy distributions of the radioactive recoil nuclei. (1) Accelerator chamber; (2) vacuum chamber; (3) collimator; (4) ring holder; (5) cassettes; (6) ionization chamber; (7) trap; (8) vacuum pump; (9) backing pump; (10) oil manometer; (11) mercury manometer; (12) trap; (13) leak; (14) gas cylinder.

The general arrangement of the apparatus for studying the angular and energy distributions of recoil nuclei is shown in Fig. 5. The cassettes 5, set at identical angles in a special holder 4, are placed in the vacuum chamber 2. In some experiments the collimator 3 is placed in front of the holder in order to reduce the background activity of the collecting films. The collimator can clearly only be used with the cassettes shown in Fig. 3. The chamber is evacuated with the backing pump 9 and filled to a specific gas pressure by means of a special leak device. In order to purify the gas used for filling, we use a liquid-nitrogen trap 12. The gas pressure in the chamber is measured with a system of oil and mercury manometers 10, 11. The accuracy with which the pressure in our experiments was measured equalled 0.1 mm Hg, i.e., 1% of the 10 mm Hg gas pressure in the chamber. The gas filling the chamber serves to slow down the recoil nuclei when measuring their range distribution. The thickness of this absorbent is between 10 and 100 $\mu g/cm^2$ (this follows from the foregoing discussion) and may be determined to an accuracy of about 1% from the value of the pressure. It is obvious enough that it would be very difficult to make or use a solid or liquid absorber of this thickness. The apparatus and gas absorber described are very simple and reliable in operation, enabling the angular and energy distribution of the recoil nuclei to be determined to high accuracy.

(b) Targets and Collecting Films. The aluminum targets in the first type of apparatus were foils 13.4 mg/cm^2 thick containing less than 0.05% of impurities. The weights of the different aluminum targets differed by no more than 0.5%. The silicon target was provided by quartz glass. The phosphorus and sulfur targets were pressed out of powder containing no more than 0.3% impurities.

In the second type of apparatus we used aluminum targets 80 $\mu g/cm^2$ thick, made by sputtering aluminum onto a triacetate substrate. The thickness of the deposited layer was determined by the activation method [44] (see Appendix I). The essence of the method is that the layer to be measured and a control sample of known thickness made from the same material are simultaneously irradiated by a beam of activating radiation under the same geometrical conditions. The thickness of the film is determined from the ratio of the sample and standard activities resulting from the irradiation.

In experiments involving the recording of Na^{24} activity, we used triacetate films 20 μ thick for capturing the recoil nuclei. In experiments with isotopes having short half lives [for example, in studying the $C^{12}(\gamma, n)C^{11}$ reaction], we used 10-μ aluminum foil. In measuring the cross sections of the photonuclear reactions $Al^{27} \rightarrow Na^{24}$, $P^{31} \rightarrow Na^{24}$, and $Co^{59} \rightarrow Mn^{56}$ we used targets in the form of discs 25.5 mm in diameter. The cobalt samples were pressed from powder containing no more than 0.5% impurity and sintered in a furnace with a hydrogen atmosphere at a temperature of about 1000°C. For monitoring the beam we used graphite targets of the same diameter as the target. The deviation of the weight of the samples from the mean was no greater than 0.5%.

(c) Apparatus for Measuring Radioactivity. The activity of the targets and the films was measured by means of six identical devices assembled into a single unit, with a common control desk and supply system. Each device consisted of a β counter (AS-1 type), a preamplifier, a counting circuit, and an electromechanical counter. The angles for recording the activity of the films was about 2π. The high voltage on the β counters was kept constant to 0.5% The construction of the apparatus provided for simultaneously checking the efficiencies of the six measuring devices by means of a radium standard.

Repeated checks showed that over long periods of time (two or three months), the efficiencies remained constant to within ±1%.

3. Measurement

The experimental part of the work consisted of measuring the effective cross sections and the angular and energy distributions of the recoil nuclei of a number of photonuclear and nuclear reactions leading to radioactive isotopes. In order to determine the effective cross section of formation for a given radioactive isotope, we must: (1) Identify and separate out the activity of the isotope; (2) determine the flux of radiation which has passed through the sample; (3) allow for the decomposition of the radioactive isotope from the instant of irradiation until the instant of measuring its activity; (4) determine the efficiency of the apparatus recording the active radiation.

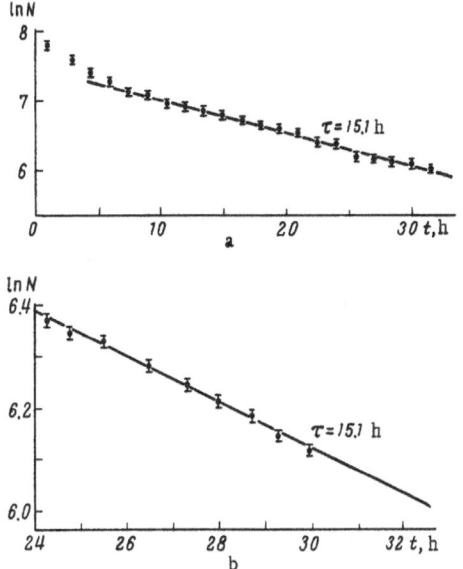

Fig. 6. (a) Decay curve of the activity of Al^{27} samples irradiated with γ-ray bremsstrahlung of maximum energy $E_{\gamma max}$ = 260 MeV; (b) decay curve of Na^{24} recoil nuclei in the collecting films after irradiating P^{31} with the same radiation.

In studying the angular and energy distributions of the recoil nuclei, there is no need to carry out the measurements indicated under (2), (3), and (4) for every experiment. It is sufficient to carry these out in one of the experiments, and in order to compare the results obtained for different angles and gas pressures in the chamber to use a monitoring sample made of the target material and occupying the same solid angle as the target in the beam. In this case, in order to take account of the decomposition of the radioactive isotope and normalize the results of the experiment to the same radiation dose, we must divide the activity of the film in which the recoil nuclei are collected by the activity of the monitoring sample, reduced to the same moment of time. If films and monitoring samples of different thicknesses are used in the experiments, then we must also introduce a correction for the different self-absorption of the measured activity.

(a) Identification of Radioactive Isotopes. Identification of the radioactive isotopes was based on the half-life periods. For this purpose we measured the activity decay curves for samples of C, Al, Mg, Si, P, S, and Co irradiated with 660-MeV protons and γ-ray bremsstrahlung with a maximum energy of 260 MeV. Figure 6a shows, as an example, the activity characteristics of aluminum samples irradiated with γ-ray bremsstrahlung of maximum energy 260 MeV. We see from the figure that, 5 to 10 h after the end of irradiation, only activity with a half-life of some 15 h remains. The activity curves of Si, Mg, P, and S have similar forms. According to tabulated data on the half-lives of radioactive isotopes in the range of atomic nuclei under consideration [45], this activity must be attributed to Na^{24} (half-life 15.01 ± 0.05 h).

In order to check whether the measured activity included any contribution from background activity due to impurities or secondary reactions, we made some special experiments with γ-ray energies near the threshold corresponding to the formation of Na^{24} from the corresponding elements. The experiments showed that, at energies below the threshold, no activity with a half-life of about 15 h existed.

The Na^{24} recoil nuclei were also identified from the half-life period. By way of example, Fig. 6b shows the decay curve of the activity of Na^{24} in the collecting films on irradiation with γ-ray bremsstrahlung of maximum energy 260 MeV.

In carbon samples, the only activity left 5-10 min after irradiation was that with a half-life of 20 min; this was identified with the C^{11} isotope formed by the $C^{12}(\gamma, n)C^{11}$ reaction.

A more complex picture is obtained on irradiating Co samples with γ-quanta. In this case, activity attributable to isotopes Co^{58}, Co^{56}, Co^{55}, and Mn^{56} occurs, and special analysis of the activity curve is required in order to separate out the desired activity of Mn^{56} (half-life 2.58 h).

(b) Measurements of Radiation Flux. Method of Allowing for the Decay in Radioactivity. The time distribution of intensity in the course of irradiation was measured by means of an ionization chamber with an ionization-current integrator. The sensitivity of the chamber was kept constant to within 3-5% during the whole irradiation session.

The integral intensity during irradiation was measured by means of the $C^{12}(\gamma, n)C^{11}$ reaction, the absolute yield of which was measured by Barber, George, and Reagan [46] as a function of the maximum bremsstrahlung energy up to 260 MeV.

The decay of the radioactive isotopes formed in the reactions studied and the decay of the isotope formed in the monitoring reaction during irradiation and measuring were taken into account by means of formulas obtained from the well-known relation for disintegration rate $dN = -\lambda Ndt$, where dN is the number of disintegrating nuclei, N is the total number of active nuclei, dt is the disintegration time, and λ is the decay constant.

For the case in which several successively irradiated and counted monitors are used during the irradiation of one target, the formula has the following form (see Appendix II):

$$\frac{N_{x0}}{N_{c0}} = \frac{N_x}{N_c} \frac{\sum\limits_{i=1}^{k} \chi_{ci} Q_{ci}}{\chi_x Q_x}.$$

Here k is the number of monitors used, N_{x0} is the total number of active nuclei formed as a result of the reaction in question on irradiating the target over a time $(0, t_k)$ (t = 0 corresponds to the beginning of the irradiation), N_{c0} is the total number of active nuclei formed as a result of the monitoring reaction in all k monitors, N_x is the number of disintegrations corresponding to the activity of the reaction in question in the range $(T_{1x}T_{2x})$, $\chi_x = e^{-\lambda_x T_{1x}} - e^{-\lambda_x T_{2x}}$, where λ_x is the decay constant of the isotope of the reaction, $Q = \int_0^{t_k} Q_0 q(t)\, dt$ is the total amount of radiation which has passed through the target, where q(t) is the time distribution of the radiation intensity, $\chi_{ci} = e^{-\lambda_c \tau_{i-1}} - e^{-\lambda_c \tau_i}$, where λ_c is the decay constant of the isotope of the monitoring reaction, and (t_{i-1}, t_i) and (τ_{i-1}, τ_i) are the times required for irradiating and recording the activity of the i-th monitor, while

$$Q_{c_i} = \int_{t_{i-1}}^{t_i} Q_0 q(t) e^{+\lambda_c t}\, dt; \quad Q_x = \int_0^{t_k} Q_0 q(t) e^{\lambda_x t}\, dt.$$

(c) Measuring the Absolute Activity of the Samples. Allowing for Self-Absorption. In order to measure the absolute activity of the samples and determine the efficiency of the apparatus, we used the method of $\beta-\gamma$ coincidences. The possibility of using the method of $\beta-\gamma$ coincidences in our case arose from the fact that the nuclei formed after radioactive decay were stable, while the geometry of the apparatus used eliminated the possibility that angular correlation of the emitted particles might affect the results.

The β-particle recording geometry of our apparatus was approximately 2π. In order to increase the counting efficiency for γ-quanta, we placed aluminum converters in front of the γ-counters.

For determining the absolute activity of Na^{24} we measured the number of coincidences between γ-quanta and electrons ($N_{\beta\gamma}$) and also the counts in the β and γ channels (N_β and N_γ). The absolute activity of the sample in terms of these quantities is $A = N_\beta N_\gamma / N_{\beta\gamma}$ [47]. In experiments with the positron activity of C^{11}, we measured the $\beta^+-\gamma$ coincidences between positrons and the γ-quanta associated with positron decay ($N_{\beta+\gamma}$) and also the counts in the $\beta^+(N_{\beta+})$ and $\gamma(N_\gamma)$ channels. The absolute activity of the sample was determined from the formula $A = N_{\beta+} N_\gamma / N_{\beta+\gamma}$. The accuracy of the absolute measurements was about 20%. The principal contribution to the error was due to statistical measuring errors.

In experiments involving recoil nuclei, in which the radiation dose passing through the target is measured with a monitoring sample made of the target material, we must know the self-absorption of the activity in question in the films and monitors in order to be able to compare the results of the experiments. We made some special measurements of the self-absorption of the Na^{24} activity in the target materials and in the triacetate films, calculating correction terms from the results so obtained.

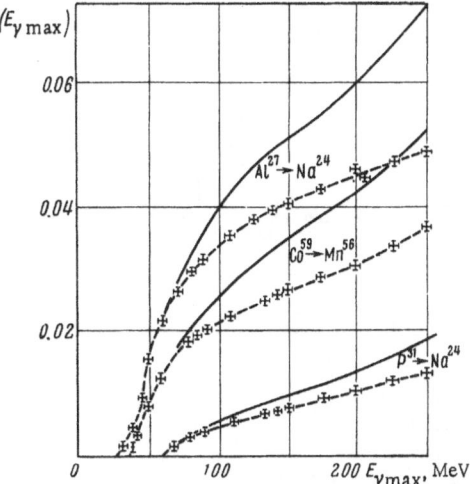

Fig. 7. Activity yields of the $Al^{27} \to Na^{24}$, $Co^{59} \to Mn^{56}$, and $P^{31} \to Na^{24}$ reactions as functions of the maximum bremsstrahlung energy $E_{\gamma\,max}$. The yields $B_x(E_{\gamma\,max})$ are given in relative units and shown as continuous curves. The broken curves drawn through the experimental points show the ratios $B_x(E_{\gamma\,max})/B_C(E_{\gamma\,max})$, where $B_C(E_{\gamma\,max})$ is the activity yield of the $C^{12}(\gamma, n)C^{11}$ reaction.

4. Yield and Effective Cross Sections of the Photonuclear Reactions $Al^{27} \to Na^{24}$, $Co^{59} \to Mn^{56}$, and $P^{31} \to Na^{24}$ for γ-Quantum Energies up to 260 MeV

We were the first to measure the yield and calculate the cross sections of the photonuclear reactions $Al^{27} \to Na^{24}$, $Co^{59} \to Mn^{56}$, and $P^{31} \to Na^{24}$ for γ-quantum energies up to 260 MeV [48]. The experiments were carried out in 1955 on the synchrotron in the P. N. Lebedev Physical Institute of the Academy of Sciences of the USSR, the maximum bremsstrahlung energy of which varied between 20 and 260 MeV and could be kept constant to ± 2%. Samples of Al, Co, and P were irradiated simultaneously and under the same geometrical conditions in a beam of bremsstrahlung for periods between 20 min and 2 h, according to the experimental conditions. In order to obtain a fair statistical accuracy and also to ensure conditions under which the background activity of the monitors and targets would not affect the results, we chose the following system of measurement.

Measurement of the activity of the monitoring samples of carbon began 100 min after the end of irradiation and lasted for 20-40 min. The activity of the Al and P samples was measured over periods of 10-14 and 35-50 h after the end of irradiation.

In experiments involving Co^{60} we had to subtract the additional activity due to the side reactions $Co^{59}(\gamma, n)Co^{58}$, $Co^{59}(\gamma, 3n)Co^{56}$, and $Co^{59}(\gamma, 4n)Co^{55}$; we therefore made three measurements of the activity of the Co samples over periods 5.5-10 h, 30-35 h, and 30-35 days after the end of irradiation. Before and after each measurement of sample activity, we measured the apparatus background and checked the efficiency of the apparatus by means of a radium standard. In all, we carried out 48 irradiations with 15 values of the maximum bremsstrahlung energy of the synchrotron. For each maximum energy the experiment was carried out at least three times.

The scatter in the results of experiments carried out with the same nominal energy value but at different times was no greater than the experimental error itself, which was mainly due to the statistical error in each count and the instability of the synchrotron energy.

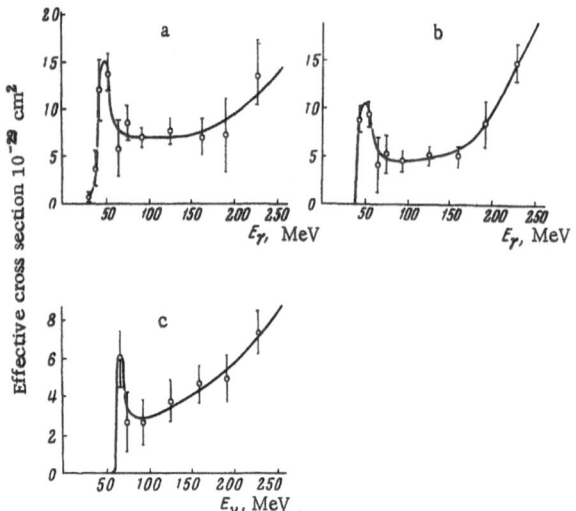

Fig. 8. Effective cross section of the reactions $Al^{27} \rightarrow Na^{24}$ (a), $Co^{59} \rightarrow$ Mn^{56} (b), and $P^{31} \rightarrow Na^{24}$ (c). E_γ = energy of the γ-quanta.

The results of the measurements are shown in Fig. 7. The x axis gives the maximum bremsstrahlung energy $E_{\gamma \max}$ and the y axis gives the activity yields of the $Al^{27} \rightarrow Na^{24}$, $Co^{59} \rightarrow Mn^{56}$, and $P^{31} \rightarrow Na^{24}$ reactions, expressed in terms of the effective cross sections of the corresponding reactions $\sigma_x(E_\gamma)$ and the bremsstrahlung spectrum $\eta(E_\gamma, E_{\gamma \max})$ as follows:

$$B_x(E_{\gamma \max}) = \int\limits_0^{E_{\gamma \max}} \sigma_x(E_\gamma)\, \eta\,(E_\gamma,\, E_{\gamma \max})\, dE_\gamma,$$

where E_γ is the energy of the γ-quanta. In addition to this, the figure shows the ratios

$$B_x'(E_{\gamma \max}) = B_x(E_{\gamma \max})/B_c(E_{\gamma \max}),$$

measured by direct experiment, where $B_c(E_{\gamma \max})$ is the yield of the $C^{12}(\gamma, n)C^{11}$ reaction.

The resultant yield curves were used to calculate the differential effective cross sections $\sigma_x(E_\gamma)$ by the "photon-difference" method [49]. For the bremsstrahlung spectrum we used that calculated by Schiff [50]. The values of $\sigma_x(E_\gamma)$ so found are shown in Fig. 8.

At the present time we know of only one paper in which the yield of the $Co^{59} \rightarrow Mn^{56}$ reaction has been measured in the energy range 160 to 320 MeV [51]. The large experimental errors (about 30%) and the small number of measurements (four values of the maximum bremsstrahlung energy) prevented the authors from calculating the cross section of the reaction. The yield of the reaction agreed with our own results within the limits of experimental error.

5. Yields of Recoil Nuclei from Thick Targets.
Range and Angular Distributions of the Recoil Nuclei

The following recoil-nuclei investigations were included in the present series [52-55].

1. With a maximum bremsstrahlung energy of 260 MeV, we measured the yield of Na^{24} recoil nuclei from thick targets of Mg, Al, Si, P, and S, at angles of 0, 90, and 180° to the beam.

2. We carried out similar measurements with Al targets for bremsstrahlung maxima of 80, 100, 150, and 200 MeV.

3. With a maximum bremsstrahlung energy of 100 MeV, we measured the yield of Na^{24} recoil nuclei from thick Al targets at angles of 0, 30, 45, 60, 75, 90, 105, 120, 135, 150, 165, and 180° to the beam.

4. We measured the differential angular distribution and integral range spectrum of Na^{24} recoil nuclei formed on irradiating thin (80 $\mu g/cm^2$) targets with bremsstrahlung having a maximum energy of 260 MeV.

5. We measured the yields of Na^{24} recoil nuclei from thick targets of Al, P, and Si irradiated with 660-MeV protons, the targets being arranged at 0, 90, and 180° to the beam.

All these were first-time measurements, except for the yield of Na^{24} recoil nuclei from Al after irradiation with 660-MeV protons [31], and so far no one has repeated them.

The experiments involving photonuclear reactions were carried out on the synchrotron of the Laboratory of Photomesonic Processes in the P. N. Lebedev Physical Institute of the Academy of Sciences of the USSR, and the experiments involving nuclear reactions in the outer beam of the synchrotron in the Laboratory of Nuclear Problems belonging to the United Institute of Nuclear Studies.

The quantity measured directly in experiments involving recoil nuclei is the activity of the collecting films. This activity is proportional to the number of nuclei escaping in the specified angular and range intervals, the parameters of these intervals being determined by the experimental geometry. In addition to this, however, the yield of recoil nuclei may be influenced by certain side effects associated with particular aspects of the method employed.

The ranges of recoil nuclei formed in high-energy nuclear reactions are very small; as a rule they are only a few hundreds of $\mu g/cm^2$. The targets, absorbers, and collecting films have thicknesses of the same order. In working with such thicknesses the following important effects may arise: 1) surface contamination of the films and targets, leading to additional absorption of the recoil nuclei; 2) diffusion of recoil nuclei from the collecting films; 3) loss of recoil nuclei from the surface layers of the collecting films as a result of mechanical damage to the films. Special test experiments showed that these effects were negligible in our measurements.

The activity of the collecting films due to the recoil nuclei N may be obtained if we subtract the intrinsic activity of the collecting films N_c from the total activity of the collecting films N_t, giving $N = N_t - N_c$. The intrinsic activity of the collecting films may be determined by placing control films identical with those used for collection in the beam of radiation under conditions in which no recoil nuclei can reach them. In experiments with thick targets, the control film (triacetate 20 μ thick) is placed between two collecting films. We found in independent experiments that the thickness of the collecting films was sufficient to absorb all the recoil nuclei. In experiments with a thin target the control film is placed immediately beyond the collecting film.

The procedure for irradiating the targets and recording their activity was as follows. The irradiation time was 10-15 h. Measurement of activity began 10-15 h after the end of irradiation and continued for 5-10 h, depending on the activity of the collecting films. In order to minimize errors associated with the intrinsic activity of the collecting films, the apparatus background, and possible instability of the apparatus, we measured the activity of the absorbing films, the control films, and the monitors, as well as the background, every 30-60 min in each of the recording devices, in turn. After introducing corrections for decay and self-absorption, the number of nuclei escaping from the target was calculated from the measured quantities. Each experiment (for a specific energy of the incident particles and a specified angle) was carried out at least two or three times. It was found that the scatter in the results was no greater than the statistical error in the number of recorded acts of disintegration.

The results of the experiments with thick targets may conveniently be expressed in the form of the relation $t = N/a_0$, where N is the number of nuclei escaping from the target (from 1 cm^2 of surface), and a_0 is the

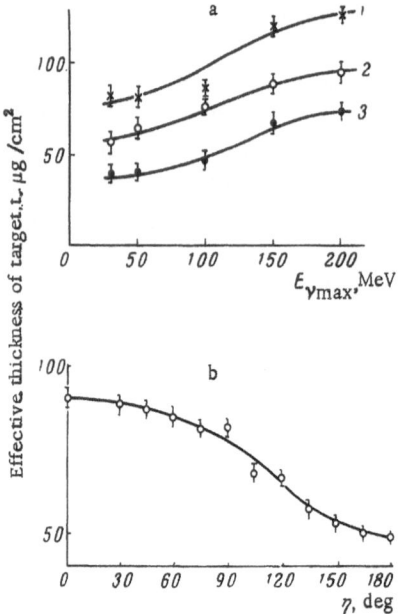

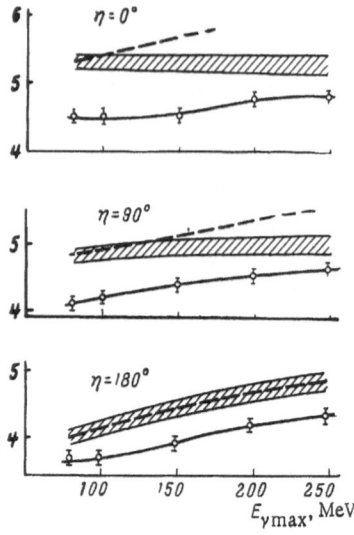

Fig. 9. Effective thickness of the target in the $Al^{27} \to Na^{24}$ reaction as a function of the maximum bremsstrahlung energy $E_{\gamma \, max}$ (a) and as a function of the inclination of the target to the beam η (b). The angle between the targets and the beam of γ-quanta is η = 0, 90, and 180° for curves 1, 2, and 3, respectively.

Fig. 10. Logarithm of the effective thickness of the target as a function of the maximum bremsstrahlung energy $E_{\gamma \, max}$ for the $Al^{27} \to Na^{24}$ reaction. Experimental results are given by the continuous line drawn through the experimental points. Calculated values based on the model of the intermediate nucleus are shown by the broken line, and those based on the cascade model by the shaded region.

specific activity of the target. The value of t is usually called the effective target thickness. It follows from definition that the effective thickness of the target is the thickness with an activity equal to the activity of the recoil nuclei escaping from the target.

Our effective thicknesses for Al, P, Si, and S targets lying at various angles to the beam and irradiated by particles of various energies are shown in Tables 1 and 2 and in Figs. 9 and 10. We remember that the position of the target relative to the beam is determined by the angle η between the normal to the target and the direction of the beam.

Figure 9a shows the effective thickness of an Al^{27} target for Na^{24} recoil nuclei as a function of the maximum bremsstrahlung energy. The targets were placed at angles of η = 0, 90, and 180° to the beam. Figure 10 shows the same results on a semi-logarithmic scale. We see from the figures that the effective thickness rises with increasing maximum bremsstrahlung energy, and in all cases t(180°) < t(90°) < t(0°).

Figure 9b shows the measured values of effective thickness for Na^{24} recoil nuclei formed on irradiating Al^{27} with bremsstrahlung having a maximum energy of 100 MeV. The measurements were made for twelve angles of the target relative to the beam. We see from the figure that the angular distribution of recoil nuclei has a considerable anisotropy, with preferential emission of recoil nuclei in the direction of the beam of γ-quanta.

Table 1 shows the effective thicknesses for Na^{24} recoil nuclei formed in the irradiation of Al, Si, P, and S targets with bremsstrahlung having a maximum energy of 260 MeV. The angles at which the targets lie relative to the beam are 0, 90, and 180°. An interesting feature of the effective thicknesses is that they diminish with increasing atomic number of the target nucleus.

Table 1. Effective Thicknesses of Targets (in $\mu g/cm^2$) for Na^{24} Recoil Nuclei, with a Maximum Bremsstrahlung Energy of 260 MeV

Target	$\eta=0°$			$\eta=90°$			$\eta=180°$		
	Experiment	Compound nucleus	Quasi-deuteron model	Experiment	Compound nucleus	Quasi-deuteron model	Experiment	Compound nucleus	Quasi-deuteron model
Al^{27}	125 ± 3	515 ± 50	220 ± 30	102 ± 3	271 ± 30	160 ± 30	72 ± 3	127 ± 15	115 ± 15
Si^{28}	80 ± 8	530 ± 50	215 ± 30	80 ± 16	200 ± 20	152 ± 30	40 ± 8	48 ± 5	115 ± 15
P^{31}	80 ± 8	680 ± 70	340 ± 40	50 ± 16	250 ± 30	160 ± 30	20 ± 8	44 ± 5	90 ± 15
S^{32}	40 ± 8	660 ± 70	340 ± 50	20 ± 8	240 ± 30	210 ± 40	8 ± 8	40 ± 5	110 ± 20

Table 2. Effective Thicknesses of Targets (in $\mu g/cm^2$) for Na^{24} Recoil Nuclei on Irradiation with 660-MeV Protons

Target	$\eta=0°$		$\eta=90°$		$\eta=180°$	
	Experiment	Cascade theory	Experiment	Cascade theory	Experiment	Cascade theory
Al^{27}	200 ± 10	300 ± 50	120 ± 5	170 ± 30	46 ± 2	70 ± 20
Si^{28}	180 ± 10	320 ± 60	63 ± 5	190 ± 40	25 ± 3	90 ± 25
P^{31}	150 ± 10	600 ± 150	127 ± 10	230 ± 40	29 ± 3	35 ± 10

Table 2 shows the effective thickness for Na^{24} recoil nuclei formed in the irradiation of Al, Si, and P with 660-MeV protons. Comparison with the results obtained in the photonuclear measurements shows that the angular distribution of the recoil nuclei in the proton experiments has a much larger anisotropy in the direction of the particle beam than in the γ-quantum experiments.

As noted earlier, our study of the $Al^{27} \rightarrow Na^{24}$ reaction with 660-MeV protons was later repeated in [31]. The experiments of [31] differed from ours in that irradiation was carried out on the inner beam of protons and the yield at $\eta = 90°$ was not measured. The effective thickness for $\eta = 180°$ agreed with our values, and only differed by 10% for $\eta = 0°$. This difference may have been due to special aspects of irradiation in the inner beam. In this connection the author's note regarding the poor reproducibility of the results for η close to 90° (resulting from fluctuations in the angle between the beam and the target) should be borne in mind.

The differential angular distributions of recoil nuclei and the range distribution of the latter were measured by means of the devices described in Sec. 3 of the present chapter. The measurements were made with a maximum bremsstrahlung energy of 260 MeV for the $Al^{27} \rightarrow Na^{24}$ reaction. The thickness of the target was 80 $\mu g/cm^2$. The pressure in the chamber was 10^{-2} mm Hg. At this pressure the thickness of the air layer between the target and the collecting films was about 0.03 $\mu g/cm^2$. The measurements were made at angles of 30, 45, 60, 90, 120, 135, and 150°. Fifteen to thirty cassettes were irradiated at once. The experiments were repeated at least two or three times for each angle. Irradiation lasted 20 to 30 h. Recording of the activity of the collecting films began 10-15 h after the end of irradiation. The procedure for measuring the activity of the collecting and control films and the monitoring samples was the same as in the experiments with thick targets.

The results of the measurements are shown in Fig. 11. The angular anisotropy is very obvious, the effective cross section for an angle of 30° to the beam being almost four times the cross section at an angle of 150°.

Measurements of the integral range spectrum were made with Na^{24} nuclei escaping in the angular range $90 \pm 7°$ with respect to the beam. Air was used as gas absorber. The pressure in the chamber was set to an accuracy of ~0.1 mm Hg. The measurements were made for five values of air pressure in the chamber, corre-

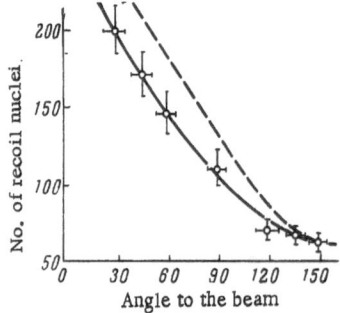

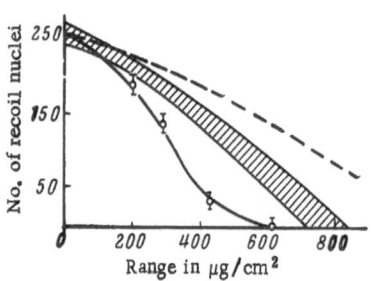

Fig. 11. Angular distribution of recoil nuclei in the $Al^{27} \rightarrow Na^{24}$ reaction for a maximum bremsstrahlung energy of $E_{\gamma max} = 260$ MeV. Notation as in Fig. 10.

Fig. 12. Integral range distribution of recoil nuclei at 90° to the direction of the beam for the $Al^{27} \rightarrow Na^{24}$ reaction with a maximum bremsstrahlung energy of $E_{\gamma max} = 260$ MeV. Notation as in Fig. 10.

sponding to the following thicknesses of air layer between the target and the collecting films: 0, 200, 280, 428, and 625 $\mu g/cm^2$. The number of simultaneously irradiated cassettes was 30-40. The procedure for irradiating and measuring the activity was the same as in measuring angular distributions.

The results of the measurements are shown in Fig. 12. We see from the figure that more than 70% of the recoil nuclei have a range smaller than 350 $\mu g/cm^2$, and the maximum range observed is smaller than 600 $\mu g/cm^2$.

CHAPTER II: DISCUSSION. COMPOUND NUCLEUS AND EVAPORATION

In the Bohr theory of the compound nucleus [56, 57], a nuclear reaction is considered as a series of successive independent processes: that of forming the compound nucleus and those associated with its decay. In forming a compound nucleus, an incident particle is absorbed by the nucleus, and its energy is distributed among the nucleons. In the decay of a compound nucleus, the excitation of the nucleus is taken up in the emission of particles. The way in which a compound nucleus disintegrates is independent of the way in which it is formed, being determined only by its state: the energy, moment of momentum, and parity.

The principal initial assumptions of the theory of the compound nucleus (absorption of the incident particle by the nucleus and independent means of formation and decay) are only satisfied if the range of the particle in the nuclear material is considerably smaller than the dimensions of the nucleus and if the excitation energy of the compound nucleus associated with one particle is considerably smaller than the energy required to separate this particle from the nucleus [58].

For incident-particle energies greater than 50 MeV, neither of these conditions is met, and the model is inappropriate. Nevertheless, it is not unreasonable to consider the experimental data at high energies from the point of view of this model. The need for such a discussion is based on two considerations. First, attempts have continued until recently to explain experimental results at high energies (including experiments with recoil nuclei) by means of the compound-nucleus model. Secondly, the results obtained may be used for calculations on the model of the intranuclear cascade: The cascade process also leads to the formation of excited nuclei, the decay of which may be used to describe an analogous decay of the compound nucleus.

In the theory of the compound nucleus in its original form, it is assumed that the particles evaporate from the nucleus successively, the emission of each successive particle not depending on the emission of the previous, and the angular distribution being isotropic in the system of the moving nucleus. In recent years there have been several papers in which the authors have considered the possibility of the simultaneous emission of several particles [59] and the anisotropy associated with this and various other factors [59, 60]. In the energy range up to 100 MeV, in which comparison with the theory of the compound nucleus is still reasonable, these effects are nevertheless very small. For higher energies, correlation becomes considerable, but at these energies the discrepancy between experiment and the theory of the compound nucleus becomes so great that allowance for these effects cannot influence the final conclusions. We therefore assumed in our calculations that the particles evaporated at different times from the nucleus and had an isotropic angular distribution in the system of the moving nucleus.

1. Mean-Square Momentum and Angular and Energy Distributions of Recoil Nuclei on the Model of the Compound Nucleus

Let us consider a motionless nucleus of mass M struck by a particle of mass m and momentum $\mathbf{P}$. After absorption of the particle, an excited state of the nucleus with mass M', momentum $\mathbf{P}$, and excitation energy E_e is formed. On evaporation of n particles with masses m_i and momenta $\mathbf{P}_i$ (in the system of coordinates of the compound nucleus), the nucleus receives momentum $\mathbf{P}_0 = \sum\limits_{i=1}^{n} \mathbf{P}_i$, and its mass becomes M_n. The total momentum of the recoil nucleus after the nuclear reaction equals $\mathbf{P} = \mathbf{P}_0 + \mathbf{P}'$, where $\mathbf{P}'$ is a part of the momentum of the incident particle. Clearly, $\mathbf{P}' = \mathbf{P} M_n / M'$, and the main difficulty lies in determining the momentum received on evaporation. Since the values and directions of momenta $\mathbf{P}_i$ may vary considerably, it is reasonable to speak of a certain momentum distribution of $\mathbf{P}_0$, characterized, in association with other parameters, by the mean-square momentum equal to $\overline{P_0^2}$.

Let us first consider the calculation of the mean-square momentum. The value of this momentum is independent of the form of the momentum distribution and is determined entirely by the kinetic energy of the nuclear reaction and also the mass and number of the evaporating particles.

The mean-square momentum of the recoil nucleus was calculated in [34, 39] on the principle of a Brownian particle, i.e., by means of the formula

$$\overline{P_0^2} = \sum_{i=1}^{n} \overline{P_i^2}. \tag{1}$$

The P_0 distribution was considered Gaussian in [39], in analogy with multiple scattering. We showed earlier [54, 55] that such calculations based on analogs were only approximate and in many cases could lead to serious errors.

In deriving relation (1) it was assumed that the momenta $\mathbf{P}_i$ were distributed isotropically in the coordinate system of the compound nucleus. Actually, each momentum $\mathbf{P}_i$ is distributed isotropically in the coordinate system of the nucleus formed after evaporation of the $(i-1)$-th particle. Allowing for the effect of the motion of the nucleus, we found (see Appendix III) that for $M/m_i > 10$ and to an accuracy of about 1%,

$$\overline{P_0^2} = [1 - \overline{\alpha}(n-1)] \sum_{i=1}^{n} \overline{P_i^2}, \tag{2}$$

where $\overline{\alpha} = \sum\limits_{i=1}^{n} \alpha_i / n$ and $\alpha_i = m_{i+1}/M_i$. We see from the formula given that relation (1) is a limiting case of relation (2) for $M_i \gg m_i$. At the same time, for $M_i \sim 10 \, m_i$, use of formula (1) instead of the more accurate formula (2) may lead to errors of 30-50%.

In the nonrelativistic case, and for evaporating particles of the same mass m_0, the mean-square momentum is expressed very simply in terms of the kinetic energy of the nuclear reaction E_K:

$$\overline{P_0^2} = \frac{[1 - \bar{\alpha}\,(n - 1)]\,E_K}{1 + \alpha_n\,[1 - \bar{\alpha}\,(n - 1)]}\,, \qquad (3)$$

where the momentum of the recoil nucleus of mass M_n is expressed in units of $\sqrt{M_n E_N / m_0}$, and E_K and E_N are the kinetic energies of the reaction and of the recoil nucleus in MeV.

The form of the angular distribution of the recoil nuclei in the laboratory system of coordinates depends on the ratio of the isotropic component of the momentum of the recoil nucleus P_0 and the momentum directed along the beam P'. The greater the value of P', the more anisotropic is the angular distribution in the forward direction.

As shown in Appendix III, the analytical expression for the angular distribution of recoil nuclei on the model of the compound nucleus has the form

$$n\,(\theta,\,P_0,\,P') = \frac{\lambda\,(x + z)^2}{z}\,, \qquad (4)$$

where $\lambda = P'/P_0$, $x = \cos\theta$, $z = \sqrt{a^2 + x^2}$, $a^2 = \lambda^{-2} - 1$, and the angle θ is reckoned from the direction of the beam.

Let us consider the energy distribution of the recoil nuclei. On evaporation of n particles with given momenta P^i, the recoil nucleus receives momentum $P_0 = -\sum_{i=1}^{n} P_i$. Clearly, the same momentum may be transferred to the recoil nucleus as a result of another combination of momenta P_1', P_2', ..., P_n'. In order to obtain the momentum distribution of the recoil nuclei, we must determine the probability of each such combination and sum over all possible variants leading to the given P_0. If, therefore, the momentum-distribution function of the evaporating particles has the form $\mathcal{P}_n\,(P_1,\,P_2,\,\ldots,\,P_n)$, then the momentum distribution of the recoil nuclei may be calculated from the formula

$$\mathcal{P}_{Nn}\,(P_0) = P_0 \int_G \mathcal{P}_n\,(P_1,\,P_2,\,\ldots,\,P_n)\,dP_1\,dP_2 \ldots dP_n, \qquad (5)$$

where the range of integration G takes account of all variants of evaporation leading to the given P_0.

We also calculated the energy distributions of the recoil nuclei on evaporation of two or three nucleons (see Appendix IV). We assumed that the evaporating nucleons had an isotropic angular distribution in the laboratory system of coordinates and took no account of the kinetic energy of the recoil nucleus in the overall balance of energy. For nuclei with atomic numbers A > 30, the error associated with these simplifications is a few percent.

For the case of two evaporating nucleons we obtained the analytical formula

$$P_2\,(\varepsilon_{N2})\,d\varepsilon_{N2} = P_0\,[2\,(1 - 2\varepsilon_1)\,\sqrt{\varepsilon_1 - \varepsilon_1^2} + \arcsin\,(1 - 2\varepsilon_1)]\,d\varepsilon_{N2}, \qquad (6)$$

where

$$\varepsilon_{N2} = [M_2 E_{N2}/(E_1 + E_2)\,m], \quad \varepsilon_1 = 1 - \sqrt{2\varepsilon_{N2} - \varepsilon_{N2}^2},$$

E_{N2} is the kinetic energy of the recoil nucleus, E_1 and E_2 are the kinetic energies of the evaporating nucleons, M_2 is the mass of the recoil nucleus, and m is the mass of a nucleon.

For the case of three evaporating nucleons, the energy distribution was found by numerical integration. The results of the calculations are shown in Fig. 13. We see from the figure that our distributions for two and three evaporating nucleons differ considerably from Gaussian. It should be noted that in calculating the energy

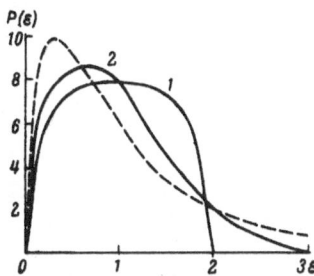

Fig. 13. Energy distributions of recoil nuclei on evaporation of two and three nucleons. (1) m = 2; (2) m = 3. Broken curve gives Gaussian distribution. The x axis gives the kinetic energy of the recoil nucleus expressed in units of $\varepsilon = E_N M_N/m E_k$. where E_N is the kinetic energy of the recoil nucleus of mass M_N, E_K is the kinetic energy of the reaction, and m is the mass of a nucleon.

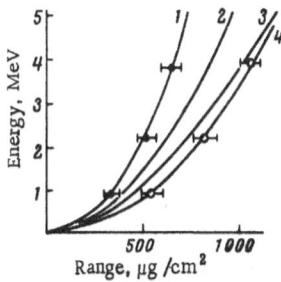

Fig. 14. Range—energy relationship for Na atoms in air and aluminum. (1) Na_{11}^{23} in air [64]; (2) Na_{11}^{24} in aluminum [31]; (3) Na_{11}^{24} in aluminum (52); (4) Na_{11}^{23} in aluminum [64].

distributions we took no account of the emission of γ-quanta and complex particles such as D, He, etc., since, in the energy range of interest, the probability of these processes was small [58, 61].

Let us now compare the computed results with the experimental values. In the experiments we measured the angular distributions of recoil nuclei from thick samples. The calculation of the angular distributions of the recoil nuclei obtained in experiments with thin targets (target thickness much smaller than the range of the recoil nucleus) requires no extra data and may be carried out on the basis of expression (4). In our experiments we measured the angular distribution of recoil nuclei in the $Al^{27} \rightarrow Na^{24}$ reaction for bremsstrahlung with a maximum energy of $E_{\gamma\,max}$ = 260 MeV. Hence, in order to compare the computed results with experiment we had to average a distribution of the form of (4) over all γ-quantum energies E_γ, allowing for the cross section $\sigma(E_\gamma)$ of the reaction and the bremsstrahlung spectrum $\eta(E_\gamma, E_{\gamma\,max})$:

$$N(\theta)\,d\theta = \int_0^{E_{\gamma\,max}} n(\theta, P_0, P') \times$$

$$\times \sigma(E_\gamma)\,\eta(E_\gamma, E_{\gamma\,max})\,dE_\gamma\,d\theta. \qquad (7)$$

The results of calculations based on formula (7) appear in Fig. 11.

2. Range – Energy Relation, Straggling, and Multiple Scattering of Heavy Ions. Range Distribution of Recoil Nuclei

For a theoretical analysis of the range distribution of recoil nuclei, in contrast to the angular distributions, we must know the range—energy relation for the corresponding ions, and also (where necessary) allow for straggling and multiple scattering. The interesting range of ion energies lies between 0 and about 10 MeV. The velocity of such ions varies during the slowing down process from a value exceeding the velocity of the atomic electrons v_0 to zero. There is at present no theory which would enable the slowing of multiply charged ions from $v > v_0$ to $v = 0$ to be explained on the basis of a single mechanism. Approximate calculations have been made for different velocity ranges, the most successful being those based on the Born approximation [62, 63], valid for $v \gg v_0$. Thus, for a correct choice of parameters, and for velocities $v \gg v_0$, the Bloch formula [63] describes the experimental data on the energy losses of multi-charged ions to an accuracy of approximately 10% [64]. For velocities $v \sim v_0$ there are no satisfactory calculations at all, while for $v < v_0$ the simplified Thomas—Fermi model of the atom is used [65, 66]. Calculations based on this model agree to an accuracy of 10-20% with experiment [64].

Experimental data on the ranges of heavy ions have up until now been very scanty and inaccurate. Attempts have nevertheless been repeatedly made to obtain semi-empirical range—energy relations for heavy ions [33, 35]. In practice, however, the use of these relations has led to substantial errors. Thus, for example, in [34], the range—energy relation for Na^{24} in Al was obtained from the semi-empirical formula of Knipp and Teller [33, 35]. Later work [52, 64] showed that the formula in question was incorrect.

The use of incorrect range—energy relations, usually underestimating the ranges of the ions, led the authors of a number of papers to erroneous conclusions. This shows the importance of knowing the correct range—energy relationships in experiments with recoil nuclei.

We measured the range of Na^{24} in aluminum and air in our experiments. In order to obtain the range—energy relation for Na^{24}, we analyzed a large number of experimental data on the ranges of ions from H to Ar [67-73]. These data were revised and reduced to ranges in Al, allowing for the variation in the stopping powers of materials at low velocities [74]. The range—energy curve for Na^{24} was obtained by extrapolating the data relating to F and Ne. The results of these calculations were first published in 1958 [52]. The following year saw the first experimental results on the ranges of Na^{23} in air (Teplova et al. [75]). The same authors measured the relative stopping powers of air and Al for several velocities of Na^{23} [64], enabling the range—energy curve for Na^{23} in air to be converted to the corresponding curve for Al.

In 1961, Crespo [31] published the measured ranges of Na^{24} in Al. In contrast to [75], in which the ranges of Na^{23} ions accelerated in a cyclotron were measured, Crespo measured the ranges of Na^{24} formed by bombarding Al with 40-MeV α particles and 10- to 20-MeV deuterons. Hence, the relation so obtained is correct if Na^{24} is the decay product of a compound nucleus.

The results obtained in [31, 75] are shown together with our own results in Fig. 14. We see from the figure that the ranges of Na^{24} found in [31] are about 20% lower than those obtained from the experimental data of [75]. The range—energy curve given by our own calculations lies between the corresponding curves of [31, 75] and closer to the experimental points of [75]. The difference between our calculations and the data of [75] lie within the limits of the corresponding computing and experimental errors. The reason for the large discrepancy between [31] and [75] is not clear. It may be that Crespo's assumption that the Na^{24} was formed only as a result of the decay of a compound nucleus was incorrect. Apparently direct processes give a considerable contribution to the formation of Na^{24}. This may lead to a reduction in the ranges of the recoil nuclei for the same incident-particle energies. In analyzing the experimental data, we used the range—energy relationship obtained in our own calculations, allowing for an error of approximately $\pm 10\%$. We should note that the experimental data of [75] and most of those of [31] lie within these limits. Thus, this error illustrates the accuracy with which the range—energy relations of ions are at present known.

The statistical spread of the ranges (straggling) may be calculated from the well-known formulas of Bethe [76] and Bohr [77]; in the ranges of present interest, this spread is some 5-10% of the mean range. Experimental data on straggling obtained by Teplova et al. [64] agree satisfactorily with the theoretical predictions. It should be noted that it is only important to allow for straggling when calculating discrete range spectra. If the spectrum is continuous and changes smoothly, as in our calculations, straggling of 5-10% changes the spectrum very little (by about 1%).

The multiple scattering of heavy ions is calculated by means of the formula [78]

$$\overline{\alpha^2} = 2\varkappa \ln (\varkappa / 2\theta_{min}), \tag{8}$$

where $\overline{\alpha^2}$ is the mean square of the angle of multiple scattering, and $\varkappa$ and θ_{min} are defined as follows:

$$\varkappa = \frac{4\pi N t z_1^2 z_2^2 e^4 (M_1 + M_2)^2}{M_1^2 M_2^2 v^4}, \tag{9}$$

$$\theta_{min} = \frac{3.8 z_2^{4/3} e^2 (M_1 + M_2)}{M_1 M_2 v^2 a_0}. \tag{10}$$

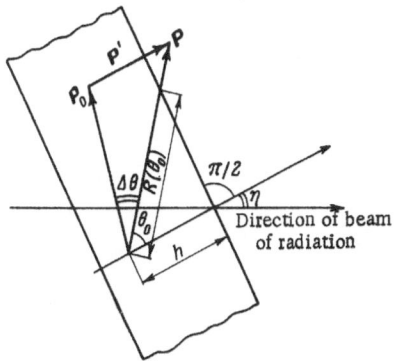

Fig. 15. To illustrate the calculation of the yield of recoil nuclei from thick targets.

Here N is the number of scattering nuclei per 1 cm^3 of material, t is the thickness at which scattering takes place, $z_1 e$ and $z_2 e$, M_1 and M_2 are the charge and mass of the incident and scattering nuclei, v is the velocity of the incident nuclei, and $a_0 = 0.529 \cdot 10^{-8}$ cm.

Calculations based on formula (8) with due allowance for the geometry of our experiments show that, in measuring the angular distribution, multiple scattering worsens the angular resolution very slightly, by about 1%. The angular distribution shown in Fig. 11 is calculated with due allowance for this effect. In addition to this, multiple scattering leads to an additional range spread of about 2%, which is considerably smaller than the range spread due to straggling.

Knowing the range−energy relation for Na^{24} ions, the energy distributions of recoil nuclei obtained in Sec. 1 of this chapter may be converted into an integral range spectrum. In doing this we must average over all γ-quantum energies (as in the case of calculating the angular distribution), allowing for the cross section of the nuclear reaction and the bremsstrahlung spectrum. The results of such a calculation are shown in Fig. 12.

3. Yield of Recoil Nuclei from Thick Targets

Most of the experimental data on recoil nuclei were obtained in thick-target experiments. In these we measured the yield of recoil nuclei from targets with thicknesses exceeding the range of the recoil nuclei, while the detector films were in close contact with the target. In order to analyze these experiments we must calculate the yield of recoil nuclei from thick samples by using their known angular and energy distributions.

Let us consider a plane target so placed that the normal to its surface makes an angle η with the direction of the beam of radiation (Fig. 15). If the range and angular distributions of recoil nuclei in a system based on the direction of the normal are expressed in the form n(R, θ'), then the number of recoil nuclei escaping from the target is determined from the expression

$$N(\eta) = \int\limits_0^\infty dR \int\limits_0^R dh \int\limits_0^{\arccos(h/R)} a_0 n(R, \theta') \frac{\sin\theta'}{2} d\theta'. \tag{11}$$

Here, a_0 is the number of recoil nuclei formed in a layer of unit thickness, h is the depth at which the recoil nucleus was formed, and the angle η is counted from the direction of the normal.

In experimental work we usually determine the relation

$$t(\eta) = N(\eta)/a_0, \tag{12}$$

which is called the effective thickness of the target at an angle η. The calculation of effective thickness is complicated and in the majority of cases can only be carried out numerically. In some cases, however, we can obtain an analytical expression. In particular, we obtained a formula for the effective thickness on the evaporation model for $\eta = 0$ and 180° (see Appendix IV):

$$t(\eta) = \frac{\lambda^{n+1}R_0}{4} \sum_{i=1}^{3} A_{in}^{(i)} \alpha_\eta(k_{in}), \tag{13}$$

where $\eta = 0°$ and $180°$;

$$k_{in} = 2i + n - 3; \quad \alpha_\eta = (b_\eta)^k - a^k; \quad b_\eta = \frac{1}{\lambda} + \cos\eta;$$

$$A_{\eta n}^{(1)} = - a^2\left[b_\eta + \frac{a^2\cos\eta}{2}\right]$$

for any $n \neq 1$, and

$$A_{\eta n}^{(1)} = 0; \quad A_{\eta n}^{(2)} = \left[b_\eta + \frac{a^2 n\cos\eta}{(n+1)}\right]$$

and

$$A_{\eta n}^{(3)} = \frac{\cos\eta\,(n+1)}{2\,(n+3)}$$

for any

$$n \geqslant 0; \quad a^2 = \lambda^{-2} - 1; \quad \lambda = (P'/\widetilde{P}_0) \leqslant 1;$$

P' is the momentum of the recoil nucleus associated with the motion of the compound-nucleus system, $\widetilde{P}_0$ is the mean-square momentum received by the recoil nucleus on evaporation of the nucleons, R_0 is the range of the recoil nucleus with momentum $\widetilde{P}_0$. In deriving the formula, it was assumed that the recoil nuclei moved in a straight line and the range−momentum relationship had the form $R = kp^n$, where n is any positive whole number. The accuracy of formula (13) is about 2-3%. The effective thicknesses for the angle $\eta = 90°$ were calculated numerically. The results of the calculations of effective thicknesses for the photonuclear reaction $Al^{27} \rightarrow Na^{24}$ are shown in Fig. 12, and those for the reactions $Si^{28} \rightarrow Na^{24}$, $P^{32} \rightarrow Na^{24}$, and $S^{32} \rightarrow Na^{24}$ are given in Table 1.

4. Comparison of Calculations with Experiment and Conclusions

The angular distributions of the recoil nuclei for $E_{\gamma\,max} = 260$ MeV calculated on the model of the compound nucleus are not very different from those obtained experimentally, as may be seen from Fig. 11. If we equate the yields at 150°, then at 45° the experimental and calculated values differ by about 30%, which is no greater than the two statistical errors.

More important results come from analyzing the integral range distributions of the Na^{24} recoil nuclei. We see from Fig. 12 that the calculated ranges are more than twice those found experimentally. Thus, the model of the compound nucleus predicts considerably greater momenta of the recoil nuclei than are found by experiment. This conclusion is confirmed by experiments with thick targets (see Fig. 10). Over the whole range of $E_{\gamma\,max}$ from 80 to 260 MeV, and for all angles η, the effective thickness calculated on the model of the compound nucleus is much greater than the experimental value. For the majority of points theory and experiment differ by factors of more than two. The most interesting fact is that even for $E_{\gamma\,max} = 80$ MeV the calculated effective thickness at angles of $\eta = 0$ and 90° exceeds the experimental value by almost a factor of two. The main contribution to the yield of recoil nuclei in this region comes from γ-quanta with energies between 50 and 70 MeV. This result shows that even in this range of energies the formation of the compound nucleus is not the decisive factor in the mechanism of the reaction studied.

Similar results obtained for reactions in Al, Si, P, and S for $E_{\gamma\,max} = 260$ MeV (see Table 1) show that the difference observed is not connected with the characteristics of any particular reaction, but is a general property of a large number of complex photonuclear reactions. In view of this conclusion it is of interest to consider the results of [29], in which the effective thicknesses were measured for Na^{24} recoil nuclei formed on irradiating Al with 60- to 340-MeV protons. In obtaining the range−energy relation for Na^{24}, the authors of this paper assumed that for proton energy of about 60 MeV, the $Al^{27} \rightarrow Na^{24}$ reaction involved the formation of a compound nucleus. We shall not need this assumption; we shall use the experimental range−energy relation shown in Fig. 14.

Table 3. Effective Thicknesses (in $\mu g/cm^2$) of an Al^{27} Target for Na^{24}_{11} Recoil Nuclei at an Angle of $\eta = 0°$

Proton energy, MeV	Experiment [34]	Theory of the compound nucleus	Cascade theory
60	460 ± 20	920 ± 120	
80	470 ± 20	940 ± 120	
140	312 ± 10	1350 ± 140	$550 \pm 100*$
340	250 ± 2	2230 ± 230	320 ± 50

---

*Calculation made for an energy of 150 MeV.

The effective thicknesses for $\eta = 0°$ may be calculated by using a formula similar to (13), but taking account of the fact that the momentum transferred to the recoil nucleus by the incident proton is greater than the mean-square momentum received by the recoil nucleus on evaporation.

For the case in which the range is proportional to the momentum (as in the case for Na^{24} in our range of energies), the formula has the very simple form

$$t(0°) = \lambda R_0, \tag{14}$$

where λ and R_0 have the same sense as in formula (13). Using (14) and the range−energy relationship illustrated in Fig. 14, we calculated $t(0°)$ for proton energies of 60, 80, 140, and 340 MeV. The results of the calculation are shown together with the experimental data of [34] in Table 3. Comparison shows that even at an energy of 60 MeV, calculation gives an effective thickness twice that of the experimental measurement, i.e., there is approximately the same discrepancy between experiment and the theory of the compound nucleus as occurred in the analysis of photonuclear reactions.

The following conclusions may be drawn from our comparison between the calculations based on the compound-nucleus model and the experimental results: (1) The model of the compound nucleus is not adequate for explaining the photonuclear and nuclear reactions studied at energies of 60 MeV or higher; (2) in order to explain the experimental data we need a new mechanism, one of the characteristics of which must be a low value of the momentum transferred to the recoil nucleus in the course of the nuclear reaction.

CHAPTER III: DISCUSSION. CASCADE THEORY

The inability of the theory of the compound nucleus to explain the experimental data for nuclear reactions at high energies is not surprising. As noted earlier (see Chapter II), the fundamental principles of this theory are only valid for low-energy particles, for which the range in the nuclear material is much shorter than the dimensions of the nucleus and the excitation energy of the compound nucleus associated with a particular particle is much smaller than the energy required to separate this particle from the nucleus.

Even for incident-particle energies of 50 MeV, the range of the particle in the nuclear material becomes comparable with the dimensions of the nucleus, the wavelength of the particle approaches the mean distance between the nucleons in the nucleus, and the energy brought into the nucleus by the particle greatly exceeds the binding energy of the nucleons in the nucleus. Hence, the interaction between high-energy particles and the nucleus involves new features, such as the partial transparency of nuclei for high-energy nucleons, the presence of a considerable number of high-energy particles among those escaping from the nucleus, anisotropy in the angular distribution of these particles, etc. At the present stage of theoretical development, all these features are satisfactorily explained by the cascade theory of nuclear reactions, based on the Heisenberg [11] and Serber [12] idea of the "quasi-free" interaction of the incident particle with the nucleons in the nucleus.

According to the cascade theory, the incident particle does not transfer all its energy to the nucleus, as assumed in the theory of the compound nucleus, but interacts with individual nuclear particles (nucleons or more complex formations), giving them part of its energy. It is assumed that this interaction only differs from interaction with free particles in the existence of a momentum distribution of the nuclear particles, the operation of the Pauli principle, and the presence of the averaged field of nucleon−nucleus interaction.

In calculations based on the cascade model, the incident particle is considered as a classical object moving on a definite trajectory. This is permissible, since the wavelength of the particle is small. Recoil nucleons which have received a considerable energy are considered analogously, and at the first stage the nuclear reaction is regarded as a cascade of paired collisions of fast nucleons with nucleons belonging to the nucleus.

Some of the nucleons which have taken part in the cascade process fly out of the nucleus, forming the experimentally observed component of high-energy nucleons, while the rest are absorbed by the nucleus, transferring their energy to the latter. The excited nucleus thus formed after the cascade passes back to the ground state by the evaporation process.

Analytical calculation of the cascade process is in principal very laborious, and at the present time it is quite impossible, since a number of the cascade characteristics (for example, the cross sections of the nucleon-nucleon collisions) have no analytical expression. Hence, in order to calculate the cascade process, we may use the method of statistical tests (Monte Carlo method), which enables us to calculate any processes the individual elements of which may be given either analytically or numerically.

Monte Carlo calculations involve a great deal of work. Wide use of this method thus only became possible recently, in view of the development of high-speed electronic computers enabling such calculations to be carried out rapidly and accurately. Such calculations provide us with data on various characteristics of nuclear reactions capable of direct comparison with experiment.

Theory and experiment are compared for two purposes; first, in order to check the validity of the fundamental assumptions of the cascade theory and, secondly, in order to obtain new information on the structure of the nucleus, which is of special importance in nuclear theory.

In order to verify the validity of the principal assumptions of the cascade theory, we must consider the results of calculations which are not too dependent on hypotheses regarding the structure of the nucleus, but are determined mainly by the mechanism of the paired nucleon-nucleon collisions. Such results include the total inelastic cross sections of nucleon-nucleus interactions, and also the angular and energy distributions of the cascade nucleons.

After we have established the validity of the original hypotheses of the cascade theory, we may proceed to analyze data sensitive to the assumed structure of the nucleus. Such data include the momentum distribution of the recoil nuclei and also various characteristics of the multicharged particles emitted in the nuclear reactions.

Unfortunately, the successive analysis of experimental data by the cascade theory had been made difficult by the absence of accurate calculations. Calculations have generally been based on a large number of crude approximations, the most serious of which are the following:

1. A considerable proportion of the calculations have been made with two-dimensional geometry.

2. In all calculations except those of Fedotov [74] for the C^{12} nucleus, the diffuse boundary of the nucleus has been ignored, the nucleus being considered as a homogeneous sphere.

3. The momentum distribution of the nucleons in the nucleus has been equated to that of a Fermi gas of nucleons at zero temperature.

4. The nucleon-nucleus potential has been approximated by a rectangular well.

5. The interaction between π-mesons and the nucleus as a whole has not been considered.

6. The reflection and refraction of nucleons and π-mesons on escape from the nucleons has been neglected.

7. In no cascade calculations except those of Fedotov [79] has the association of nucleons been taken into account.

A detailed analysis of all these factors and an estimate of their possible influence on the results of calculations are given in [80]. We shall later give special consideration to questions of association and the interaction of cascade nucleons with associations, but at the moment we shall confine attention to a brief note on certain aspects.

First of all let us consider the question of the density distribution of nuclear matter. From experiments on fast-neutron scattering [1] and a variety of other data [81] we find that the nucleus does not have a sharp boundary, but that its density falls off gradually at the edges. Allowance for the diffuse boundary of the nucleus should increase the transparency of nuclei and may lead to a change in the total inelastic-interaction cross sections.

The results of cascade calculations may be considerably affected by the reflection and refraction of the cascade nucleons on escaping from the nucleus. Since reflection is more important for low-energy nucleons, the number of such nucleons escaping is reduced. The rise in the number of absorbed nucleons leads to a rise in the excitation energy. The angular distributions of the cascade nucleons also change as a result of refraction.

There is little doubt that calculations in which these and various other effects are ignored can only be regarded as extremely rough. This makes it hard to analyze experimental data and reduces the value of the conclusions drawn. In view of this, high-speed computer calculations of the cascade process were carried out in the P. N. Lebedev Physical Institute of the Academy of Sciences of the USSR in 1961-1963, a large proportion of the approximations listed being removed [14]. The aim of the calculations was to verify the cascade theory, disregarding the association of nucleons in the nucleus. The calculations were carried out in three-dimensional geometry, using relativistic conservation laws. The Si_{14}^{28}, $(AgBr)_{41}^{95}$, and Au_{79}^{197} nuclei were considered for irradiation by 150, 340, and 660-MeV protons. The calculations allowed for: the density distribution of nuclear matter derived from fast-neutron scattering experiments, the interaction between the π-meson and the nucleus as a whole, and the refraction and reflection of nucleons and π-mesons on escape from the nucleus.

In contrast to the earlier calculations, several variants of the momentum distribution of the nucleons in the nucleus were considered, including the Fermi and Gauss distributions. Calculations of the photonuclear cascade were carried out. In the calculations of proton-induced cascades, data relating to the transparency of the nucleus and the total inelastic-interaction cross sections were obtained, together with data relating to the angular and energy distributions of the cascade nucleons, the angular and energy distributions and excitation energies of the nuclei formed after the cascade, the distributions of nucleons knocked-on in the cascade over the volume of the nucleus, the relation between the excitation energy and the number of knocked-on nucleons, and a large amount of other data.

In this paper we shall only give detailed consideration to data relating to recoil nuclei. As regards the rest of the results, we shall confine ourselves to setting out the principal data, which will be analyzed separately in later papers.

The calculations thus executed showed that the total effective cross sections for the inelastic interaction of protons with nuclei, and also the angular and energy distributions of the fast nucleons, were well explained by the cascade theory. The calculations also showed that these characteristics of nuclear reactions depended little on specific hypotheses regarding the spatial and momentum structure of the nucleus, being mainly determined by the mechanism of the nucleon—nucleon collisions.

The agreement between theory and experiment shows that at large energies the interaction of the incident high-energy nucleon with the nucleus is in fact determined by a cascade of quasi-free collisions with individual nucleons of the nucleus, and thus confirms the validity of the basic fundamentals of the cascade theory.

In view of this result, the data relating to recoil nuclei which, as will shortly be shown, fail to agree with experiment, are of special interest. Before passing on to an analysis of these data, however, we must give more detailed consideration to the basic principles and data underlying the calculations.

1. General Arrangement and Initial Data of the Nuclear-Cascade Calculations

1. The cascade was considered in three-dimensional space, and the collisions were treated as relativistic, allowing for the formation of mesons. Neutrons and protons were distinguished by charge and collision cross section, but their masses were regarded as equal ($mc^2 = 938.85$ MeV); the π-mesons were also distinguished by charge sign (+, −, or 0), while their mass was considered identical and equal to $m_\pi c^2 = 137$ MeV.

2. For the density distribution of nuclear matter, we used a distribution similar to that obtained in experiments with fast electrons, but not identical with this. In analyzing experiments on fast-electron scattering, the density distribution of nuclear matter is approximated by continuous functions which only give a zero density at infinity. The cross section of a nucleus with such a density should, strictly speaking, also be infinite. It is very inconvenient to work with this kind of density in Monte Carlo calculations: the larger the cross section of the nucleus, the lower the density at the edges, the larger the number of nucleons flying over the edges, and the larger the number of nucleons flying right through the interaction nucleus. This will lead to a large number of "blank" cascades and to an unnecessary load on the computer. In our calculations we therefore used a "truncated" density distribution of nuclear matter

$$\rho\,(r) = \frac{\rho_1}{1 + e^{\frac{r-c}{z}}} \quad \text{at} \quad r \leqslant R$$

and

$$\rho\,(r) = 0 \quad \text{at} \quad r > R.$$

(15)

The values of parameters c and z were taken from [1]. The "cutoff" (truncation) radius R, taken as the radius of the nucleus, was calculated on the principle that $\rho(R)/\rho(0) = 0.1$. For nuclei with A = 28, 95, and 197, the radii R equalled 4.35, 6.16, and 7.54 F.

3. The potential of nucleon−nucleon interaction was taken in the form of a rectangular well of radius R and depth determined in the same way as in [13].

4. The total collision cross sections of the incident nucleons and π-mesons with nucleons belonging to the nucleus were obtained by averaging the cross sections corresponding to nucleon−nucleon and meson−nucleon collisions over the types of collision. The cross sections of nucleon−nucleon collisions for incident-proton energies smaller than 4 GeV were taken as in [13]. For energies greater than 4 GeV we used the data given in [82]. The cross sections for meson−nucleon collisions were taken from [13].

5. In calculating the range of nucleons in nuclear matter we considered the inhomogeneous density of the nucleus. We used the following formula for the range X_N corresponding to random number N:

$$\int_0^{x_N} \rho\,(t)\,dt = -\frac{1}{\sigma_{av}} \ln \frac{\sqrt{(N_0 - N)(N_0 - N + 1)}}{N_0},$$

(16)

where N = 100 is the number of equally probable intervals of range and $\rho(t)$ and σ_{av} are determined as in paragraphs 2 and 4. Equation (16) was solved numerically.

6. After determining the collision coordinates r, θ, and φ, we checked whether $r \gtrless R$. If r > R, it was considered that the particle had not experienced a collision inside the nucleus. If r < R, it was considered that a collision with one of the nucleons in the nucleus had taken place.

7. We considered two types of momentum distributions of the nucleons in the nucleus, of the Fermi and "truncated" Gauss types, respectively, with the same mean-square momentum value (18 MeV). The energy at which the Gaussian distribution was cut off equalled 49 MeV.

8. We considered eight forms of collision. For nucleon—nucleon collisions these were: elastic, in-elastic with the formation of one π-meson, and inelastic with the formation of two mesons. For π-meson—nucleon collisions they were: elastic, absorption, and scattering with charge exchange or the formation of one or two mesons. The relative probabilities of collisions of different kinds, the cross sections, and the scattering and angular distributions of the particles formed as a result of the collisions were taken as in [13]. The absorption of the π-mesons was calculated on the model of pair correlations (pn pair).

9. After calculating a collision, we checked whether the collisions were allowed by Pauli's principle.

10. After collision a nucleon was considered to have been absorbed if its kinetic energy was less than the average of the depth of the potential well (for protons and neutrons) and the Coulomb barrier.

11. A particle not experiencing a collision inside the nucleus was considered as having escaped from the nucleus if its energy were greater than the sum of the energies of the centrifugal potential and the potential of the nucleus. The momentum of the particle outside the nucleus was calculated with due allowance for refraction at the boundary of the nucleus. A particle with a kinetic energy insufficient to overcome the potential of the nuclear forces and the centrifugal potential was considered as having been reflected. The angle of reflection was equated to the angle of incidence. Further motion of the particle was considered analogously to the motion of a primary nucleon.

12. For escaping π-mesons the depth of the potential well was taken as 25 MeV, in accordance with [83]. Reflection and refraction were considered as for nucleons.

13. After the calculation of each cascade, the momentum balance was set up and the momentum of the recoil nucleus found. The kinetic energy of the recoil nucleus and its direction relative to the beam were found from the momentum.

14. The excitation energy of the residual nucleus was found from the energy balance, with due allowance for the kinetic energy of the recoil nucleus.

15. In the course of the calculation, the collision points of the cascade nucleons with the nuclear nucleons were recorded if the nuclear nucleon changed its momentum by an amount greater than 6 MeV$^{\frac{1}{2}}$ in these collisions. *

16. As a result of the calculation we obtained the following quantities: the mean number of particles of each sort escaping in a single cascade, the angular and energy distributions of all the escaping particles, the mean excitation energy of the residual nucleus, the relation between the excitation energy and the number of collision points recorded, the relation between the number of collision points recorded and the number of escaping nucleons, the distribution of collision points over the volume of the nucleus, and the excitation energy of the residual nucleus in cascades involving a definite number of escaping nucleons.

The data indicated in paragraphs 4-7 were not recorded in calculations carried out earlier. Thus, our calculations differ from the earlier work in that the volume of resultant information is greater as well as the original data more accurate.

Finally, let us consider the principal simplifications made in our calculations.

1. The potential of the nuclear forces was taken as a rectangular well. This form of potential is not the best approximation. Closer to reality is a diffuse potential, the parameters of which might be obtained, for example, by analyzing data on the elastic scattering of nucleons at nuclei from the point of view of the optical model [81]. Numerical calculations showed, however, that the introduction of a diffuse potential had little effect on the reflection and refraction of cascade nucleons, and hence could have little serious effect on the final calculations.

---

*The momentum is measured in units of $\sqrt{E_k}$, where E_k is the kinetic energy of the nucleon.

2. Allowance for the dependence of the nucleon–nucleus potential on the velocity of the nucleon may be more important [80, 81]. Unfortunately, data on the form of this relationship are not yet reliable. Use of the real part of the optical potential is not sufficiently justified, since the real potential in the optical model is only concerned with elastic processes, while the nuclear potential in the cascade model also leads to the absorption of some of the cascade nucleons, i.e., to inelastic processes. Hence, the dependence of potential on velocity was not considered in our calculations.

3. The absorption of π-mesons was considered a two-nucleon process in our calculations. This form of π-meson absorption is clearly not the only one, especially for light nuclei and low energies [84]. However, there are no reliable experimental data in respect of other processes for the nuclei and energy ranges of present interest. On the other hand, the main bulk of our calculations were carried out for primary particle energies for which meson formation played no serious part. Hence, a refinement allowing for many-particle absorption of π-mesons will not produce any great change in the results.

2. Photonuclear Cascade

From the point of view of cascade theory, high-energy photonuclear reactions only differ from the cascade process just considered in respect of the primary interaction of the photon with the nucleus. The model usually employed to describe this interaction is the quasi-deuteron model of Khokhlov [85] and Bethe and Levinger [86], based on the fact that for high-energy γ-quanta the nucleons in the nucleus may be regarded as free. Yet the laws of conservation of energy and momentum forbid the transfer of the momentum and energy of the γ-quantum to a free, noninteracting nucleon (the momentum of the γ-quantum is smaller than that of a nucleon with the same kinetic energy), so that the γ-quantum cannot be absorbed by such a nucleon. In order to satisfy the conservation laws, several nucleons in the nucleus must participate in the absorption. It is assumed in the quasi-deuteron model that absorption by a pair of neighboring nucleons plays the major part in this.

Since the main contribution in the energy range considered comes from the dipole absorption of photons, and the proton-proton pair has no dipole moment, we have only to consider the proton–neutron pair. It may be shown [86] that the wave function of such a pair is similar to that of a deuteron. Hence, in calculating the absorption of a photon we may use experimental data relating to the photodisintegration of a deuteron, so long as we remember the motion of the "quasi-deuteron" in the nucleus.

The first calculations of a photonuclear cascade were those which we carried out in 1956 and 1957; the results were published in [53-55]. Later the calculations were repeated with a number of refinements, and the statistical accuracy of the results was greatly increased.

Let us consider the original data and the arrangement of the subsequent calculations.

1. Calculations were carried out for a nucleus with A = 28 and γ-quantum energies of 100, 150, and 200 MeV. We took due notice of the radial density distribution of nuclear matter (see p. 140). We considered that the probability of the absorption of a γ-quantum depended on the distance from the center of the nucleus and was proportional to the density of nuclear matter.

2. The absorption of a γ-quantum was calculated in analogy with that of a π-meson in the calculation of a nuclear cascade. The escape angle of the proton with respect to the beam of γ-quanta was found from the differential cross section for the photodisintegration of a deuteron:

$$\frac{d\sigma}{d\Omega} = k(A + \sin^2\theta')(1 + 2B\cos\theta'). \tag{17}$$

The values of coefficients A and B for various energies were obtained from the data presented in [87].

3. The motion of the nucleons absorbing the γ-quantum was calculated as in the case of a nuclear cascade.

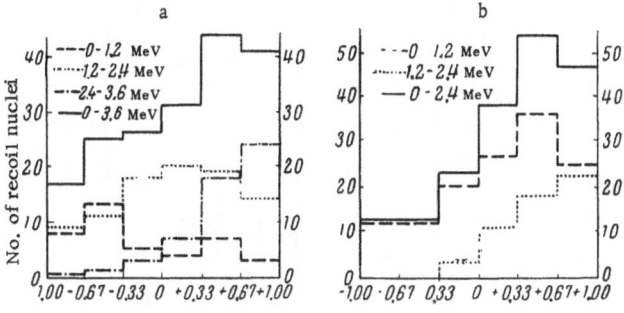

Cosine of angle relative to the beam

Fig. 16. Angular distributions of recoil nuclei formed on irradiating Si^{28} nuclei with 100-MeV γ-quanta. (a) For final nuclei three or four atomic units from the initial nucleus ($\Delta A = 3\text{-}4$); (b) for final nuclei with $\Delta A \neq 3\text{-}4$.

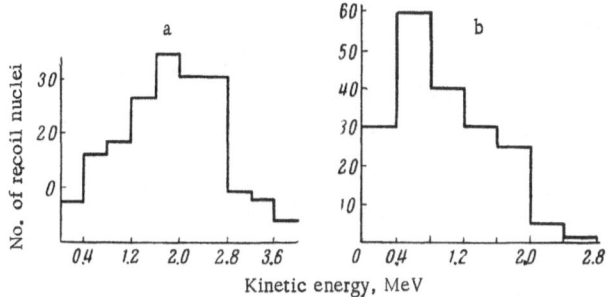

Kinetic energy, MeV

Fig. 17. Energy distribution of recoil nuclei formed on irradiating Si^{28} nuclei with 100-MeV γ-quanta. (a) For final nuclei with atomic numbers three or four units smaller than that of the target ($\Delta A = 3\text{-}4$); (b) for final nuclei with $\Delta A \neq 3\text{-}4$ and $\Delta A \neq 7\text{-}8$.

4. Some 500 cascades were considered in all for each value of photon energy; the angular and energy distributions of the escaping particles were obtained, as well as the momenta and excitation energy of the residual nuclei formed as a result of the cascade.

3. Results of Calculations and Comparison with Experiment

Photonuclear Reactions. In each of the cascades calculated, we recorded the number and type of cascade nucleons escaping from the nucleus, and the excitation energy and momentum of the nucleus formed after the cascade. From these data we found the atomic number and charge of the final nuclei formed after the cascade and plotted a momentum distribution for each of these.

Analysis of the results showed that the form of the momentum distribution of the recoil nuclei varied extremely slowly with the total number of nucleons escaping from the target nucleus, and for a given number of escaping nucleons was practically independent of their type. Hence, in order to increase the statistical accuracy in analyzing the reactions $Al^{27} \rightarrow Na^{24}$ and $Si^{28} \rightarrow Na^{24}$, we used overall data relating to final nuclei with atomic numbers three or four units smaller than that of the target nucleus ($\Delta A = 3\text{-}4$). In the same way, in

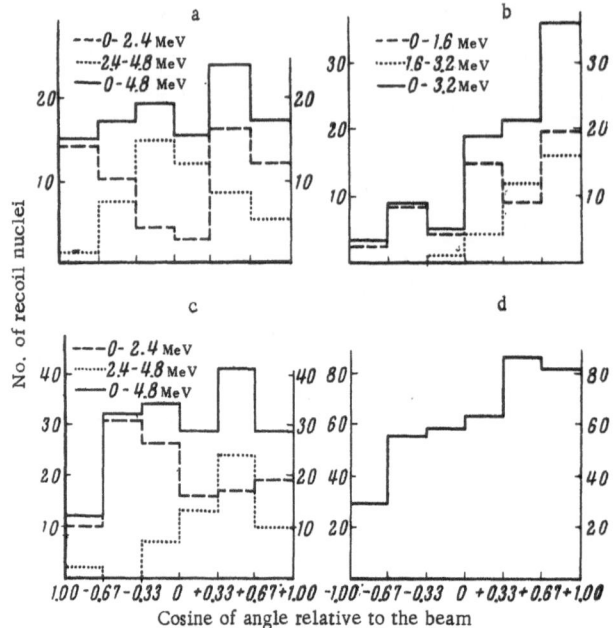

Fig. 18. Angular distributions of recoil nuclei formed on irradiation of Si^{28} with 150-MeV γ-quanta for final nuclei with atomic numbers differing from that of the original nucleus by $\Delta A = 3$-4 (a), $\Delta A = 7$-8 (b), $\Delta A \neq 3$-4 and $\Delta A \neq 7$-8 (c), and for all final nuclei together (d).

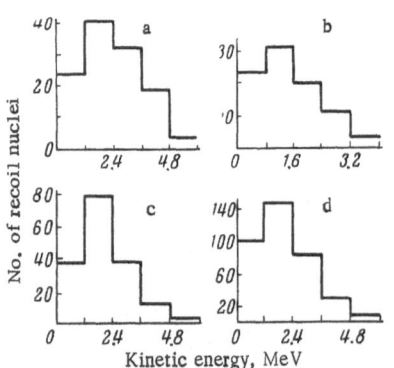

Fig. 19. Energy distribution of recoil nuclei formed on irradiating Si^{28} with 150-MeV γ-quanta for final nuclei with atomic numbers differing from that of the initial nucleus by $\Delta A = 3$-4 (a), $\Delta A = 7$-8 (b), $\Delta A \neq 3$-4 and $\Delta A \neq 7$-8 (c), and for all final nuclei together (d).

analyzing the reactions $P^{31} \rightarrow Na^{24}$ and $S^{32} \rightarrow Na^{24}$, we used overall data for final nuclei seven or eight atomic units away from the initial nucleus ($\Delta A = 7$-8).

Figures 16-21 show the angular and energy distributions of recoil nuclei for γ-quantum energies of 100, 150, and 200 MeV for various groups of final nuclei ($\Delta A = 3$-4, $\Delta A = 7$-8, $\Delta A \neq 3$-4, $\Delta A \neq 7$-8, and ΔA arbitrary). In constructing the distributions we only considered the momentum received in the cascade. The momentum transferred to the recoil nucleus on evaporation was calculated separately from the excitation energy in accordance with the results of Sec. 1 of Chapter II. The calculations showed that for $\Delta A = 3$-4 the momentum transferred to the recoil nucleus on evaporation was quite small, contributing in all something like 5% of the momentum received by the nucleus in the cascade. Hence, in plotting the range and angular distributions of the recoil nuclei (Fig. 22), this momentum was not considered. The error associated with this approximation was 2% in the range distributions and 1% in the angular distributions.

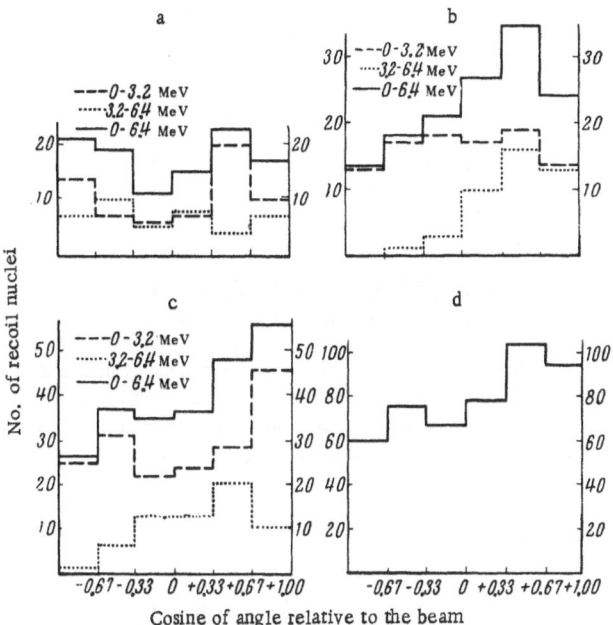

Fig. 20. Angular distributions of recoil nuclei formed on irradiating Si[28] with 200-MeV γ-quanta for reactions with $\Delta A = 3\text{-}4$ (a), $\Delta A = 7\text{-}8$ (b), $\Delta A \neq 3\text{-}4$ and $\Delta A \neq 7\text{-}8$ (c), and all final nuclei together (d).

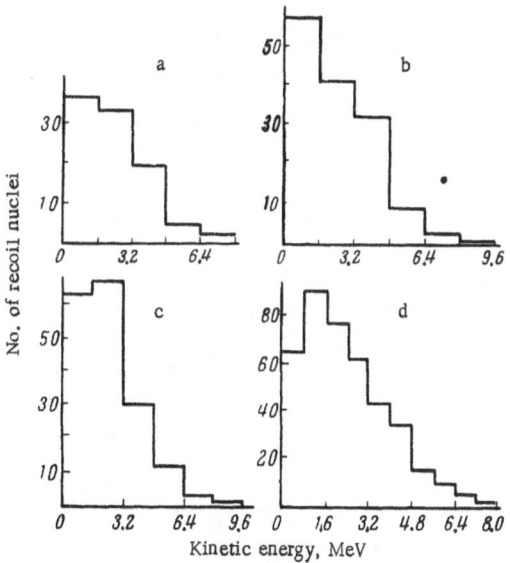

Fig. 21. Energy distribution of recoil nuclei formed on irradiation of Si[28] with 200-MeV γ-quanta for reactions with $\Delta A = 3\text{-}4$ (a), $\Delta A = 7\text{-}8$ (b), $\Delta A \neq 3\text{-}4$ and $\Delta A \neq 7\text{-}8$ (c), and for all final nuclei together (d).

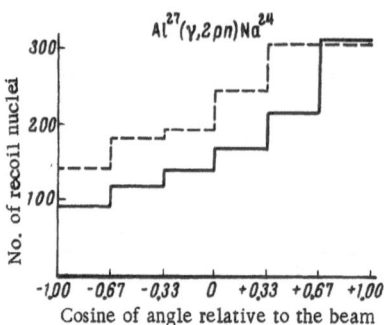

Fig. 22. Angular distributions of Na²⁴ recoil nuclei formed on irradiating aluminum with bremsstrahlung having a maximum energy of $E_{\gamma\,max}$ = 260 MeV. Continuous line represents our own experimental data and the broken line the results of a calculation on the quasi-deuteron model.

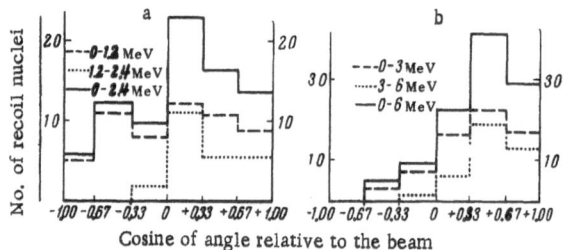

Fig. 23. Angular distributions of recoil nuclei formed on irradiating Si²⁸ with 660-MeV protons for reactions ΔA = 3-4 (a) and for $\Delta A \neq$ 3-4 (b).

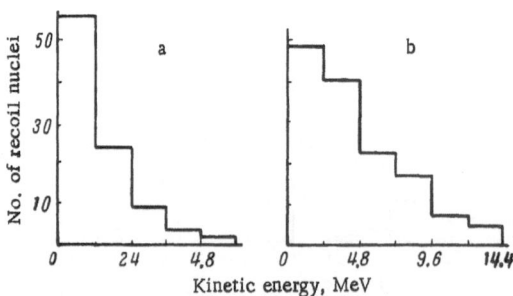

Fig. 24. Energy distributions of recoil nuclei formed on irradiating Si²⁸ with 660-MeV protons for reactions with ΔA = 3-4 (a) and $\Delta A \neq$ 3-4 (b).

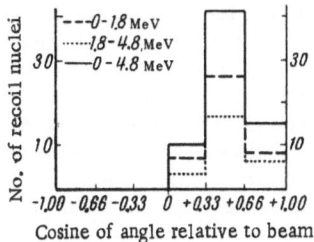

Fig. 25. Angular distributions of recoil nuclei formed on irradiating Si^{28} with 150-MeV protons for reactions with ΔA = 3-4.

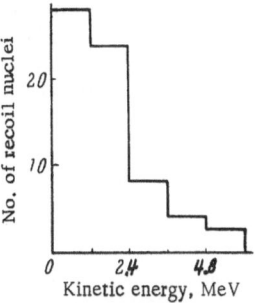

Fig. 26. Energy distributions of recoil nuclei formed on irradiating Si^{28} with 150-MeV protons for reactions with ΔA = 3-4.

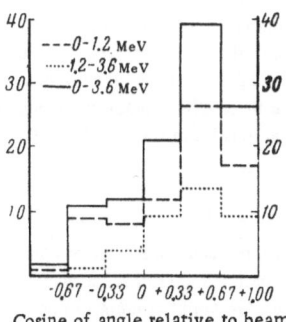

Fig. 27. Angular distributions of recoil nuclei formed on irradiating Si^{28} with 340-MeV protons for reactions with ΔA = 3-4.

The effective thicknesses relating to the various reactions studied were calculated with proper allowance for the momentum received on evaporation (see Fig. 10 and Table 1). In calculating the angular and energy distributions of recoil nuclei and the effective thicknesses, we averaged over all the γ-quantum energies. For γ-quantum energies of 50-70 MeV, the momenta of the recoil nuclei were calculated on the model of the compound nucleus, since (as shown by corresponding calculations) at these energies the cascade process led to the formation of a compound nucleus in the overwhelming majority of cases (more than 90%). The errors associated with the results of the calculations incorporated the statistical accuracy of the cascade calculations and the errors already existing in the range—energy relationship of the Na^{24} ions.

Comparison between theory and experiment showed that the cascade theory (like the theory of the compound nucleus) gives overestimated kinetic energies of the recoil nuclei. In the majority of cases the discrepancy exceeds three or four statistical errors.

Nuclear Reactions. It would be interesting to discover whether the discrepancy just mentioned is a characteristic of photonuclear reactions or whether it constitutes a general property of nuclear reactions of all kinds. For this purpose we calculated the momentum distributions of recoil nuclei formed on irradiating Si$^{28}_{14}$, (AgBr)$^{95}_{41}$, and Au$^{197}_{79}$ nuclei with 150-, 340-, and 660-MeV protons.

The results of the calculations relating to Si^{28} and 660-MeV protons are shown in Figs. 23 and 24. Using these results, we calculated the effective thicknesses for the reactions Al^{27} → Na^{24}, Si^{28} → Na^{24}, and P^{31} → Na^{24} for 660-MeV incident protons. The cascade-theory calculations (allowing for evaporation) are compared with our experimental results [52] in Table 2. We see from the table that, as in the case of photonuclear reactions, calculation gives effective thicknesses two or three times too high.

Figures 25-28 show the angular and energy distributions of recoil nuclei formed on irradiating Si$^{28}_{14}$ with 150- and 340-MeV protons. Using these data, and considering the momentum transferred on evaporation, we calculated the effective thicknesses for the reaction Al^{27} → Na^{24} for proton energies of 150 and 340 MeV. The results are compared with experiment [34] in Table 3. The comparison shows that, as for 660-MeV protons, calculation gives effective thicknesses several times too high.

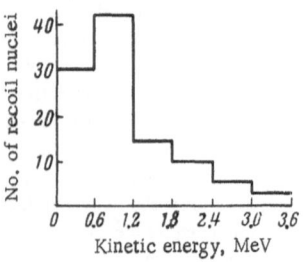

Fig. 28. Energy distributions of recoil nuclei formed on irradiating Si^{28} with 340-MeV protons for reactions with $\Delta A = 3$-4.

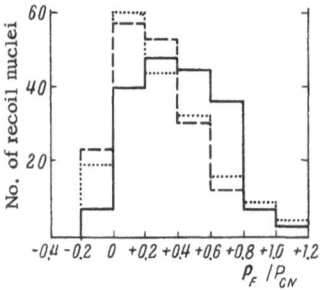

Fig. 29. Distribution of recoil nuclei in respect of the component of momentum parallel to the beam. Continuous curve represents our own calculations for 660-MeV protons and A = 95. Broken and dotted lines represent the results of [34] for 460- and 940-MeV protons and A = 100. The x axis gives the momentum component parallel to the beam P_F in proportions of the momentum received on formation of the compound nucleus P_{CN}. The y axis gives the number of recoil nuclei with a momentum component within the corresponding range of cosines.

It would have been interesting to carry out a similar comparison with heavier nuclei. Analysis of experimental data for A > 30, however, is made difficult by the absence of sufficiently reliable information on the range–energy relation for the corresponding ions. Nevertheless, some indication of an analogous discrepancy between theory and experiment may also be obtained for this range of nuclei.

We considered the results of Ostroumov's paper [37]. In addition to other quantities, Ostroumov determined the mean projection of the momentum of the recoil nuclei on the direction of the beam for recoil nuclei formed on irradiating heavy nuclei in photo-emulsion with 660-MeV protons. This ratio was found to be $(P_F/P_{CN}) = 0.25 \pm 0.004$. In Porile's calculations for the Ru^{100} nucleus with 460- and 940-MeV incident protons the values found for the same ratio were 0.30 ± 0.03 and 0.26 ± 0.03. In our own calculations $(P_C/P_{CN}) = 0.41 \pm 0.3$. This value exceeds experiment by a factor of 1.6 and Porile's value [42] by a factor of 1.4.

Figure 29 shows the distributions of recoil nuclei in terms of the component of momentum parallel to the beam, including our own values calculated for A = 95 and 660-MeV protons and those of Porile [42] for A = 100 and 460- and 940-MeV protons. We see from the figure why our values of the component of momentum parallel to the beam are larger than those of Porile [42]: Our distributions have been displaced in the direction of higher recoil-nucleus energies. We should remember that, in contrast to our own calculations, those of Porile [42] omitted the reflection and refraction of nucleons on escaping from the nucleus as well as a number of other effects. The comparison given in Fig. 29 shows that neglect of these effects leads to considerably different characteristics of the recoil nuclei.

4. Interaction with Correlated Groups of Nucleons

The results of the present investigation show that the cascade theory, not considering the association of nucleons in the nucleus, seriously contradicts experimental data. The kinetic energies of recoil nuclei calculated by this theory are much greater than those observed experimentally. We might attempt to remove this discrepancy by means of some other theory of high-energy nuclear reactions. Another theory discussed in the literature together with the cascade principle is the "tube" theory [88], which predicts low kinetic energies of the recoil nuclei. Detailed analysis shows, however, that in the energy range of present interest this theory encounters serious disagreement with experimental data [89]. At the same time the cascade theory gives a good representation of the most important aspects of nuclear reactions [89-91]. Hence, in our view, the search for methods of eliminating the observed discrepancy should take place within the framework of the cascade theory.

Let us examine several possible modifications of the cascade theory bringing it into agreement with experiments on recoil nuclei. As mentioned earlier, the momentum received by the recoil nucleus in the cascade process is made up of a momentum P_0 (which belonged to the knocked-on nucleons in the nucleus), a momentum ΔP_e transferred by the nucleons on escape from the nucleus, and a momentum ΔP_r received by the nucleus on the absorption and reflection of nucleons. A reduction in the value of any of these momenta may lead to a fall in the total momentum of the recoil nucleus.

The value of the momentum P_0 is determined by the momentum distribution of nucleons in the nucleus. In our calculations we used momentum distributions with a mean-square momentum equal to about 18 MeV, which agrees closely with the values obtained in the study of the elastic scattering of protons [92] and in experiments with p-n coincidences [93]. Hence this momentum cannot be reduced by anything very much without running contrary to these experimental data.

The values of momenta ΔP_e and ΔP_r are given by the depth of the nucleon—nucleus interaction potential. The nucleon—nucleus potential, however, is directly linked to the momentum distribution of the nucleons in the nucleus and the energy for removing the last nucleon. Thus, if we consider the potential as independent of the velocity of the nucleon,* the values of moments ΔP_e and ΔP_r also cannot be changed in any substantial way.

Thus, the discrepancy between theory and experiment is apparently due, not to an incorrect choice of the momentum distribution of the nucleons and the depth of the nuclear potential, but to some other factors, one of which may be the failure to allow for the association of nucleons in the nucleus.

We shall show that allowance for association may lead to a reduction in the momenta of the recoil nuclei. In fact, if there are correlated groups of nucleons in the nucleus, then, when an incident nucleon interacts with one of such groups, the recoil nucleus receives a momentum equal to the sum of two momenta: (1) the momentum which it had at the instant at which the incident nucleon interacted with the group P_{on} (since the nucleus as a whole is stationary, this momentum is equal and opposite to that of the corresponding group in the nucleus); and, (2) the momentum transferred to the nucleus of removal of the group in question (ΔP). For high energies of the incident particles and low binding energies between the group and the nucleus, the momentum $\Delta P < P_{on}$ and the total momentum of the recoil nucleus will be determined mainly by the momentum P_{on}. If the latter is small (momentum correlation of the nucleons in the nucleus), then the total momentum of the nucleus is also small. Thus, information regarding the spatial and momentum structure of the nuclei may be obtained from the momentum distributions of the recoil nuclei.

How does a group become separated from the nucleus? Three processes by which nucleon groups may be emitted from the nucleus are being discussed in current literature: (1) the elastic scattering of incident nucleons at the group [21, 22]; (2) the inelastic interaction of an incident nucleon with the group; and (3) the removal of the group from the nucleus as a result of the breaking of nuclear bonds. The processes indicated by (2) and (3) were first put forward in our own earlier papers [29, 55].

It is hard to assess the relative contributions of these processes just at the moment. The calculation of a cascade in which all these processes were considered would thus be fairly reasonable. By comparing the results of such calculations with experimental data on recoil nuclei, we should be able to obtain information on the various characteristics of nucleon associations: the probabilities of their existence in the nucleus, the binding energies, and the momentum distributions. In order to carry this program out we shall need further research in both the experimental and theoretical fields.

In the experimental field we shall have to pass on from measuring effective thicknesses to measuring differential angular and energy distributions of the recoil nuclei. The method proposed in the present paper may find wide application in this. We shall also have to carry out a great deal of work on direct measurements of range—energy relations for ions of all masses, including the very heaviest.

---

*We discussed the velocity—potential relationship earlier (Sec. 1 of this chapter), but unfortunately have no reliable information on its form. The question of allowing for this effect thus remains open.

In the theoretical field we shall have first to show that a reasonable energy and spatial correlation of the nucleons in the nucleus (for example, in the form of light nuclei such as D, He^3, and Li^7) is sufficient to ensure the escape of all the nucleons of the group in question from the nucleus, at least in the majority of cases. It will also have to be shown that for the spatial correlation chosen the nucleons of the group escaping from the nucleus will, in the majority of cases, not interact with the residual nucleus (such an interaction would, as a rule, lead to the transfer of a large amount of momentum to the nucleus). To this must be added the urgent necessity of establishing a firmer basis for the actual phenomenon of association, together with a number of other problems presented to nuclear theory by research into high-energy nuclear reactions.

In conclusion, the author wishes to express his sincere thanks to Professor P. A. Cherenkov for constant interest and for directing this research. The author also thanks his colleagues for assistance in the work: K. V. Kosarev, R. A. Latypova, T. D. Kruglov, and A. Duisebaeva.

The author is extremely grateful for valuable comments by Professor N. A. Perfilov, Doctor of Physico-Mathematical Sciences V. S. Barashenkov, Master of Physico-Mathematical Sciences V. I. Ostroumov, V. N. Mekhedov, O. V. Lozhkin, G. A. Leksin, and V. M. Mal'tsev.

APPENDICES

Appendix I. Activation Method of Measuring the Thicknesses of Thin Films and Foils

The methods of measuring the thickness of thin films at present used in experimental technology are be-set by serious failings. The widely used method in which the thickness is determined from the range of α par-ticles of known energy [94, 95] is only useful for comparatively thick films (500-3000 $\mu g /cm^2$) not having a substrate. Optical methods [97, 98] enable us to measure thicknesses from 1 $\mu g/cm^2$, but the film must be fairly transparent and there must be no substrate.

In the present investigation we used the activation method of measuring the thickness of thin films; this was based on artificial radioactivity. The method enables us to measure thicknesses from 1 $\mu g/cm^2$ and may be used for measuring thin films with thick substrates. The only condition which has to be observed in this method is that an easily measured induced radioactivity must be developed in the film or foil by means of some form of radiation.

The method is as follows. The film to be measured is irradiated simultaneously and under the same geometrical conditions as a standard sample of known thickness (made of the same material) in a beam of activating radiation. The thickness of the film is determined from the ratio of the activities developed in the film and standard as a result of radiation.

The procedure for determining film thickness by the activation method is the following. A sample about 10 cm^2 in area and a control sample of the same material are prepared from the film; the control sample should satisfy the following conditions: (1) It must be absolutely similar to the film sample in all di-mensions save thickness; (2) its thickness must be much smaller than its other dimensions; (3) the mass of the sample must be sufficient for its thickness to be determined by weighing. Conditions 2 and 3 may easily be satisfied at the same time by taking a sample 0.1 mm thick. The weight of such a sample about 10 cm^2 in area is greater than 10 mg for the majority of materials used in film preparation, and may be measured to an accuracy of about 0.1%.

The control sample and the film are fixed closely together and placed in a beam of activating radiation ($protons$, neutrons, γ-quanta), the form, energy, and flux of which must be sufficient to produce a considerable induced activity. After irradiation, the activities of the film and control sample are measured under identical conditions.

---

*By measuring the energy losses of monoenergetic α particles in thin films, thicknesses of about 1 $\mu g/cm^2$ [96] may be determined.

If all these conditions are satisfied, the radiation flux passing through the film and control sample is the same; on measuring the activities of the film A_f and standard A_s, and knowing the thickness of the standard t_s, the thickness of the film t_f is determined as

$$t_f = t_s \alpha A_f / A_s. \tag{I.1}$$

Here α is a correction for the different self-absorption of the induced activity in the film and control sample. The value of this correction may easily be determined. For this we must measure the activity A of samples of different thickness $\tilde{t}$ and plot the relation $a(\tilde{t}) = A/\tilde{t}$ as a function of $\tilde{t}$. For small $\tilde{t}$ (inaccessible to measurement), $a(t)$ is obtained by extrapolation. Clearly $\alpha(t_f, t_s) = \alpha(t_f)/\alpha(t_s)$. Since t_f is unknown, calculation by formula (I.1) is based on the method of successive approximations: we first estimate t_f roughly and evaluate α, which enables us to obtain a better value of t_f, etc. Usually the first approximation is adequate, since self-absorption in the thin films is very small. In cases in which the loading of the counters is considerable, the correction for self-absorption in measuring the activities of the films and sample must be supplemented by a correction for the different degree of counting loss.

The accuracy of the proposed method is determined mainly by the statistical accuracy with which A_f is measured. The radiation fluxes obtained with contemporary physical apparatus enable us to measure film thickness from a few $\mu g/cm^2$ to an accuracy of 1-4%. In cases in which A_f is small, owing to a small value of t_f or insufficient radiation intensity, several films may be irradiated simultaneously in order to improve the statistical accuracy.

The following comments should be made:

1. If two identical irradiating and measuring systems are not available, then A_f and A_s may be measured on the same one, provided that the decay curves of the active film material are well known. In this case a factor allowing for decay must be introduced on the right-hand side of formula (I.1).

2. If high-energy particles are used for activating the films and standards, then the effect of knocking activated recoil nuclei out of the film and standard must be considered [52]. For this purpose the irradiated film and standard must be placed between two foils capturing the escaping recoil nuclei. The thickness of these foils must be sufficient to absorb the most energetic recoil nuclei (say 1-2 mg/cm^2). In choosing the material for the foils we must remember that its activity in the radiation flux must differ considerably from that of the film material (in absolute magnitude, energy, or half-life period).

In order to check the intrinsic activity of the collecting foils, identical foils not touching the films or standard must be irradiated under the same conditions. In the case considered, $A_f = A_f' + A_{fo.f}' - A_{fo}$ and $A_s = A_s' + A_{fo.s}' - A_{fo}$, where A_f' and A_s' are the activities of the film and standard in question, $A_{fo.f}'$ and $A_{fo.s}'$ are the activities of the collecting foils touching the film and standard, respectively, and A_{fo} is the intrinsic activity of the collecting foils.

3. The proposed method is also applicable for measuring the thicknesses of thin films deposited on a substrate. In this case, however, the material of the substrate is subject to the same requirements as the material of the collecting foils (paragraph 2). In the presence of a substrate, $A_f' = A_f'' + A_{fo.f}' - A_{fo} - A_{sub}$, where A_s' is the activity of the film with the substrate, A_{sub} is the intrinsic activity of the substrate, and $A_{fo.f}'$, A_{fo}', and A_s are defined as in paragraph 2.

The method in question was used to measure thin aluminum films (about 80 $\mu g/cm^2$) deposited on a triacetate substrate. The activating radiation was the 260-MeV bremsstrahlung of the synchrotron in the Physical Institute of the Academy of Sciences. As induced activity for calibrating the films we used the activity (half-life period 15.01 h) resulting from the reaction $Al^{27} \rightarrow Na^{24}$. The activation of the deposited films was sufficient to determine their thickness to an accuracy of about 4%.

In the table presented below, we show the reactions which may be used for the activation of the most widely used films.

Film material	Nuclear reactions	Final product	Half-life period
Organic compounds . .	$C^{12}(\gamma, n)$, $C^{12}(p, pn)$	C^{11}	20.2 min
Aluminum.	$Al^{27}(\gamma, 2pn)$, $Al^{27}(p, 3pn)$	Na^{24}	15.01 h
Nickel	$Ni^{58}(\gamma, n)$, $Ni^{58}(p, pn)$	Ni^{57}	36.4 h
Copper	$Cu^{65}(\gamma, n)$, $Cu^{65}(p, pn)$	Cu^{64}	12.8 h
	$Cu^{63}(\gamma, n)$, $Cu^{63}(p, pn)$	Cu^{62}	9.7 min

Appendix II: Correction for the Decay of Radioactivity in the Case of the Successive Irradiation of Several Monitors

In our measurements of the effective cross sections of photonuclear reactions, the target-irradiation time was much greater than the half-decay period of the activity of the monitoring samples. Hence, in order to measure the radiation flux, we used several successively irradiated monitors, the activity of which was measured in the same sequence as they were irradiated.

Let us find an expression for the decay correction applicable to such irradiation and recording conditions. As zero-point for time we take the beginning of the irradiation process. Let the i-th monitor be irradiated for a time interval (t_{i-1}, t_i) and its activity be recorded in the interval (τ_{i-1}, τ_i). If the time distribution of radiation intensity has the form $Q_0 q(t)$, the decay constant of the monitoring activity λ_c, and the number of active nuclei formed in the monitor by unit quantity of radiation a_c, the number of nuclei decaying in the time interval (τ_{i-1}, τ_i) will equal

$$N_{ci} = \chi_{ci} Q_{ci} a_c, \qquad (\text{II}.1)$$

where

$$\chi_{ci} = e^{-\lambda_c \tau_{i-1}} - e^{-\lambda_c \tau_i}, \quad Q_{ci} = \int_{t_{i-1}}^{t_i} Q_0 q(t) e^{\lambda_c t} dt.$$

If k monitors are used in all, the total number of activity disintegrations in all the monitors during the activity recording period will equal

$$N_c = \sum_{i=1}^{k} a_c \chi_{ci} Q_{ci}. \qquad (\text{II}.2)$$

The total number of active nuclei formed in all the monitors is

$$N_{co} = a_c Q, \qquad (\text{II}.3)$$

where

$$Q = \int_{0}^{t_k} Q_0 q(t) dt.$$

In the same way, for a target x, irradiated for time $(0, t_k)$, the number of nuclei disintegrating in time (T_{1x}, T_{2x}) is

$$N_x = \chi_x a_x Q_x. \qquad (\text{II}.4)$$

Here, $\chi_x = e^{-\lambda_x T_{1x}} - e^{-\lambda_x T_{2x}}$; λ_x is the decay constant of the target activity, $\quad Q_x = \int\limits_0^{t_k} Q_0 q\,(t)\, e^{\lambda_x t}\, dt$,

and a_x is the number of active nuclei formed by unit quantity of radiation during the irradiation of the target.

The total number of active nuclei formed in the target is

$$N_{xo} = a_x Q. \tag{II.5}$$

Using relations (II.2) and (II.5), we obtain

$$\frac{N_{x0}}{N_{c0}} = \frac{N_x}{N_c} \cdot \frac{\sum\limits_{i=1}^{k} \chi_{ci} Q_c}{\chi_x Q_x}. \tag{II.6}$$

Appendix III: Mean-Square Momentum and Angular Distribution of Recoil Nuclei on the Compound-Nucleus Model

The mean-square momentum acquired by the recoil nucleus on evaporation of nuclear particles was calculated in [34, 39] in the same way as that of a Brownian particle:

$$\overline{P_0^2} = \sum_{i=1}^{k} \overline{P_i^2}, \tag{III.1}$$

where P_i is the momentum of the i-th evaporating particle and k is the total number evaporated. This kind of analogy-based calculation, however, is only very rough, and in a number of cases may lead to serious error. More detailed study of the question demands due allowance for the motion of the nucleus produced by the evaporation of all the previous particles from within it. Let us consider a stationary nucleus with mass M' excited to an energy E_e, from which k particles with momenta P_i and masses m_i evaporate. According to the law of the conservation of momentum, the momentum received by the recoil nucleus equals

$$\mathbf{P}_0 = -\sum_{i=1}^{k} \mathbf{P}_i. \tag{III.2}$$

Let us express this equation in scalar quadratic form and average over the directions and absolute magnitudes of the vectors

$$\overline{P_0^2} = \sum_{i=1}^{k} P_i^2 + \sum_{i \neq j}^{k} (\mathbf{P}_i, \mathbf{P}_j). \tag{III.3}$$

We see that the true expression for $\overline{P_0^2}$ differs from (III.1) by the additional term $\delta = \sum\limits_{i \neq j}^{k} (\mathbf{P}_i, \mathbf{P}_j)$. In order

to estimate δ, let us represent each momentum $\mathbf{P}_i$ in the laboratory system of coordinates in the form of its isotropic component $\mathbf{P}_{i0}$, i.e., the momentum received by the i-th nucleon in the system of the moving nucleus, and the momentum $\mathbf{P}_{i1}$ associated with the motion of the nucleus and produced by the evaporation of all the preceding nucleons .

Plainly, $\mathbf{P}_{i1} = -\sum\limits_{l=1}^{i-1} \alpha_l \mathbf{P}_l$, where $\alpha_l = m_{l+1}/M_l$, where M_l is the mass of the nucleus after the

evaporation of the l-th particle. Putting $\alpha_i = \bar{\alpha} = \sum\limits_{l=1}^{k} \alpha_l/k$ in δ for simplicity, remembering that $(\mathbf{P}_{i0},$

$\mathbf{P}_{j0}) \equiv 0$, and neglecting terms containing α^2, we obtain

$$\delta = \sum_{i \neq j}^{k} (\mathbf{P}_i, \mathbf{P}_j) = -2\alpha \sum_{i=1}^{k} (k-i)\, \overline{P}_i^2.$$

Neglecting the change in the temperature of the residual nucleus, we may consider that in the correction $\overline{P}_i^2 = \overline{P}_j^2$ (i, j = 1, 2, . . . , k). Then

$$\delta = -2\overline{\alpha} \sum_{i=1}^{k} (k-i)\, \overline{P}_i^2 = -\overline{\alpha}\,(k-1) \sum_{i=1}^{k} \overline{P}_i^2,$$

$$\overline{P}_0^2 = [1 - \overline{\alpha}\,(k-1)] \sum_{i=1}^{k} \overline{P}_i^2. \tag{III.4}$$

Formula (III.4) for k = 3 was first used in [54, 55]. For $\overline{\alpha}$ = 0, formula (III.4) transforms into relation (III.1). However, for $\overline{\alpha} \approx 0.1$ ($M_l / m_{l+1} \approx 10$), the error associated with the use of (III.1) may reach 30-50%.

The values of $\overline{P}_0^2$ may be expressed in terms of the kinetic energy of the reaction, using the law of conservation of energy. In the nonrelativistic case, for the evaporation of particles of equal mass m,

$$\overline{P}_0^2 = \frac{[1 - \overline{\alpha}\,(k-1)]\, E_{\mathrm{K}}}{1 - \alpha_{\mathrm{K}}\,[1 - \overline{\alpha}\,(k-1)]}\,, \tag{III.5}$$

where the momentum of the recoil nucleus with mass M_K is expressed in units of $\sqrt{M_K E_N / m}$, and E_K and E_N are the kinetic energies of the reaction and the recoil nucleus in MeV. According to the evaporation model, the momentum of the final nucleus $\mathbf{P}$ is made up of the momentum $\mathbf{P}'$ received on evaporation and distributed isotropically, and the momentum $\mathbf{P}_0$ constituting part of the momentum of the incident particle $\mathscr{P}$ and directed along the beam. Clearly, $\mathbf{P}' = (M_K / M')\mathscr{P}$.

The angular distribution of recoil nuclei in the laboratory system for given $\mathbf{P}_0$ and $\mathbf{P}'$ may be determined in the following way. In the system of the moving nucleus and in the angular range $\{\theta', \theta' + d\theta'\}$, the number of particles escaping is $N(\theta')d\theta' = (\sin \theta' d\theta'/2) = -(d \cos \theta'/2)$ (normalized to one particle in an angle of 4π). The angles θ' are related to angles θ in the laboratory system of coordinates by the relation $\theta' = \theta + \Delta\theta$, where $\sin \Delta\theta = \lambda \sin \theta$, and $\lambda = P'/P_0$. Expressing θ' in terms of θ, we obtain $N(\theta, P_0)\, d\theta = [\lambda(x + z)^2/z][\sin \theta d\theta/2]$, where $x = \cos\theta$, $z = \sqrt{a^2 + x^2}$, and $a^2 = \lambda^{-2} - 1$. Thus, the angular distribution of recoil nuclei in the laboratory system of coordinates for given P_0 and P' has the form

$$n\,(\theta,\, P_0,\, P') = \frac{\lambda\,(x+z)^2}{z}\,. \tag{III.6}$$

It is the distribution n(θ, P_0, P') averaged over all P_0 which is to be compared with experiment. Numerical calculations showed, however, that the distribution averaged over P_0 (for the distribution functions of current interest — see Appendix IV) coincided to within 2 or 3% with the angular distribution plotted for $\overline{P}_0^2$. If, therefore, the required accuracy is no better than 2-3%, we may replace the distribution averaged over P_0 by the distribution obtained for $\overline{P}_0^2$.

The absolute value of the momentum of the recoil nucleus in the laboratory system of coordinates (as may be seen from simple geometric considerations) is given by the expression

$$P - P'\,(x + z). \tag{III.7}$$

Appendix IV: Energy Distributions of Recoil Nuclei
on the Evaporation of Two and Three Nucleons

Let us use the following notation:

E_K = total kinetic energy of the reaction;

E_i (i = 1, 2, 3) = kinetic energies of the evaporating nucleons;

P_i (i = 1, 2, 3) = momenta of the evaporating nucleons (in units of $\sqrt{m_i E_i / m}$, where m_i = mass of particle, E_i = kinetic energy of particle, and m = mass of nucleon);

$E_{20} = E_1 + E_2$ = sum of kinetic energies of the first and second nucleons;

E_{Ni} (i = 2, 3) = kinetic energy of the recoil nucleus after the evaporation of two (i = 2) and three (i = 3) nucleons;

$E_{Ni} = M^{(i)} E_{Ni} / m$ (i = 2, 3), where $M^{(i)}$ is the mass of the recoil nucleus after the evaporation of the i-th nucleon;

P_{Ni} (i = 2, 3) are the momenta of recoil nuclei after the evaporation of two (i = 2) and three (i = 3) nucleons, expressed in units of $P_{Ni} = \sqrt{M^{(i)} E_{Ni} / m} = \sqrt{E_{Ni}}$;

θ_i (i = 1, 2) are the angles between the directions of the momenta P_1 and P_2 (i = 1) and P_{N_2} and P_3 (i = 2).

Let us write down the laws of the conservation of momentum and energy (nonrelativistic case). After evaporation of the first and second nucleons

$$P_{2N}^2 = P_1^2 + P_2^2 + 2P_1 P_2 \cos \theta_1 \tag{IV.1}$$

or

$$E_{2N} = E_1 + E_2 + 2\sqrt{E_1 E_2} \cos \theta_1. \tag{IV.1'}$$

After evaporation of the third nucleon

$$P_{3N}^2 = P_{2N}^2 + P_3^2 + 2P_{2N} P_3 \cos \theta_2 \tag{IV.2}$$

or

$$E_{3N} = E_{2N} + E_3' + 2\sqrt{E_{2N} E_3} \cos \theta_2. \tag{IV.2'}$$

Neglecting the kinetic energy of the recoil nucleus, we obtain

$$E_1 + E_2 + E_3 = E_N, \tag{IV.3}$$

$$E_1 + E_2 = E_{20}. \tag{IV.4}$$

If $\mathscr{P}(E_2, E_3) dE_2 dE_3$ is the probability that the energies of nucleons 2 and 3 lie in the ranges $(E_2, E_2 + dE_2)$ and $(E_3; E_3 + dE_3)$, * then, for isotropic evaporation of the particles in the laboratory system, the probability that the recoil nucleus receives kinetic energy in the range E_{3N} to $E_{3N} + dE_{3N}$ equals

$$\mathscr{P}(E_{3N}) dE_{3N} = \int_G \mathscr{P}(E_2, E_3) d\Omega_1 d\Omega_2 dE_2 dE_3. \tag{IV.5}$$

The range of integration G is chosen in such a way that integration covers all variants of evaporation leading to energy E_{3N}, and

---

*In view of (IV.3), the energy of nucleon 1 is not an independent variable.

$$d\Omega_i = \frac{\sin\theta_i \, d\theta_i \, d\varphi_i}{4\pi} \qquad (i = 1, 2). \tag{IV.6}$$

Let us find $\mathscr{A}(E_2, E_3)$. The probability of the evaporation of a nucleon with kinetic energy E_i from a nucleus with temperature T (neglecting the change in temperature on evaporation and the Coulomb barrier) may be written in the form [61]:

$$\mathscr{P}(E_i)\, dE_i = \frac{E_i}{T^2} e^{-\frac{E_i}{T}} \, dE_i. \tag{IV.7}$$

Clearly,

$$\mathscr{P}(E_2, E_3)\, dE_2 \, dE_3 = \prod_{i=1}^{3} \mathscr{P}(E_i)\, dE_i = \frac{e^{-\frac{E_K}{T}}}{T^2} \prod_{i=1}^{3} E_i \, dE_i = \mathscr{P}_0 \prod_{i=1}^{3} E_i \, dE_i. \tag{IV.8}$$

Let us determine the range G. We can show that for given P_1 and P_2 (or P_{2N} and P_3) the probability of obtaining a given energy E_{2N} (or E_{3N}) does not depend on the angles θ_1 and θ_2. In fact, according to (IV.6), $d\Omega_i \sim -d\cos\theta_i$ (i = 1, 2), but it follows from (IV.1') that

$$d\cos\theta_1 = \frac{dE_{2N}}{2\sqrt{E_1 E_2}}. \tag{IV.9}$$

Hence,

$$d\Omega_1 \sim -\frac{dE_{2N}}{2\sqrt{E_1 E_2}}. \tag{IV.10}$$

Analogously,

$$d\Omega_2 \sim -\frac{dE_{3N}}{2\sqrt{E_{2N} E_3}}. \tag{IV.11}$$

Introducing (IV.10) and (IV.11) into (IV.5), integrating with respect to φ_1 and φ'_1, and collecting all the constants into a coefficient $\mathscr{P}'_0$, we obtain

$$\mathscr{P}_3(E_{3N})\, dE_{3N} = \mathscr{P}'_0 \, dE_{3N} \int_{G'} \frac{E_3 E_2 (E_{20} - E_3)}{\sqrt{E_1 E_2 E_3 E_{2N}}} \, dE_2 \, dE_3 \, dE_{2N}. \tag{IV.12}$$

Let us determine the limits of integration with respect to E_2. According to (IV.1'), as $\cos\theta_1$ varies from +1 to −1, E_{2N} varies from

$$E_{2N\,min} = (\sqrt{E_1} - \sqrt{E_3})^2 \tag{IV.13}$$

to

$$E_{2N\,max} = (\sqrt{E_1} + \sqrt{E_3})^2. \tag{IV.14}$$

Since $E_{2N\,min} \le E_{2N} \le E_{2N\,max}$, we have

$$(\sqrt{E_1} - \sqrt{E_3})^2 = E_{2N\,min} \le E_{2N}. \tag{IV.15}$$

Thus,

$$0 \le E_{2N\,min} = (\sqrt{E_1} - \sqrt{E_3})^2 \le E_{2N}. \tag{IV.16}$$

From (IV.16) and (IV.4) we find the limits of integration with respect to E_2:

$$E_2^{(1)} = \frac{E_{20}}{2},$$ (IV.17)

$$E_2^{(2)} = \frac{E_{20} \pm \sqrt{2E_{20}E_{2N} - E_{2N}^2}}{2}.$$ (IV.18)

It follows from (IV.18) that $2E_{20}E_{2N} - E_{2N}^2 \geq 0$, i.e., $E_{2N} \leq 2E_{20}$.

Integrating with respect to E_2, we obtain the energy spectrum of the recoil nuclei after the evaporation of two nucleons

$$\mathscr{P}_2(E_{2N})\,dE_{2N} = \mathscr{P}_{20}''\,dE_{2N} \int_{E_2^{(1)}}^{E_2^{(2)}} \sqrt{E_2(E_{20} - E_2)}\,dE_2 =$$

$$= \mathscr{P}_{20}' \left[2(E_{20} - 2E_2^{(2)}) \sqrt{E_2^{(2)}(E_{20} - E_2^{(2)})} + E_{20}^2 \arcsin \frac{E_{20} - 2E_2^{(2)}}{E_{20}} \right] dE_2$$ (IV.19)

where all the constants are brought into the coefficients $\mathscr{P}_{20}'$ and $\mathscr{P}_{20}''$. For plotting this distribution it is convenient to use the dimensionless units

$$\varepsilon_{2N} = \frac{E_{2N}}{E_{20}}$$ (IV.20)

and

$$\varepsilon_1 = \frac{E_2^{(2)}}{E_{20}},$$ (IV.21)

which, on the basis of (IV.18), are related by

$$2\varepsilon_1 = 1 \pm \sqrt{2\varepsilon_{N2} - \varepsilon_{N2}^2}.$$ (IV.22)

In these units,

$$\mathscr{P}_2(\varepsilon_{N2})\,d\varepsilon_{N2} = \mathscr{P}_{20}\left[2(1 - 2\varepsilon_1)\sqrt{\varepsilon_1 - \varepsilon_1^2} + \arcsin(1 - 2\varepsilon_1)\right] d\varepsilon_{2N}.$$ (IV.23)

We note that in view of (IV.4), $E_2 \leq E_{20}$, i.e., $\varepsilon_1 \leq 1$.

Putting (IV.19) into (IV.5) and using (IV.23), we obtain the following integral expression for the energy spectrum of the recoil nuclei after the evaporation of three nucleons:

$$\mathscr{P}_3(E_{3N})\,dE_{3N} = \mathscr{P}_0'\,dE_{3N} \int_{G''} \frac{E_3 E_{20}^2}{\sqrt{E_{2N}E_3}}\, \mathscr{P}_2(\varepsilon_{2N})\,dE_{2N}dE_3.$$ (IV.24)

The range of integration G'' is formed by the variables E_3 and E_{2N} only.

The limits of integration with respect to E_{2N} are found in an analogous way to (IV.17) and (IV.18), being

$$E_{2N}^{(1)} = E_3,$$ (IV.25)

$$E_{2N}^{(2)} = (\sqrt{E_3} + \sqrt{E_{3N}})^2.$$ (IV.26)

The limits of integration with respect to E_3 equal

$$E_3^{(1)} = 0, \qquad \text{(IV.27)}$$

$$E_3^{(2)} = E. \qquad \text{(IV.28)}$$

Considering (IV.25)-(IV.28), and introducing the dimensionless units $\varepsilon_3 = E_3/E_K$, $\varepsilon_{2N}' = E_{2N}/E_K$, $\varepsilon_{3N} = E_{3N}/E_K$, $\varepsilon_{2N} = E_{2N}/E_{20} = \varepsilon_{2N}'/(1 - \varepsilon_3)$, we may write expression (IV.24) in the form

$$\mathcal{P}_3\left(\varepsilon_{3N}\right) d\varepsilon_{3N} = \mathcal{P}_{30} \, d\varepsilon_{3N} \int_0^1 d\varepsilon_3 \int_{\varepsilon_3}^{(\sqrt{\varepsilon_3}+\sqrt{\varepsilon_{3N}})^2} \frac{\sqrt{\varepsilon_3}\,(1 - \varepsilon_3)^2\,\mathcal{P}_2\left(\varepsilon_{2N}\right)}{\sqrt{\varepsilon_{2N}'}} \, d\varepsilon_{2N}'. \qquad \text{(IV.29)}$$

We integrated formula (IV.29) numerically.

Appendix V: Yield of Recoil Nuclei from Thick Targets. Effective Thickness

Let us find the yield of recoil nuclei from targets with thicknesses exceeding the maximum range of the recoil nuclei. Let the target (a plane sample) be so placed that the normal to its surface makes an angle η with the direction of the beam of radiation (see Fig. 15). Then, in a system of coordinates in which the polar angle θ' is reckoned from the direction of the normal, the number of recoil nuclei escaping from the target is given by the expression

$$N(\eta) = \int_0^\infty dR \int_0^R dh \int_0^{\text{arc cos}(h/R)} a_0 n\left(R, \theta'\right) \frac{\sin \theta'}{2} \, d\theta', \qquad \text{(V.1)}$$

where a_0 is the number of recoil nuclei formed in a layer of unit thickness, $n(R, \theta')$ is the distribution of recoil nuclei with respect to range R and angle θ', and h is the depth of the target at which the recoil nucleus was formed.

The usual experimentally determined quantity is the ratio

$$t\left(\eta\right) = \frac{N\left(\eta\right)}{a_0}, \qquad \text{(V.2)}$$

which is called the effective thickness of the target at an angle η for the given reaction. The effective thickness is the thickness of target material in which a number of recoil nuclei equal to the number actually escaping is formed. In order to calculate the effective thickness, we must know the range and angular distributions of the recoil nuclei.

In calculations based on the intranuclear-cascade model, this distribution can only be obtained in the form of histograms. The effective thickness can therefore only be calculated numerically from formula (V.2). With the compound-nucleus model, an expression for the effective thickness can in a number of cases be obtained in analytical form. Let us consider a case in which the range—momentum relationship of the recoil nucleus has the form $R \sim p^n$ (n being any positive whole number) and $\eta = 0$ and $180°$.

Using the expressions for the absolute momentum of the recoil nucleus and its angular distribution given in Appendix III, (III.7) and (III.6), we may write the expression for the effective thickness in the form

$$t\left(\eta\right) = \int_0^{R\left(\eta\right)} dh \int_\eta^{\theta_0 + \Delta\theta_0} \frac{\sin \theta}{2} \, d\theta. \qquad \text{(V.3)}$$

Here η takes two values, 0 and 180°. The remaining symbols have the following meanings (see Fig. 15); θ is the angle between P_0 and the direction of the beam; P_0 is the momentum received by the recoil nucleus in the system of the compound nucleus; h is the depth in the target at which the recoil nucleus is formed; $R(\eta)$ is the range of the recoil nucleus at an angle η; θ_0 is the extreme angle at which a recoil nucleus formed at depth h and having range $R(\theta_0)$ can escape from the target; R_0 is the range of the recoil nucleus with momentum P_0.

These quantities are related to each other and to the quantities in (V.3) by the following equations: $\sin \Delta \theta_0 = \lambda \sin \theta_0$, $\lambda = P'/P_0$, P' being the momentum of the recoil nucleus associated with the motion of the compound-nucleus system, and P being the momentum of the recoil nucleus in the laboratory system of co-ordinates; $R(\theta_0) = R_0(\sqrt{a^2 + x^2} + x)^n \lambda^n$, $a^2 = \lambda^{-2} - 1$, $x = \cos \theta_0 = h/R(\theta_0)$. After integrating, we obtain

$$t(\eta) = \frac{\lambda^{n+1} R_0}{4} \sum_{i=1}^{3} A_{\eta n}^{(i)} \alpha_\eta (k_{in}),$$

(V.4)

where

$$k_{in} = 2i + n - 3, \quad \alpha_\eta (k) = (b_\eta)^k - a^k, \quad b_\eta = \lambda^{-1} + \cos \eta, \quad A_{\eta n}^{(1)} = -a^2 \left[b_\eta + \frac{a^2 \cos \eta}{2} \right]$$

for any $n \neq 1$, and

$$A_{\eta n}^{(1)} = 0, \quad A_{\eta n}^{(2)} = \left[b_\eta + \frac{a^2 n \cos \eta}{(n+1)} \right]$$

and

$$A_{\eta n}^{(3)} = \frac{(n+2) \cos \eta}{2(n+3)}$$

for any $n \geq 0$.

Calculations based on formula (V.4) were first published in August 1958 [52]. In 1961, the author of [31] cited a contribution of Winsberg [99] printed in September 1958 and giving approximate formulas for calculating $t(\eta)$. * The results of calculations based on these formulas coincide with ours for the limiting case $\lambda = 0$ while for $\lambda \simeq 1$ there are slight discrepancies not exceeding ~2-3%.

In order to verify the validity of our analytical formula, we made some numerical calculations of $t(\eta)$ for $\eta = 0$ and 180°. The results were the same.

In order to obtain effective thicknesses on the evaporation model, we must average over all momenta P_0. Numerical calculations showed, however, that, for n = 0, 1, 2, $\overline{t(P_0)} = t(\sqrt{\overline{P_0^2}})$ to an accuracy of about 2 or 3%. This considerably simplifies calculations, so that, instead of averaging over all P_0, we may simply use the effective thickness obtained for the mean-square value of momentum. We also note that formulas (V.4) take no account of multiple scattering and straggling of the recoil nuclei. These effects are small and have no influence on the results.

LITERATURE CITED

1. R. Hofstädter, Phys. Rev., 28:214 (1956); Usp. Fiz. Nauk, 63:696 (1957).
2. M. Riou, Rept. on the Internat. Sympos. on Direct. Interactions and Nuclear Reaction Mechanisms, Padua, September 3-8, 1962. New York-London (1962).
3. N. A. Perfilov, Rep. on the Internat. Conf. on High-Energy Physics and Nuclear Structure, CERN, February 25-March 1, 1963. Geneva (1963).
4. J. Combe, Nuovo Cim. Suppl., 3(2):182 (1956).
5. V. I. Ostroumov, N. A. Perfilov, and R. A. Filov, Zh. Eks. i Teor. Fiz., 39:105 (1960).
6. V. P. Dzhelepov and V. I. Moskalev, Dokl. Akad. Nauk SSSR, 110:539 (1956).
7. T. Coor, D. A. Hill, W. F. Hornyak, L. W. Smith, and G. Snow, Phys. Rev., 98:1369 (1955).
8. G. P. Millburn, W. Birnbaum, W. E. Grandall, and L. Schecter, Phys. Rev., 95:1268 (1954).
9. G. Bernardini, E. T. Booth, and S. Y. Lindenbaum, Phys. Rev., 85:826 (1952).
10. J. Combe, J. Phys. Radium, 16:445 (1955).

---

*The author neglected terms smaller than λ^2.

11. W. Heisenberg, Kosmische Strahlung. Berlin (1943).

12. R. Serber, Phys. Rev., 72:1114 (1947).

13. N. Metropolis, R. Bivins, M. Storm, Y. M. Miller, G. Friedlander, and A. Turkevich, Phys. Rev.,
     110:185 (1958).

14. F. P. Denisov, T. D. Kruglova, R. A. Latypova, Yu. A. Oratovskii, and P. A. Cerenkov, FIAN* Preprint
     A-3 (1962).

15. G. B. Zhdanov, V. M. Maksimenko, M. I. Tret'yakova, and M. N. Shcherbakova, Zh. Eks. i Teor. Fiz.,
     37:620 (1959).

16. T. Bowen, Y. Hardi, G. T. Reynolds, G. Tagliaferri, A. E. Werbronck, and W. H. Moore, Phys. Rev.,
     119:2041 (1960).

17. V. S. Barashenkov, V. A. Belyakov, Wang Shu-fen, V. V. Glagolev, N. Dolkhazhaev, L. F. Kirillova, R. M.
     Lebedev, V. M. Mal'tsev, P. K. Markov, K. D. Tolstov, E. N. Tsyganov, M. G. Shafranova, and Yao
     Ch'ing-Hsieh, Atomnaya Energiya, 7:376 (1959).

18. P. E. Hodgson, Nucl. Phys., 8:1 (1958).

19. A. P. Zhdanov et al., Korpuskularfotografie, IV. Munich (1963).

20. T. J. Gooding, J. Igo, and I. F. Hausen, Phys. Rev., 131:337 (1963).

21. P. Cüer and A. Samman, J. Phys. Radium, 19:13 (1958).

22. H. Gauvin, M. Lefort, and X. Tarrage, Phys. Rev. Letters, 4:215 (1963).

23. N. A. Perfilov, O. V. Lozhkin, and V. P. Shamov, Usp. Fiz. Nauk, 70:3 (1960).

24. V. I. Ostroumov and R. A. Filov, Zh. Eks. i Teor. Fiz., 37:643 (1959).

25. V. I. Ostroumov, N. A. Perfilov, and R. A. Filov, Zh. Eks. i Teor. Fiz., 39:105 (1960).

26. Yu. R. Arifkhanov, M. M. Makarov, N. A. Perfilov, and V. P. Shamov, Zh. Eks. i Teor. Fiz., 38:1115
     (1960).

27. M. M. Makarov, Zh. Eks. i Teor. Fiz., 44:463 (1963).

28. O. V. Lozhkin, N. A. Perfilov, and Yu. P. Yakovlev, Dokl. Akad. Nauk SSSR, 151:826 (1963).

29. F. P. Denisov, K. V. Kosareva, and P. A. Cerenkov, In: Transactions of the All-Union Conference on
     the Peaceful Uses of Atomic Energy, Izd. Akad. Nauk UzbSSR, Tashkent (1961); F. P. Denisov, E. V.
     Kuz'min, and Yu. A. Oratovskii, FIAN Preprint A-4 (1962); F. P. Denisov, Tr. Fiz. Inst. Akad. Nauk
     SSSR, 22:129 (1964).

30. N. Sugarman, M. Campos, and K. Wielgoz, Phys. Rev., 101:388 (1956).

31. V. P. Crespo, Lawrence Radiation Laboratory, Univ. of California, Berkeley, California, UCRL-9183.

32. N. Hintz, Phys. Rev., 86:1042 (1952).

33. J. Knipp and E. Teller, Phys. Rev., 59:659 (1941).

34. S.-C. Fung and I. Perlman, Phys. Rev., 87:623 (1952).

35. J. H. M. Brunings, J. K. Knipp, and E. Teller, Phys. Rev., 60:657 (1941).

36. S.-C. Fung and A. Turkevich, Phys. Rev., 95:176 (1954).

37. V. I. Ostroumov, Zh. Eks. i Teor. Fiz., 32:3 (1957).

38. Y. K. Böggild, K. Y. Brostrom, and T. Lauritsen, Kgl. Danske vid. selskab. Mat.-fys. medd.

39. N. I. Borisova, M. Ya. Kuznetsova, L. N. Kurchatova, V. N. Mekhedov, and L. V. Chistyakov, Zh. Eks.
     i Teor. Fiz., 37:366 (1959).

40. C. B. Fulmer, Phys. Rev., 108:1113 (1957).

41. E. W. Baker and S. Katcoff, Phys. Rev., 117:1353 (1960).

42. N. T. Porile, Phys. Rev., 120:572 (1960).

43. F. P. Denisov and V. E. Kolesov, Pribory i Tekhn. Eksp., 2:34 (1958).

44. F. P. Denisov, Pribory i Tekhn. Eksp., 1:155 (1963).

45. D. Strominger, Y. M. Hollander, and G. T. Seaborg, Rev. Mod. Phys., 30:585 (1958).

46. W. C. Barber, W. D. George, and D. D. Reagan, Phys. Rev., 98:73 (1955).

47. I. V. Estulin, Radioactive Radiations. Fizmatgiz (1962), p. 199.

48. A. N. Gorbunov, F. P. Denisov, and V. A. Kolotukhin, Zh. Eks. i Teor. Fiz., 38:1084 (1960).

*The abbreviation FIAN refers to the P. N. Lebedev Physics Institute of the Academy of Sciences of the USSR.

49. L. Katz and A. G. E. Cameron, Can. J. Phys., 29: 518 (1951).
50. L. I. Schiff, Phys. Rev., 83: 252 (1951).
51. R. L. Wolke and N. A. Bonner, Phys. Rev., 102: 530 (1956).
52. L. V. Volkova and F. P. Denisov, Zh. Eks. i Teor. Fiz., 35: 538 (1958).
53. F. P. Denisov and P. A. Cerenkov, Zh. Eks. i Teor. Fiz., 35: 544 (1958).
54. F. P. Denisov and P. A. Cerenkov, Proc. Conf. on Nuclear Phys., Paris (1959), p. 676.
55. V. A. Balitskii and F. P. Denisov, In: Transactions of the Second All-Union Conference on Low- and Medium-Energy Nuclear Reactions. Izd. Akad. Nauk SSSR (1962), p. 450.
56. N. Bohr, Nature, 137: 344 (1936).
57. N. Bohr and F. Kalckar, Kgl. Danske vid. selskab. Mat.-fys. medd., 14: 10 (1937).
58. J. Blatt and V. Weisskopf, Theoretical Nuclear Physics. Wiley, New York (1952).
59. A. Tomasini, Nuovo Cim., 6: 404 (1957).
60. T. Ericson and V. Strutinski, Nucl. Phys., 8: 284 (1958).
61. K. Y. Le Couteur, Proc. Phys. Soc., 65A: 718 (1952).
62. H. A. Bethe, Ann. Phys., 5: 325 (1930).
63. F. Bloch, Ann. Phys., 16: 285 (1933).
64. Ya. A. Teplova, V. S. Nikolaev, I. S. Dmitriev, and L. N. Fateeva, Zh. Eks. i Teor. Fiz., 42: 43 (1962).
65. E. Fermi and E. Teller, Phys. Rev., 72: 399 (1947).
66. O. B. Firsov, Zh. Eks. i Teor. Fiz., 36: 1517 (1959).
67. P. M. S. Blackett and D. S. Lees, Proc. Roy. Soc., 134: 658 (1932).
68. N. Feather, Proc. Roy. Soc., 141: 194 (1933).
69. W. W. Eaton, Phys. Rev., 4: 48, 921 (1935).
70. R. L. Antony, Phys. Rev., 50: 726 (1936).
71. J. I. Carthy, Phys. Rev., 53: 30 (1938).
72. G. A. Wrenshall, Phys. Rev., 57: 1095 (1940).
73. H. L. Reynolds, D. W. Scott, and A. Zucker, Phys. Rev., 95: 671 (1954).
74. S. K. Allison and S. D. Warshaw, Rev. Mod. Phys., 25: 779 (1953).
75. Ya. A. Teplova, V. S. Nikolaev, I. S. Dmitriev, and L. N. Fateeva, Izv. Akad. Nauk SSSR, Ser. Fiz., 23: 894 (1959).
76. M. S. Livingston and H. A. Bethe, Rev. Mod. Phys., 9: 245 (1937).
77. N. Bohr, Kgl. Danske vid. selskab. Mat.-fys. medd., 18: 8 (1948).
78. N. F. Mott and H. S. W. Massey, The Theory of Atomic Collisions (2nd ed.). Oxford (1949).
79. P. I. Fedotov, Dissertation, Leningrad. Izd. Akad. Nauk SSSR (1961).
80. N. A. Perfilov, O. V. Lozhkin, and V. I. Ostroumov, Nuclear Reactions Due to High-Energy Particles. Izd. Akad. Nauk SSSR (1962).
81. L. Elton, Nuclear Sizes. Oxford University Press (1961).
82. V. S. Barashenkov and Huang Nien-Ning, Zh. Eks. i Teor. Fiz., 36: 1395 (1959).
83. B. A. Nikol'skii, L. P. Kudrin, and S. A. Ali-Zade, Zh. Eks. i Teor. Fiz., 32: 48 (1957).
84. A. I. Varfolomeev, Tr. Fiz. Inst. Akad. Nauk SSSR, 22: 101 (1964).
85. Yu. K. Khokhlov, Dissertation, FIAN (1955).
86. J. S. Levinger, Phys. Rev., 84: 43 (1951).
87. Constitution of the Atomic Nucleus. [Collection of Russian translations], IL (1959).
88. E. L. Feinberg, Usp. Fiz. Nauk, 58: 193 (1956).
89. V. S. Barashenkov, V. M. Mal'tsev, and E. K. Mihul, Nucl. Phys., 24: 642 (1961).
90. V. S. Barashenkov et al., Nucl. Phys., 14: 522 (1960).
91. V. S. Barashenkov, A. Ya. Boyadzhiev, I. A. Kulyukina, and V. M. Mal'tsev, Joint Institute for Nuclear Research Preprint R-1341. Dubna (1963).
92. J. B. Cladis, W. N. Hess, and R. J. Moyer, Phys. Rev., 87: 425 (1952).
93. A. Wathenberg, A. S. Odian, P. S. Steinand, and H. Wilson, Phys. Rev., 104: 1710 (1956).
94. W. H. T. Davidson, J. Scient. Instrum., 34: 418 (1957).
95. G. M. Osetinskii and M. V. Savenkov, Pribory i Tekhn. Eksp., 6: 115 (1959).

96. S. A. Baranov, V. M. Kulakov, A. G. Zelenkov, and V. M. Shatinskii, Zh. Eksp. i Teor. Fiz., 43 : 795 (1962).

97. R. O. Lane and D. I. Zaffarano, Phys. Rev., 94 : 203 (1954).

98. I. S. Dmitriev, Ya. A. Teplova, V. S. Nikolaev, and L. N. Fateeva, Pribory i Tekhn. Eksp., 6 : 131 (1959).

99. L. Winsberg, Chem. Div. Semi-Annual Rep. UCRL-8618, 44 (1958).

A STUDY OF THE ANGULAR DISTRIBUTION OF FRAGMENTS IN THE PHOTOFISSION OF URANIUM ISOTOPE U²³³*

N. M. Kulikova, N. V. Nikitina, and N. V. Popov

The angular anisotropy of the fragments produced in the photofission of nuclei with odd atomic weights was studied in [1, 2]. A dipole character was assumed for the absorption of the γ-quanta, and in all cases the yield ratio $I(\theta = 90°)/I(\theta = 0°)$ was measured, θ being the angle relative to the direction of the γ-ray beam. In view of the controversial role of the quadrupole absorption of γ-quanta in the photofission process, it was considered desirable to check the isotropy of the distribution for such nuclei over a wide angular range.

The FIAN (Physical Institute of the Academy of Sciences) 30-MeV synchrotron was used to measure the angular distribution of fission fragments produced by irradiating uranium-238 with x-rays having a maximum energy of $E_{max} = 12$ MeV. The fission fragments were detected in photo-emulsions charged with uranium of the following composition: 97.6% uranium-233, 2.4% uranium-238, 0.01% thorium-232. During irradiation, the photographic plates were placed along the x-ray beam at a distance of 60 cm from the synchrotron target.

In studying the angular distributions of fission fragments, the method of charged photo-emulsions has the following advantages:

1. A better angular resolution than other methods; the direction of escape of the fragments being determined to an accuracy of about 1°

2. The possibility of observing the flight direction of the fragments without any serious error resulting from the scattering of fragments at nuclei

3. The absence of randomness in the fragments recorded

4. The observation of fragments escaping at different angles under completely identical conditions of irradiation and detection

5. The absence of any possibility of recording false acts of fission, such as may occur when working with ionization chambers and counters, owing to the superposition of α-particle pulses, or (in the case of x-ray irradiation) groups of electrons

The method of charging and developing various photo-emulsions when working with uranium-238 and thorium-232 was described in [3-7]. In the present investigation, we used emulsions of the Ilford-D1 type, 100-μ thick. Table 1 shows the method used by the authors for emulsions of this type in earlier work on the angular distribution of fragments produced by the photofission of uranium-238 [8]; the method is taken over (with slight changes) from [3, 4].†

---

*This paper describes research work carried out in 1958 and 1959 in the Photonuclear-Reaction Laboratory.
† According to the data of [3], the method given for the uranium charging and subsequent treatment of the emulsion also gives good results for Ilford-C_2, -E1, and -B2 plates.

Table 1. Method of Treating the Photo-Emulsion when Charging with Uranium

Treatment stage	Uranium-238			Uranium-233		
	Solution	T, °C	t, min	Solution	T, °C	t, min
Charging the Photo-Emulsion with Uranium						
Saturating with distilled water	—	18-22	30	—	22-25	45
Charging the uranium	7% uranyl acetate or 4% sodium uranyl acetate	18-22	30	0.5% sodium uranyl acetate	18-22	15*
Washing in distilled water	—	18-22	3	—	18-22	1
Saturating with alcohol containing glycerin	2% glycerin in ethyl alcohol	18-22	15	2% glycerin in ethyl alcohol	18-22	5
Drying	—	18-22	~1 h	—	~30	12-18
Time required to prepare the emulsion for irradiation, starting from the uranium-charging state			~1 h, 50 m			~40
Developing the Photo-Emulsion after Irradiation						
Preliminary soaking in an alkaline medium	3% soda	18-22 / 5-6	20 / 10	1% soda	18-22 / 5-6	15 / 5
Soaking in developer	2 g para-amino-phenol, 25 g sodium sulfite (a.w.), water to 500 ml	5-6	30	2 g para-aminophenol, 25 g sodium sulfite (a.w.), water to 500 ml	5-6	25
Dry development	—	18-22	10-25†	—	18-22	15†
Time required prior to stop-bath			~1 h, 20 m			~1 h
Stop-bath	2% acetic acid	7-8	15			
Washing in distilled water	—	7-8	20			
Fixing by decantation	40% sodium hypo-sulfite	7-8 / 18-22	60 / Until fully fixed	As for uranium-238		
Washing	—					
Drying	—					

*On inspecting the plates, it was found that the uranium was not distributed evenly enough over the thickness of the emulsion; the charging time should be raised to 20 or 25 min.

†Depends on the batch of emulsion and the conditions of irradiation.

Table 2. Angular Distribution of Fission Fragments Obtained from U^{233}
on Irradiation with X-Rays of Maximum Energy 12 MeV

$\Delta\theta$, deg	Relative yield per unit solid angle*	$\Delta\theta$, deg	Relative yield per unit solid angle*
0—15	0.97±0.03	45—60	0.97±0.03
15—30	1.03±0.03	60—75	1.02±0.03
30—45	0.99±0.03	75—90	1.01±0.03

*As unit we take the yield obtained by the method of least squares
for the distribution I(θ) = const.

If the thickness of the emulsion is greater than 100 μ, all the soaking times should be correspondingly increased. In this way, plates with emulsion thicknesses up to 200 μ were used in [8]. The note in [6] regarding the impossibility of using photo-emulsion thicker than 50 μ for uranium charging was apparently inspired by an unfortunate choice of uranium salt. The development prescription presented in Table 1 gave very sharp tracks of the fission fragments and scarcely noticeable tracks of α particles, which greatly aided inspection of the plates.

On charging with uranium-238, the photographic density of the plates was almost entirely determined by the dose of γ radiation received during irradiation. In the case of the uranium-233 isotope, for which the probability of an α disintegration is about $3 \cdot 10^4$ times greater than for uranium-238, the use of the charged-photo-emulsion method is complicated by the very heavy background of α particles. In order to avoid this difficulty, it is above all important to minimize the duration of those stages of emulsion treatment which correspond to the buildup of the α-particle background. The duration of the uranium-charging process (from the instant of placing the plate in the solution containing the uranium salt) was accordingly reduced to about 40 min and the duration of the initial stages of development prior to the stop-bath to about 1 h. Together with the time for irradiating the plates with x-rays (equal to 20 min), this gave a period of approximately 2 h. The α-particle background was further weakened by reducing the concentration of the sodium uranylacetate solution with which the experiments were carried out. Table 1 shows suitable conditions for charging and developing the emulsion in the case of uranium-233; these conditions gave plates quite suitable for inspection.

The plates so obtained were inspected in MBI-2 microscopes with an objective of 60× and an ocular of 5×; the angles were measured with an ocular of 10×. For greater accuracy of measurements, only tracks of fission fragments inclined at a maximum of 15° to the plane of the plate were recorded. For the 5860 tracks selected in this way, the angle θ between the direction of the escaping fragments and the direction of the incident x-ray beam was measured in each case. Since the fragmentation point was not fixed, the angles θ and ($\pi - \theta$) were not distinguished, and only the angular distribution [I(θ) + I($\pi - \theta$)] was determined.

The angular distribution of fission fragments obtained from U^{233} on irradiation with x-rays having E$_{max}$ = 12 MeV is given in Table 2.

The relative yields per unit solid angle are given. The distribution obtained is corrected for tracks missed because of having escaped from the emulsion (between 4.2 and 5.5% for the range of θ in question). The correction for the anisotropy of the angular distribution of fission fragments arising from U^{238} (present to the extent of 2.4%) was less than 0.2%. The number of fissions due to background neutrons around the accelerator was checked by monitoring plates placed at the same distance from the synchrotron, but outside the x-ray beam, during the irradiation period. Inspection of these plates showed that neutron-induced fission contributed no more than 1%.

The resultant angular distribution was isotropic to within ∓2% over the whole range of angles θ. This coincides with the results of [2] carried out by the ionization-chamber method, in which the yield ratio I(θ = 90°)/I(θ = 0°) was measured. The observed isotropy in the angular distribution of fragments resulting from the photofission of uranium-233 agrees with the predictions of Bohr's theory [9] for the fission of nuclei with odd "A."

In conclusion, the authors offer their sincere thanks to L. E. Lazareva for constant attention and interest in the work.

LITERATURE CITED

1.   E. I. Winhold and I. Halpern, Phys. Rev., 103:990 (1956).
2.   L. Katz, A. P. Baerg, and F. Brown, in the book: Second U. N. Internat. Conf. on the Peaceful Use of Atomic Energy. Geneva (1958), p. 200; A. P. Baerg, P. H. Bartholomew, F. Brown, L. Katz, and S. B. Kowalski, Canad. J. Phys., 37:1418 (1959).
3.   G. E. Belovitskii and T. A. Romanova, Report of the P. N. Lebedev Physics Institute of the Academy of Sciences of the USSR (1951); T. A. Romanova and L. D. Chikil'dina, Pribory i Tekhn. Eksp., No. 5, p. 88 (1962).
4.   N. A. Perfilov and N. S. Ivanova, Zh. Eks. i Teor. Fiz., 29:551 (1955).
5.   H. Faissner and F. Gönnenwein, Z. Phys., 153:257 (1958).
6.   B. Forkman and S. A. E. Johansson, Nucl. Phys., 20:136 (1960).
7.   H. G. de Carvalho, A. G. da Silva, and J. Goldemberg, Nuovo Cim., 19:1131 (1961).
8.   B. P. Bannik, N. M. Kulikova, L. E. Lazareva, and V. A. Yakovlev, Zh. Eks. i Teor. Fiz., 33:53 (1957); A. I. Baz', N. M. Kulikova, L. E. Lazareva, N. V. Nikitina, and V. A. Semenov, in the book: Transactions of the Second International Conference on the Peaceful Use of Atomic Energy. Geneva, 1958, Vol. 1. Atomizdat (1959), p. 362.
9.   O. Bohr, in the book: Transactions of the International Conference on the Peaceful Use of Atomic Energy. Geneva, 1955, Vol. 2. Fizmatgiz (1958), p. 175.

SHOWER-TYPE GAMMA SPECTROMETERS, THEORY AND CALCULATION OF THE PRINCIPAL CHARACTERISTICS

V. F. Grushin and E. M. Leikin

SEC. 1. INTRODUCTION

The majority of experimental methods of determining the energy of γ quanta are based on the measurement of the energy distribution of secondary particles created in matter by the γ radiation. Prominent among these are methods based on using a "thick" radiator and measuring the energy of the particles arising in this [1]. An important feature of the γ spectrometer with a thick radiator is the high recording efficiency. As regards resolving power, however, these spectrometers are inferior to those with thin radiators, because of the multitude of processes taking place in the actual radiator and the consequent spread in the evolution of energy. In order to reduce the intrinsic width of the γ-spectrometer line in the low-energy range, we may either limit the number of processes constituting the main contribution to the formation of the line, or else ensure conditions such that none of the secondary particles should leave the radiator, i.e., conditions ensuring complete absorption. As the energy of the γ quanta recorded increases, the character of the processes taking place in the radiator becomes much more complex, so that complete absorption is in fact the only method of ensuring a minimum intrinsic line breadth for a thick-radiator γ spectrometer.

In the high-energy range, the absorption of a γ quantum in matter is accompanied by the development of an electron—photon shower. Hence, the condition of complete absorption for high-energy γ spectrometers (or "shower" γ spectrometers) is equivalent to the requirement that no secondary particles of the shower should escape from the radiator.

The measurement of energy by a shower γ spectrometer falls into several stages. The first is the conversion of the primary γ quantum into an electron—photon shower. The number of charged particles in the shower N is proportional to the energy of the γ quantum, E_γ; this cannot be measured directly, however, but only by way of secondary effects produced by these particles.

It is well known that the principal secondary effect is ionization; most of the energy losses of the charged component of the shower may be attributed to this. Unfortunately, however, the collection of the "free" charges arising in the considerable volume of the solid radiator is almost impossible. Hence, the energy evolution in a thick radiator is usually measured by reference to the intensity of the electromagnetic radiations accompanying the retardation of the particles in matter. The emission of electromagnetic radiation by the shower particles is (generally speaking) a second-order effect compared with the ionization which they produce. Nevertheless, the use of scintillation, or Čerenkov, light reduces the problem of obtaining information regarding the energy liberated in the radiator to one of measuring the intensity of this light. Thus, the second stage of energy measurement is associated with the generation of ν photons by each shower particle; the intensity of the light emitted by all the shower particles $N\nu$ is also proportional to E_γ.

167

For recording the light, sensitive light detectors (photoelectron multipliers, or FEM) are usually employed. The FEM-intensity measurement is preceded by collecting or focusing the light into the detector (stage 3a), which takes place with an efficiency g. The actual intensity measurement falls into a number of stages: the conversion of the photons into photoelectrons, with an efficiency ψ (stage 3b), the collection (focusing) of the photoelectrons into the multiplying system, with an efficiency ρ (stage 3c), and, finally, the multiplication of the electrons, with an amplification factor M (stage 4). Only after completion of the fourth stage do we obtain the signal characterizing E_γ; the value of this may be measured by standard methods to the appropriate accuracy. In other words, the generation of a charge Q in the anode (plate) circuit of the FEM completes the multistage process of transforming the original γ quantum of energy E_γ into a quantity convenient for measurement. The whole chain of successive transformations may be written as follows:

$$E_\gamma \overset{1}{\to} N \overset{2}{\to} N\nu \overset{3a}{\to} N\nu g \overset{3b}{\to} N\nu g\psi \overset{3c}{\to} N\nu g\psi\rho \overset{4}{\to} Q, \tag{1}$$

where $Q = N\nu g\psi\rho M$ (in units of the charge on the electron).

Since the transformation of one quantity into the other is a stochastic process, each of these is a random quantity characterizing the energy of the γ quantum at some stage of the measuring process. When monoenergetic γ quanta impinge on the radiator of the spectrometer, the output of the system will yield signals having a certain statistical spread. The problem of theory is to find a distribution function of the output signals which will characterize the line shape of the shower γ spectrometer for a definite energy of the γ radiation. In any particular case we must know the first two moments of the distribution function characterizing the magnitude of the signal and the energy spread of the apparatus. Such problems may be solved by using the methods of the theory of branch processes [2, 3].

Thus, the principle of operation of the apparatus as a spectrometer is based on the fact that, at each stage of measuring the energy of the γ quantum, the quantity E_γ is characterized by the total number of particles of a particular generation. This means that, in the mathematical apparatus of the theory, finding the distribution function of the output signal $\Omega(Q)$ reduces to constructing the distribution of the sum of a random number of independent random components. It is usually assumed that all the terms have identical distributions, i.e., that, for any pair of stages, all the "primary" particles generate "secondary" particles with identical distributions. We shall assume that the first two moments (mathematical expectancy and dispersion) exist in all the distributions with which we shall be dealing subsequently.

SEC. 2. CALCULATION OF THE MATHEMATICAL EXPECTANCY AND DISPERSION. ENERGY RESOLUTION

The problem of finding the mathematical expectancy and the dispersion of a random quantity arising as a result of a branching process may be solved without knowing the specific form of the distribution function of this quantity, if we know the corresponding characteristics of the individual stages. Here we shall use the formalism of generating functions.

In the case of the mechanism of energy measurement described above in (1), the probability generating function has the form [4]:

$$f_{SS}[\zeta] = f_1[f_2[f_{3a} \cdots f_4[\zeta] \cdots]], \tag{2}$$

where f_i are the probability generating functions of the individual stages. This way of writing $f_{SS}|\zeta|$, however, fails to reflect a number of important characteristics of the apparatus. In any practical apparatus, the assumption that the distributions of the "secondary" particles created by different "primary" particles are identical is not satisfied in a number of the stages listed. In addition to causes of a technical character (optical inhomogeneity of the radiator, inhomogeneity of the quantum yield over the surface of the FEM photocathode, nonuniformity of the focusing of the photoelectrons from various parts of the photocathode into the multiplying section of the FEM), which, in an ideal case, are capable of elimination, there are other causes which constitute matters of principle. These include the dependence of light-collecting factor g on the point of the radiator x and wavelength λ, and also the dependence of the quantum yield ψ of the photocathode on λ.

In order to give a proper representation of these relationships and so eliminate the limiting features underlying the theory, we should have to present the corresponding event, for example, the light-focusing, in the form of a set of incompatible events with a given distribution of a random value, and then, using the additivity of the probabilities of these events, carry out a summation (averaging) over this distribution in the expression for $f_{SS}|\zeta|$. The formal operation of averaging must be carried out in accordance with the specific physical averaging mechanism on which the principle of operation of the apparatus is based.

In the present case of the shower γ spectrometer, the averaging of the light-collection and quantum yield is effected in each act in which the energy of an individual γ quantum is measured. Mathematically, this operation may be carried out by integrating over the volume of the radiator, allowing for the distribution of light sources $L(x)$, and over the spectral range, allowing for the emission spectrum $S(\lambda)$. It must here be remembered that averaging over the spectrum takes place at the stages of the collection and conversion of the emitted light, and averaging over the volume takes place at the stage of shower development. This means that, in expression (2) for $f_{SS}|\zeta|$, averaging over the volume of the radiator must be carried out with respect to f_2, and averaging over the spectrum with respect to f_3. Thus [5]:

$$f_{SS}\ [\zeta] = f_1\Big[\ \int L(\mathbf{x})\,d^3x f_2\big[\int S(\lambda)\,d\lambda f_3\,[f_4\,[\zeta]]\big]\Big]. \tag{3}$$

In this formula, $L(\mathbf{x})$ coincides with the shower function $N_\Gamma\,(E_\gamma,\ E,\ x)$, which characterizes the average number of charged particles with energy greater than E,* at a depth x in the shower produced by a γ quantum of energy E_γ.

In formula (3), the three stages 3a, 3b, and 3c are combined into one with common index 3. This is permissible because in each of these stages the random quantity can only take two values: 1 with a probability g, ψ, ρ, and 0 with a probability $1-g$, $1-\psi$, $1-\rho$. The corresponding generating functions are linear in ζ, so that the generating function of these three stages, $f_3\,[\zeta]$, retains the structure of each of the distributions. In view of what has been said, averaging over λ is of a trivial nature; it leads to a simple replacement of $g\psi$ by the quantity, obtained as a result of integration over the spectrum. As regards averaging over the volume, this leads to an important consequence to be discussed later.

Calculating the mathematical expectation $\overline{Q}$ from (3), we obtain

$$\overline{Q} = \int N_\Gamma\,(E_\gamma,\ E,\ x)\,dx \int S(\lambda)\,g(\lambda,\ x)\,\psi(\lambda)\,\rho M\,d\lambda. \tag{4}$$

In the case of the Čerenkov spectrometer $S(\lambda)g(\lambda,x)$ is expressed by the Frank–Tamm formula for a medium with absorption [6]

$$S(\lambda)\,g(\lambda,\ x) = \Big(1 - \frac{\mathrm{Re}\,\varepsilon(\lambda)}{(v/c)^2\,|\,\varepsilon(\lambda)\,|^2}\Big)\,\frac{1}{\lambda^2}\,e^{-(v/c\lambda)\,\mathrm{Im}\,\varepsilon(\lambda)R(x)}, \tag{5}$$

where v/c is the velocity of the particle in terms of the velocity of light, $\varepsilon(\lambda)$ is the dielectric constant of the medium, and $R(x)$ is a quantity characterizing the path of the light through the medium to the detector.

The function $\psi(\lambda)$ has a sharp maximum. Hence, $g(\lambda,\ x)$ may be taken out from under the integral with respect to $d\lambda$ at some point λ_1 in the neighborhood of the maximum. Applying analogous considerations to N_Γ and taking $g(\lambda_1,\ x)$ from under the integral sign at a point x_1 close to the maximum of N_Γ, we have

$$\overline{Q} \sim g(\lambda_1,\ x_1) \int N_\Gamma\,(E_\gamma,\ E,\ x)\,dx. \tag{6}$$

---

*Energy E corresponds to the threshold of formation of the electromagnetic radiation.

The integral of N_Γ extended to infinity represents the zeroth moment, or the total range of the charged particles up to the radiation threshold $P_\Gamma (E_\gamma, E)$. In order to evaluate this, we may use the Tamm—Belen'kii formula [7]:

$$P_\Gamma (E_\gamma, E) = \frac{E_\gamma}{\beta} \left\{ \xi e^\xi \int_\xi^{\xi_\gamma} \frac{e^{-s}}{s^2}\, ds - \frac{\xi}{\xi_\gamma^2} (1 - e^{-\xi_\gamma + \xi}) \right\} = \frac{E_\gamma}{\beta}\, \Phi\, (\xi_\gamma, \xi), \qquad (7)$$

where $\xi_\gamma = 2.29 E_\gamma /\beta$ and β is the critical energy of the radiator material.

Let us introduce a quantity $\alpha(E_\gamma, T)$ characterizing the proportion of the range of the charged particles in the shower produced by a γ quantum of energy E_γ, and being contained by the distance T traveled in the radiator; $\alpha(E_\gamma, T)$ is the ratio of the integral in formula (6) to $P_\Gamma(E_\gamma, E)$, i.e.,

$$\alpha (E_\gamma, T) = \frac{\int_0^T N_\Gamma (E_\gamma, E, x)\, dx}{P_\Gamma (E_\gamma, E)} . \qquad (8)$$

In order to complete the separation of the factors depending on the energy of the recorded γ quantum (E_γ) in $\overline{Q}$, we use the expression for $g(\lambda_1, x_1)$ in formula (5), putting $\mathrm{Im}\, \varepsilon (\lambda) = k(\lambda)$, $v/c \approx 1$, and $R(x_1) \sim x_1$. Since the position of the maximum of the shower x_1 varies $\sim \ln(E_\gamma/\beta)$, then $g \sim e^{k(\lambda_1)\ln(E_\gamma/\beta)}$ or $g \sim (E_\gamma/\beta)^{k(\lambda_1)}$. Thus,

$$\overline{Q} \sim (E_\gamma / \beta)^{1+k(\lambda_1)}\, \alpha (E_\gamma, T)\, \Phi\, (\xi_\gamma, \xi). \qquad (9)$$

The expression so obtained enables us to draw a number of conclusions regarding the relation between the readings of the shower γ spectrometer, i.e., the value of $\overline{Q}$, and the energy of the quantum recorded.

In all cases of practical interest, the threshold of radiation emission E is normally quite low ($\xi \leq 10^{-2}$), so that $\Phi(\xi_\gamma, \xi) \approx \Phi(\xi_\gamma, 0) = 1$. In addition to this, in view of the condition of complete absorption, $\alpha(E_\gamma, T)$ will be close to unity over a large part of the operating range of the apparatus, which is ensured by choice of appropriate radiator dimensions. Near the lower limit of the working range, the development of the shower is completed at a comparatively shallow depth, and the main part of the radiator (calculated for complete absorption of a shower) operates as an absorber for the most penetrating component of the shower, which corresponds to the minimum in the absorption cross section of γ quanta. As energy increases, the dimensions of the region in which a shower develops increase in proportion to $\ln E_\gamma$. Hence, the proportion of particles leaving the radiator, i.e., the quantity $\alpha(E_\gamma, T)$, will depend little on the energy E_γ.

We see from relation (9) that the condition of complete absorption is necessary but not sufficient in order to ensure linearity of the readings of the apparatus. A deviation from linearity may take place over the whole working range of the apparatus owing to unsatisfactory transparency of the radiator. With increasing energy, there is a displacement in the maximum of the shower (which moves deeper into the radiator), so that the conditions of light collection alter. If the medium is characterized by high transparency, i.e., $k(\lambda_1) \ll 1$, then the deviations from linearity will not be serious. We should note that especially rigorous demands are made on the transparency of spectrometers operating over a wide energy range, since, in this case, the values of E_γ/β at the top and bottom of the energy range differ considerably from one another. Deviations from linearity due to insufficient transparency are in opposition to those associated with the variation of $\alpha(E_\gamma, T)$. A certain amount of compensation between these effects was apparently observed in [8].

As a second important characteristic of shower γ spectrometers, we usually employ not the dispersion, but the relative mean-square fluctuation η_{SS}^2 related to the dispersion by

$$\eta_{SS}^2 = \frac{D(Q)}{(\overline{Q})^2} . \qquad (10)$$

The quantity η_{SS}^2 is similar in significance to the energy resolution used in practice, δ_{SS}^2, this being defined as the ratio of the half-width of a "monochromatic" line (the total width at half height) to the quantity characterizing the position of the line maximum (in the same units). In cases in which the shape of the line is Gaussian:

$$\delta_{SS}^2 = 5.54 \eta_{SS}^2. \tag{11}$$

Calculating η_{SS}^2 from expression (3) in accordance with (10), we obtain [5]:

$$\eta_{ss}^2 = \eta_1^2 + \frac{\eta_3^2}{\overline{N}} + \frac{\eta_3^2 + 1}{\overline{N}} \left(\eta_2^2 - \frac{1}{\nu} \right) + \frac{1 + \eta_4^2}{\overline{n}} , \tag{12}$$

where $\overline{N}$, $\overline{\nu}$, and $\overline{n}$ are the respective mathematical expectations of the number of shower particles, the number of photons generated by an individual shower particle, and the number of photoelectrons falling into the multiplying system of FEM, in which $\overline{N} = \alpha(E_\gamma, T) \, P_\Gamma(E_\gamma, 0) = \alpha(E_\gamma, T) \, E_\gamma / \beta$, and $\overline{n} = <N\nu g\psi\rho> \sim \alpha(E_\gamma, T)$. $(E_\gamma / \beta)^{1 + k(\lambda_1)}$.

The quantities η_i^2 are the relative mean-square fluctuations of the individual stages. Thus, formula (12) expresses the fluctuations of pulses at the output of the spectrometer in terms of the fluctuations associated with particles of each generation. The intensity of the light emitted by an individual shower particle may, on average, be considered identical for all particles and normal fluctuations may therefore be ascribed to this. Hence, formula (12) may be simplified, assuming the form

$$\eta_{ss}^2 = \eta_1^2 + \frac{\eta_3^2}{\overline{N}} + \frac{1 + \eta_4^2}{\overline{n}} . \tag{13}$$

It should be noted that, if the spectrometer radiator is such that the light yield depends on its point of origin (as tends to occur in scintillation spectrometers owing to nonuniform activation of the phosphors), then this fact will lead to results completely analogous to those associated with accounting for the distribution of light collection.

For an FEM with identical stages, the quantity η_4^2 is determined entirely by the first two moments of the distribution of the secondary emission coefficient σ [9]; on the assumption that $\overline{\sigma} > 1$:

$$\eta_4^2 = \frac{D(\sigma)}{\overline{\sigma}(\overline{\sigma} - 1)} . \tag{14}$$

Quite often in practice, the intensity of the light scintillation is measured by not one but several FEM with, generally speaking, different η_{4i}^2. In these cases, the output signal is obtained by summing the Q_i from the various FEM. The summation should be carried out in such a way that the spread in the value of $Q = \Sigma Q_i$ should be a minimum, i.e., with due allowance for the ratios $\eta_{4i}^2 / \overline{n}_i$ associated with the different FEM. As shown in [10], in order to ensure a minimum spread of Q values, the amplitudes of the signals entering the summing system should be inversely proportional to the squares of the resolutions of the corresponding FEM. In the particular case in which one FEM is replaced by several "equivalent" FEM, the use of this rule leaves the previous resolving power of the spectrometer unaltered.

The final formulas (12) and (13) lead to a number of conclusions regarding special features in the behavior of the energy resolution of the shower γ spectrometer. These show that the resolution of the shower γ spectrometer is only limited by the fluctuations accompanying the development of the shower, η_1^2. If the conditions of complete absorption are satisfied, we may expect that, in the limit, all factors restricting the energy resolution of such apparatus will be removed.

This reveals the difference in principle between the shower γ spectrometer and the ordinary scintillation γ spectrometer intended for work with γ quanta of comparatively low energies. The reason for this difference is rooted in the mechanism of measuring energy with these two types of apparatus. As already noted, in the shower γ spectrometer, the evolution of energy embraces the whole radiator, as a result of which the value of the light yield is averaged at each act of light recording. In the ordinary γ spectrometer, owing to the local character of the evolution of energy, the light yield over the volume of the crystal is only averaged as a result of recording many scintillations. Owing to the absence of the first stage, the formula for η^2_{OS} (ordinary spectrometer) analogous to (12) takes the form [11]:

$$\eta^2_{OS} = \eta^2_3 + (\eta^2_3 + 1)\left(\eta^2_2 - \frac{1}{\nu}\right) + \frac{1+\eta^2_4}{n}.$$ (15)

Thus, the resolution of the ordinary scintillation γ spectrometer is always limited by the scatter introduced by the collection of the light.

SEC. 3. LINE SHAPE OF THE SHOWER GAMMA SPECTROMETER

It was indicated in the preceding discussion that the line shape of the shower γ spectrometer was described by the distribution of the sum of a random number of random terms. We shall use a model in which the averaging (summing) of the light yield takes place in the actual process of shower development. Then, the distribution obtained as a result of the automatic averaging of the light yield by the shower, and characterizing the probability that the sum of the quantities g takes the value G, has the form [12]

$$F(G) = \sum_{N=0}^{\infty} \varphi_N \chi^{(N)}(G),$$ (16)

where φ_N is the distribution of the number of charged particles in the shower N, and $\chi^{(N)}(G)$ is an N-fold composition of the density of the light-yield distribution $\chi(g)$, based on the law

$$\overset{(i)}{\chi(y)} = \int \overset{(i-1)}{\chi(x)} \chi(y-x)\, dx.$$ (17)

In order to find the functions φ_N and χ_g, we must in general know the solution to the fluctuation problem of cascade theory, i.e., the distribution $w_\Gamma(E_\gamma, N, \mathbf{x})$, which characterizes the probability of the appearance of a given number of cascade particles with the required properties at the point $\mathbf{x}$ in the case of a shower produced by a γ quantum of energy E_γ. Here, $\varphi_N = \int w_\Gamma(E_\gamma, N, \mathbf{x})d^3x$, and $\chi(g)$ is equivalent (in view of the direct relation between g and $\mathbf{x}$) to the distribution $N_\Gamma(E_\gamma, \mathbf{x}) = \sum_{N=0}^{\infty} N w_\Gamma(E_\gamma, N, \mathbf{x})$, i.e., to the ordinary cascade curve (the index Γ and argument E_γ are omitted in φ_N and χ_g). Unfortunately, theory will not allow us to obtain a general expression for $w_\Gamma(E_\gamma, N, \mathbf{x})$ in cases of practical interest.

As regards the distribution of the number of photoelectrons coming from an individual shower particle and falling into the multiplying system of the FEM, this may be obtained in explicit form in view of our knowledge of the distribution functions of each of the intermediate stages (light emission, conversion of light into photoelectrons, and passage of the photoelectrons into the multiplying system of the FEM). For this purpose we must construct a probability generating function for these three stages: $f[\zeta] = f_2[f_{3b}[f_{3c}[\zeta]]]$; this takes the form $f[\zeta] = e^{\overline{\nu}\psi\rho(\zeta-1)}$. Thus, for the number of photoelectrons from an individual shower particle we obtain a Poisson distribution with parameter $\overline{q} = \overline{\nu}\psi\rho$.

As shown in [9], the distribution for the final stage (electron multiplication) does not in general have any analytical expression.[*] In our case, this is not a very serious matter, for the following reason. The limit theorem for the sum of a random number of random terms [13] asserts that a sufficient condition for the asymptotic normality of the distribution of the sum is the asymptotic normality of the number of its terms. As shown earlier, the distribution of the number of photoelectrons from an individual shower particle passing into the multiplying system of the FEM is described by the Poisson law, i.e., it is asymptotically normal, with a

_____

[*] The distribution is Gaussian if $\overline{\sigma} \gg 1$ and φ_0 is the Poisson distribution. In practice, however, such assumptions cannot be justified.

mathematical expectation and dispersion equal to $\bar{q}$. Hence, the distribution of pulses from an individual shower particle at the output of an FEM with r identical stages will also be asymptotically normal, with a mathematical expectancy $\omega = \bar{q}(\bar{\sigma})^r$ and a dispersion $\Delta = \bar{q}(\bar{\sigma})^{2r}[1 + D(\sigma)/\bar{\sigma}(\bar{\sigma}-1)]$.

The applicability of the asymptotically normal distribution may be estimated quantitatively on the basis of the fact that semi-invariants* of the third and higher orders, i.e., $\varkappa_m = 0$ for $m \geq 3$, are absent from the normal distribution. Bringing the semi-invariants to the same scale, i.e., dividing them by $D^{m/2}$, where D is the dispersion of the distribution, we obtain for the n-fold composition describing the distribution of the sum of n independent terms:

$$\frac{\varkappa_m}{D^{m/2}} < n^{1-m/2}. \tag{18}$$

Relation (18) shows that, as the number of terms increases (this number being $\bar{q} = \bar{\nu}\psi\rho$ in the present case), the normalized semi-invariant of the third order, equal to the third central moment or the asymmetry of the distribution, falls most slowly, about $(\bar{q})^{-\frac{1}{2}}$. Thus, the smallness of the quantity $(\bar{q})^{-\frac{1}{2}}$ may be considered as a criterion for the closeness of the distribution of the composition to a normal distribution.

The discussion just presented enables us to draw a number of important conclusions. First of all, it is easy to see that in the shower γ spectrometer the condition for the applicability of the asymptotic distributions is satisfied for practically all stages, beginning with the generation of light by the shower. In fact, even in the Čerenkov spectrometer, the value of $\bar{q}$ may reach 10^2 for energies of $E_\gamma \approx 100$ MeV, so that $\varkappa_3/D^{3/2} < 0.1$. Hence, in general, we may write the following expression for the distribution of Q:

$$\Omega(Q) = \frac{1}{\sqrt{2\pi\Delta}} \int F(G) e^{-\frac{(Q-G\omega)^2}{2G\Delta}} \frac{dG}{G^{1/2}}. \tag{19}$$

Near the lower boundary of the working range, N is small, and hence for the first two stages asymptotic distributions are invalid. In general, the line shape of the γ spectrometer is described by a composition of the distribution obtained as a result of the internal averaging of the light yield by the shower, F(G) [Eq. (16)], with a Gaussian distribution for the remaining stages. Use of the asymptotic distributions for all the stages becomes practically possible for energies of $E_\gamma \approx 0.5$-1 GeV. In this case, $(\overline{N})^{-\frac{1}{2}} < 0.1$, the asymmetry of the distribution of light yield is considerably weakened, and we may expect that, on satisfying the conditions of complete absorption, the distribution $\Omega(Q)$ will be close to a normal distribution, the dispersion of which may be obtained from formula (13).

If the condition of complete absorption is broken, then the line shape may differ considerably from normal since, as indicated by formulas (16) and (19), the character of $\Omega(Q)$ will be determined by fluctuations associated with the development of the shower, on condition that the first term in (13) predominates.

We must also note another fact which may prove important in the practical use of shower γ spectrometers. Let us suppose that the formation of electron—photon showers in the radiator of the spectrometer is preceded by the generation therein of γ quanta resulting from some reaction such as the annihilation of antineutrons. In this case, simple correlation of formulas (13) and (19) shows that there will be a "weakening" of the fluctuations of the shower, and the line shape $\Omega(Q)$ will have the form of a composition of a distribution characterizing the process of generating γ quanta with a normal distribution describing the apparatus line shape of the system for the case in question.

---

*We remember that the semi-invariant of order m, $\varkappa_m$, is defined as the m-th order derivative at the point x = 0 of the logarithm of the generating function of moment $W(x) = \Sigma\varphi_k e^{kx}$, i.e., $\varkappa_m = V^m(0)$, where $V(x) = \ln W(x)$.

SEC. 4. NUMERICAL CALCULATION OF THE CHARACTERISTICS
OF ČERENKOV GAMMA SPECTROMETERS

The theory developed in Secs. 2 and 3 enables us to calculate the principal characteristics of shower γ spectrometers. Of special interest is the calculation of distribution F(G) (16), since this enables us to find the value of δ_3 characterizing the spread introduced by the light-collection process $\eta_3^2/\overline{N}$.

Calculations of F(G) for a Čerenkov γ spectrometer with a radiator of lead glass (of various thickness and transparencies) and γ-quantum energies from 50 to 1000 MeV [14] are presented below. As φ_N we used the distributions calculated in [15] by the Monte Carlo method for a radiator of lead glass. The distributions $\chi(g)$ were calculated from cascade curves for copper calculated by the method of moments in [16]. In the case of a Čerenkov spectrometer, the directional properties of the light enable us quite simply (see p. 169) to estimate the contribution of the light reaching the end of the radiator from which the light is collected. The calculations were carried out for radiators of two types of lead glass: Corning glass 8392 (or SF-5) and TF-1 glass. These glasses were similar in chemical composition and differed only as regards spectral transmission. Table 1 shows some of the characteristics of these glasses. The effective short-wave boundaries of the region of transparency (λ_{lim}) cut off the Čerenkov-radiation spectrum in the high-frequency range. Owing to the difference in λ_{lim} there is a difference in the average number of light quanta $\bar{\nu}$ created in these media by an individual shower particle in the range of spectral sensitivity of the FEM. In addition to this, the table shows the absorption coefficients of Čerenkov radiation in the range in question (k), reduced to a single radiation wavelength (about 2.4 cm). The choice of these glasses for calculation was governed by the fairly large number of spectrometers in which glasses of these types are used.

Calculations of the F(G) distributions were carried out by the Monte Carlo method on an electronic computer. Tables of probabilities $\Phi(N > N_i)$ and $X(g > g_k)$ were fed into the machine. Then a "draw" of $\Phi(N > N_i)$ values was made, and a specific value of N was chosen from the result of this. The number of terms used in the subsequent summing of the light yield was determined accordingly. The value of the sum G was found as a result of an N-fold "draw" of $X(g > g_k)$ values and the selected values of g were stored. Since the distribution of the discrete random quantity N may be interpreted as a distribution of the overall range of the shower particles, the program of calculations included linear interpolation, which was used in selecting the values of N and g. Each distribution was obtained as the result of 10^4 tests. The results of the calculations are shown in Figs. 1 and 2.

The results obtained enabled us to calculate the energy distribution of the Čerenkov γ spectrometer, δ_{SS}^2 (see p. 171). It follows from formula (13) that

$$\delta_{SS}^2 = \delta_{13}^2 + \delta_4^2, \qquad (20)$$

where $\delta_{13} = \sqrt{\delta_1^2 + \delta_3^2}$ is the resolution corresponding to the distribution which arises after the light yield has been averaged by the shower, δ_1 characterizes the contribution of fluctuations in the number of shower particles, and δ_3 is the scatter introduced in collecting the light. The value of δ_{13} may be found from our calculated F(G) distributions. In view of the fact that the distribution is normal for the remaining stages, the value of δ_4 may be expressed in terms of the relative mean-square fluctuations of the FEM (η_4^2) and the average number of photoelectrons arising from an individual shower particle and falling into the multiplying system ($\bar{n}$), namely:

$$\delta_4^2 = 5.54 \cdot \frac{1 + \eta_4^2}{\bar{n}}. \qquad (20')$$

It was assumed in the calculations that $\bar{n} = 0.1\bar{\nu}$ and $1 + \eta_4^2 = 2$, which approximately corresponds to the parameters of Soviet-manufactured FEM of the FEU-24 or FEU-49 types with intrinsic resolutions of 6% [17]. The results are shown in Fig. 3. The broken lines correspond to the use of extrapolated δ_3 values. For comparison, Fig. 4 shows the δ_1 obtained from the data of [15]. The δ_1 curves illustrate the effect of the escape of particles from the radiator; this is expressed as an increase in the fluctuations of the number of shower particles with increasing energy E_γ. We see from the figure that the escape of particles plays a decisive part in

Table 1

Type of glass	$\lambda_{\lim}$, Å	$\bar{\nu}$, quant	k, (rad. u.)$^{-1}$
Corning glass 8392 (SF-5)	4200	400	0.05
TF-1	3900	500	0.015

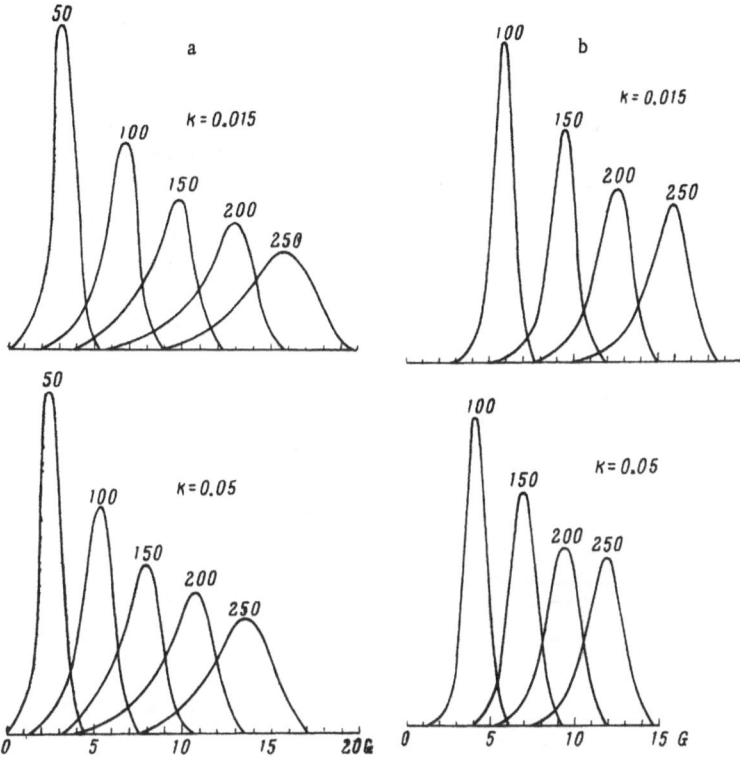

Fig. 1. Distribution obtained as a result of the averaging of the light yield due to the shower. Thickness of radiator: (a) 7; (b) 10 rad. u. Numbers of curves give the energies of the γ quanta in MeV.

the behavior of δ_1 in the case of a radiator 7 rad. u. thick, introduces a noticeable contribution to the behavior of δ_1 in the case of a radiator 10 rad. u. thick, and has a certain influence on the value of δ_1 in the neighborhood of E = 1000 MeV, even in the case of a radiator 15 rad. u. thick.

Comparison of Figs. 3 and 4 shows that, in the case of a radiator 7 rad. u. thick, the relation between δ_{SS} and E_γ reproduces the behavior of δ_1 to some considerable extent. It should be noted that the effect of escaping particles begins to make itself felt, generally speaking, at lower energies in the case of spectrometers with radiators of the more transparent glass than in the case of those with the less-transparent glass.

Figure 5 shows more detailed data characterizing the contribution of various stages to the value of δ_{SS} for a radiator 10 rad. u. thick. The weak energy dependence of δ_{SS} for the radiator with k = 0.015 and E_γ > 200 MeV (the result of the opposing behavior of δ_1 and δ_4 in this energy range) is very noticeable.

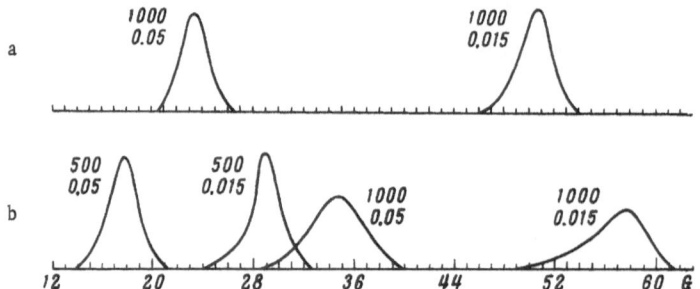

Fig. 2. Distributions obtained as a result of the averaging of the light yield due to the shower. Thickness of radiator: (a) 20; (b) 15 rad. u. Upper figures on the curves give the energies of the γ quanta in MeV; lower figures give the k values.

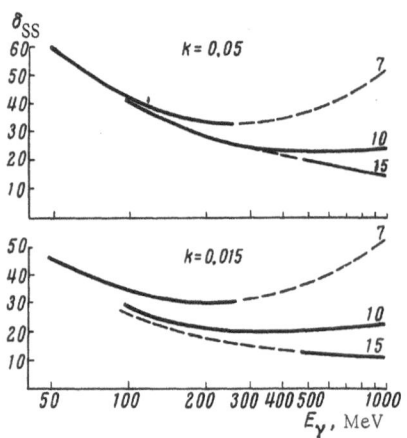

Fig. 3. Energy resolution of Čerenkov γ spectrometers, δ_{SS}, as a function of the energy of the γ quanta. Numbers on curves give the thickness of the radiators in rad. u.

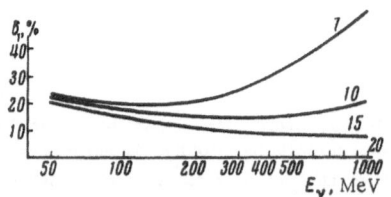

Fig. 4. Contributions of fluctuations in the number of shower particles δ_1 as a function of the energy of the γ quanta. Notation as in Fig. 3.

Figure 6 shows the variation of δ_{SS} and δ_i with radiator thickness T at energy $E_\gamma = 1000$ MeV. On increasing the thickness from 10 to 15 rad. u., the decisive factor in the behavior of δ_{SS} is a fall in δ_1 due to the reduction in the escape of particles from the radiator. We see from the figure that at large thicknesses the contribution of δ_1 plays no decisive role, and that the energy resolution of the spectrometer can only be improved by increasing the dimensions of the radiator if the transparency of the radiator is high enough. This conclusion is also illustrated by the data presented in Fig. 3 for radiators 10 and 15 rad. u. thick in the energy range $E_\gamma < 300$ MeV, where the role of particle escape is relatively small. This characteristic may be explained by the fact that δ_1 and δ_3 have a tendency, generally speaking, to fall with increasing thickness, in the first case as a result of the creation of conditions for more complete absorption, and in the second because of the relative equalization of the path of the light rays. On the other hand, an increase in the thickness of the radiator leads to an increase in the path of the light, and this is accompanied by a fall in the intensity of the light reaching the detector and a corresponding rise in δ_4. This leads to a weaker dependence of δ_{SS} on thickness, and in the case of a radiator with inadequate transparency may completely annul the gain in resolution expected with increasing thickness.

Comparison of the contributions of the various stages, i.e., a comparison of the quantities δ_1, δ_3, δ_4, enables us to draw certain conclusions regarding the influence of the constructional features of the apparatus on the line shape of the Čerenkov spectrometer. In the distributions F(G), calculated for radiators 7

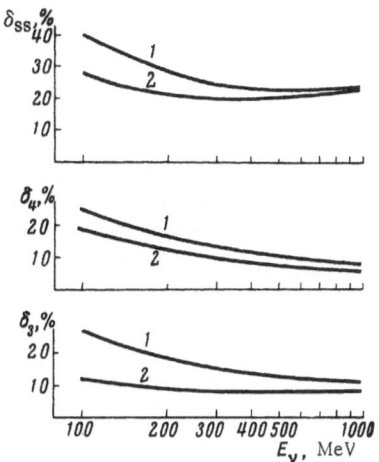

Fig. 5. Energy dependence of the contributions of various stages to the energy resolution of a Čerenkov spectrometer with a radiator 10 rad. u. thick. (1) k = 0.05; (2) k = 0.015.

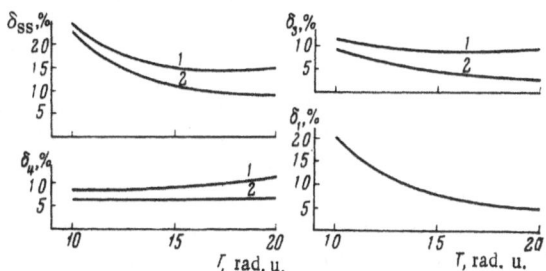

Fig. 6. Contributions of various stages to the energy resolution of the Čerenkov spectrometer at energy E_γ = 1000 MeV as functions of radiator thickness. (1) k = 0.05; (2) k = 0.015.

Table 2

E_γ, MeV	T, rad. u.	k = 0.05	k = 0.015
500	15	$\delta_3 \approx 2\delta_1$	
1000	15	$\delta_3 \approx \delta_1$	$\delta_1 \approx 2\delta_3$
1000	20	$\delta_3 \approx 2\delta_1$	

and 10 rad. u. thick (see Fig. 1), and corresponding to energies E_γ = 200 and 250 MeV, there is a marked influence of particle escape on the form of the curve. This influence appears in the asymmetry of the distribution, which is the more considerable, the smaller the thickness and the greater the energy. However, on comparing radiators of the same thickness and different transparency, one can hardly fail to note that in the case of the less transparent radiator the curves are more symmetric. This feature may be explained by the fact that, for the radiator with the higher transparency, δ_1 is about double δ_3, i.e., the fluctuations in the number of shower particles play the main part, while for the radiator with the lower transparency δ_1 is approximately equal to δ_3, as a result of the increase in the scatter introduced in the collection of the light.

Especially clear from this point of view are the data for E_γ = 500 and 1000 MeV (see Fig. 2). In this case, $\bar{N}$ reaches approximately 70-80, and the distribution of the light yield is practically normal. Table 2 shows the

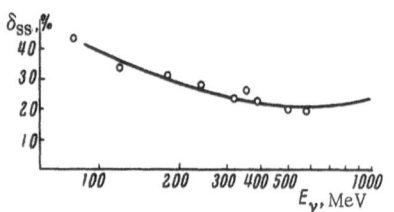

Fig. 7. Results of calculations and calibration data of a Čerenkov spectrometer. Continuous line: calculation; points: calibration data.

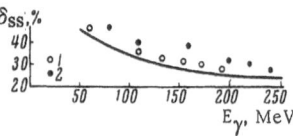

Fig. 8. Results of calculations and calibration data of Čerenkov spectrometers. Continuous line: calculation; (1), (2): calibration data of spectrometers with radiators 10 and 12 rad. u. thick, respectively.

relation between δ_1 and δ_3 for radiators 15 and 20 rad. u. thick. In the case of T = 15 rad.u., the escape of particles leads to a considerable asymmetry in the shape of φ_N, especially for E_γ = 1000 MeV. For the radiator with the higher transparency, the contribution of the scatter in light yield is smallish and the character of the composition F(G) is determined by the shape of φ_N. In the case of the radiator with the lower transparency, the considerable contribution of the normal distribution of light yield leads to a symmetric form of F(G).

In the case of a radiator 20 rad. u. thick, particle escape does not play a decisive part, and the form of the F(G) distribution is close to normal, even in the case of the more transparent radiator. Thus, the results obtained show that the Gaussian line shape of the Čerenkov spectrometer may sometimes be the result of low transparency of the radiator, which, as it were, "maximizes" the fluctuations associated with the escape of particles from the radiator.

Despite the extensive literature containing descriptions of Čerenkov spectrometers, our results can only be compared with published experimental data relating to the characteristics of such systems in a limited number of cases. This is because the number of spectrometers with suitable constructional parameters is low. In addition to this, in many cases the resolution was measured for a single energy or over a very narrow energy range. We therefore selected, for comparison, data relating to three Čerenkov spectrometers set up by different groups in the P. N. Lebedev Physical Institute of the Academy of Sciences of the USSR.

Figure 7 presents data relating to a spectrometer with a radiator made of TF-1 glass 10 rad.u. thick and of high transparency, and a light detector in the form of a single FEU-49 [18]. This apparatus was calibrated over a wide energy range, from 60 to about 600 MeV. The computed values of δ_{SS} were obtained with due allowance for the proportion of the area of the light-collecting end covered by the photocathode of the FEM (approximately 0.3). Figure 8 shows data relating to two spectrometers with glass of the same type but 10 [19] and 12 rad.u. thick. The light detectors in these systems were FEU-24 photomultipliers (seven and eight in the two cases, respectively), the proportion of the area of the end of the radiator covered by the photocathode of the FEM being approximately 0.5 in each case. These spectrometers were calibrated from 60 to 240 MeV. We see from the figures that the calculations correctly represent the variation of δ_{SS} with energy E_γ. The low resolution (as compared with calculation) of the spectrometers containing several FEM (see Fig. 8) may be explained by an inferior selection of amplitudes (see p. 171) when summing the amplitudes of the pulses from the several FEM,* and also to the slightly lower transparency of the batch of glass from which the radiator 12 rad.u. thick was made.

SEC. 5. CONCLUSIONS

The principle underlying the operation of such forms of apparatus as spectrometers, based on the proportionality of the total number of secondary particles of each generation to the energy of the γ quantum, enables us to apply the mathematical apparatus of the theory of branch processes to an examination of the characteristics of such systems. Adequate consideration, however, demands a modification of the theory in order

---

*In calibrating the spectrometer with eight FEM, this selection was not made.

to allow for the mechanism of energy measurements in the shower spectrometer. The mathematical expectation of the output signal (9) calculated in this paper demonstrates the proportionality of the readings of the apparatus to the energy of the γ-quantum recorded and indicates the possible causes of departures from linearity.

The formulas for the energy resolution of the shower γ spectrometer illustrate the main difference between this apparatus and the ordinary scintillation γ spectrometer. The weakening of the fluctuations of light yield in these formulas is due to the fact that in the shower γ spectrometer the energy evolution embraces the whole radiator, leading to an averaging out of the light yield in the course of shower development.

The employment of limit theorems for obtaining the sum of a random number of random terms enables us to secure a general expression for the line shape of the shower γ spectrometer; this is given as formula (19). This expression has the structure of a composition of a distribution obtained as a result of the averaging of the light yield brought about by the shower itself and a normal distribution characterizing the output signal created by a single shower particle.

The results of numerical calculations based on formulas (16), (20), and (20') illustrate the effect of the transparency of the radiator on the characteristics of Čerenkov spectrometers. The effect of the escape of cascade particles from the radiator begins to appear, generally speaking, at lower energies in the case of a radiator made of more transparent glass than in the case of one made of less transparent glass. In addition to this, the energy resolution of the apparatus can only be achieved by increasing the radiator dimensions if the glass is sufficiently transparent. Finally, a symmetrical line shape may sometimes be the result of low transparency in the radiator.

The authors are indebted to R. A. Latypova for setting up the program of calculations and also A. S. Belousov for information on the parameters and the results of calibrating the Čerenkov spectrometer.

LITERATURE CITED

1. Experimental Nuclear Physics, Ed. by E. Segre, Vol. III. Wiley, New York (1959).
2. M. Bartlett, Introduction to the Theory of Random Processes. [Russian translation], IL (1958).
3. V. Feller, Introduction to the Theory of Probabilities and Its Application (second ed.,) Wiley, New York (1957).
4. I. J. Good, Proc. Cambridge Philos. Soc., 45 : 360 (1949).
5. E. M. Leikin, Pribory i Tekhn. Eksp., No. 1, p. 56 (1964).
6. P. Budini, Nuovo Cim., 10 : 236 (1953).
7. S. Z. Belen'kii, Avalanche Processes in Cosmic Rays. Moscow-Leningrad, Gostekhizdat (1948).
8. G. Gatti et al., Rev. Sci. Instrum., 32 : 949 (1961).
9. P. M. Woodward, Proc. Cambridge Philos. Soc., 44 : 404 (1948).
10. V. M. Zapevalov and E. M. Leikin, Physics Institute of the Academy of Sciences Report (1958).
11. E. Breitenberger, Progr. Nucl. Phys., 4 : 56 (1955).
12. V. F. Grushin and E. M. Leikin, Pribory i Tekhn. Eksp., No. 3, p. 33 (1964).
13. H. Robbins, Bull. Amer. Math. Soc., 54 : 1151 (1948).
14. V. F. Grushin, R. A. Latypova, and E. M. Leikin, Pribory i Tekhn. Eksp., No. 5, p. 40 (1965).
15. T. Yamagata, Thesis, Univ. of Illinois (1956).
16. I. P. Ivanenko and B. E. Samosudov, Zh. Eks. i Teor. Fiz., 35 : 1265 (1958).
17. V. V. Matveev and A. D. Sokolov, Photomultipliers in Scintillation Counters. Gosatomizdat (1962).
18. A. V. Kutsenko, V. P. Maikov, and V. V. Pavlovskaya, Pribory i Tekhn. Eksp., No. 4, p. 38 (1964).
19. V. F. Grushin, V. A. Zapevalov, and E. M. Leikin, Pribory i Tekhn. Eksp., No. 2, p. 27 (1960).

DETERMINATION OF THE SHOWER EFFICIENCY
OF SCINTILLATION DETECTORS

V. N. Bolotov, M. I. Devishev, and V. M. Knyazev

The scintillation method has now become widespread in practical physical experiments. The use of spark (scintillation) detectors for studying phenomena due to high-energy particles requires a determination of such fundamental characteristics of the apparatus as shower efficiency.

By shower efficiency we mean the probability of the recording of each particle in a flux of N particles simultaneously passing through the apparatus. The shower efficiency Q may be determined experimentally from the relation

$$Q = n/N, \tag{1}$$

where N is the flux of particles passing through the apparatus, and n is the number of particles recorded by the system. In this case we must know precisely the number of particles which have passed through the working space of the apparatus. For example, relation (1) may be used when the apparatus is irradiated by a beam with a known number of particles which may be varied over a wide range.

The creation of such beams or the use of apparatus with a 100% shower efficiency for determining Q is in practice beset by a number of difficulties. In the Wilson chamber, the "memory" is much greater than that of scintillation detectors, while the generation and absorption of particles in the comparatively thick walls of the chamber may also take place. The hodoscope resolves particles traveling through distances comparable with the diameter of Geiger counters or neon tubes; in scintillation counters the recording of the number of particles depends considerably on the ionizing method and the direction of each particle.

In this paper we consider a method of determining Q, using only scintillation detectors. For this purpose the detector to be studied is placed between analogous detectors and the showers are recorded in each of these. The number of scintillations lying in the same straight line in the upper and lower detectors (cross-linked scintillations) enables us to determine the number of particles which have passed through the apparatus under test. The proportion of this flux recorded in the middle detector is determined from the number of scintillations arising from trajectories found in both upper and lower detectors. We assume that the detectors are independent of each other; this requires that each should have a separate electrical supply. The value of Q may thus be determined using either electron−photon showers or accelerator beams containing particles of unknown number, spatial distribution, and time distribution. It is simplest to use this method for determining the shower efficiency of scintillation chambers with large gaps [1] (Fig. 1) in which particle tracks appear.

When using electron−photon showers generated in lead, three chambers are sufficient to determine Q. In this the scattering, generation, and origin of particles in the chamber electrodes are fully taken into account. For this case,

$$Q_2 = \frac{n_{123}}{3}, \tag{2}$$

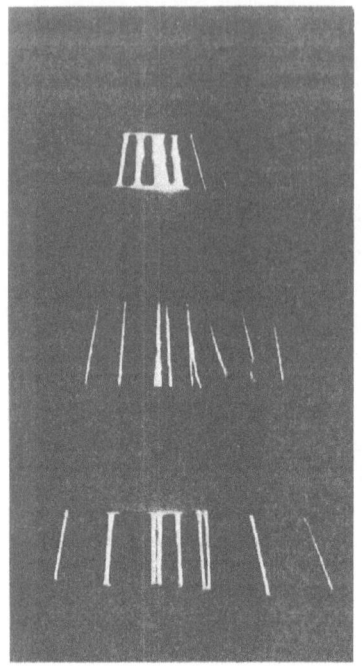

Fig. 1

where n_{13} is the number of pairs of scintillations in the upper and lower chambers (number of "cross-linked" scintillations), n_{123} is the number of "cross-linked" scintillations in all three chambers, and Q_2 is the shower efficiency of the middle chamber.

The use of this method for determining the Q of multiple-plate chambers with small gaps, independent of one another, in which the coordinate of the particles is measured instead of the trajectory, requires four analogous sections

$$Q_3 = \frac{n_{1234}}{n_{124}}, \qquad (3)$$

where Q_3 is the shower efficiency of the third section, n_{1234} is the number of "cross-linked" trajectories in the four chambers, and n_{124} is the number of "cross-linked" trajectories in three chambers. If the direction of the beam is fixed, or it is known that the whole beam has passed through three sections, then three sections are sufficient:

$$Q_2 = \frac{n_{123}}{n_{12}}, \qquad (4)$$

where Q_2 is the shower efficiency of the second section, n_{123} is the number of trajectories in the three sections "cross-linked" as regards beam direction, and n_{12} is the number of similarly "cross-linked" trajectories in two sections. Relations (2)-(4) are valid when Q = const. If Q is a function of certain parameters, for example the inclination of the particles α, or the number of particles N [Q = Q(α, N)], then the relations in question are valid for finite ranges $\Delta\alpha_i$ and ΔN_i, where Q_i = const. [For determining the form of function Q(α, N) due choice of the ranges of variation of parameters $\Delta\alpha_i$, ΔN_i is an important factor.] Contraction of the ranges ($\Delta\alpha_i$, ΔN_i) is continued until the relationship $Q_i(\Delta\alpha_i, \Delta N_i)$ no longer varies. Relations (2)-(4) are valid inside the ranges so obtained. The resultant values of Q_i may be used for constructing the form of the function Q(α, N). The form of Q(α) in [2] was determined in the same way.

By way of an illustration of this method, we may present a determination of the shower efficiency of air counters used in a Nor-Amberde installation [3]. The apparatus consists of three sectioned scintillation counters 1300 cm^2 in area situated in a magnetic field at a distance of 75 cm from one another. For each recorded particle passing through the scintillation telescope, the number of scintillations in the upper, middle, and lower scintillation counters (n_1, n_2, n_3) and the number of trajectories n_{123} in the telescope were determined. Cases of low and high telescope loading (corresponding to the passage of nucleons and a wide atomic beam [ShAL], respectively) were considered. The experimental results appear in the table.

In order to determine the shower efficiency from these data, let us consider the following equations:

Type of event in telescope	$\overline{n}_1$	$\overline{n}_2$	$\overline{n}_3$	$\overline{n}_{123}$
Protons (1)	2.75	2.40	2.10	0.60
ShAL (2)	3.90	3.80	3.80	1.15

$$
\begin{aligned}
Q_1(N_1) &= n_1, \\
Q_2(N_2) &= n_2, \\
Q_3(N_3) &= n_3, \\
Q_1 Q_2 Q_3 N_{123} &= n_{123},
\end{aligned}
\qquad (5)
$$

where Q_1, Q_2, and Q_3 are the shower efficiencies of the upper, middle, and lower scintillation counters for recording N_1, N_2, and N_3, while N_{123} is the number of particles which have passed through the scintillation telescope. In order to determine the lower limit of the product of Q_1, Q_2, and Q_3, we shall consider $N_{123} = N_3$. For each type of selection, the number of scintillations varies only slowly from chamber to chamber, so that we may put

$$Q_1\,(N_1) = Q_2\,(N_2) = Q_3\,(N_3) = Q. \tag{5a}$$

Hence,

$$Q = \sqrt{\frac{n_{123}}{n_3}}. \tag{6}$$

Thus, for each type of selection $Q_1 = 0.6$, $Q_2 = 0.6$, which proves the validity of proposition (5a).

Using the same example, let us consider how to estimate the errors in determining the shower efficiency. The statistical error is found from the binomial distribution for n_{123} and N_3 [4]:

$$\delta Q_1 = \frac{1}{2}\sqrt{\frac{1 - Q_8^3}{n_3}} = 0.03. \tag{7}$$

Another source of error is associated with the inaccuracy in determining the value of n_{123}, which depends on the accuracy of localization of the particle in the counter. The number of false trajectories may be determined, for a uniform distribution of incident particles over the area of the detector, from the relation

$$\delta n_{123} = \frac{Q^3 N \pi \Delta^2}{S},$$

where Δ is the accuracy of localization of a trajectory in the telescope, S is the area of the detector, and N is the flux passing through the detector. Thus we find δQ_2 as

$$\delta Q_2 = \frac{1}{2}\frac{\delta n_{123}}{n_{123}} = \frac{1}{2}\frac{N Q^3 \Delta^2 \pi}{n_{123} S} = \frac{1}{2}\frac{\Delta^2 \pi}{S}.$$

In our case, $\delta Q_2 = 0.0001$, and this may subsequently be neglected. Since we actually determine the lower limit of Q from (5), we may estimate the upper limit of shower efficiency from data relating to the recording of single particles. The recording efficiency of a single particle is $q = 0.70 \pm 0.01$ [3]. Thus, $\delta Q_3 = q - Q$. Finally, the expression for determining Q has the form

$$Q = \sqrt{\frac{n_{123}}{n_3}} \begin{array}{l} +(\delta Q_3 + \delta Q_1) \\ -\delta Q_1 \end{array} = 0.6 \begin{array}{l} +0.13 \\ -0.03 \end{array}.$$

This method of determining the shower efficiency enables us to find Q quite simply without employing auxiliary apparatus of known efficiency, by using beams of arbitrary composition.

In conclusion, let us consider the characteristics of other methods of determining shower efficiency. In [5, 6] the shower efficiency was determined by comparing the form of the experimental and theoretical cascade curves; this only gave a qualitative estimate of Q.

The method used in [7, 8] was based on observing a single particle in parallel scintillation chambers placed one over the other. This method of determining Q is less universal, since it demands an exact knowledge of the number of particles which have passed through the detector, and also an increase in the number of detectors when studying the Q(N) relationship.

The authors wish to thank Professor A. I. Alikhanyan for constant interest in the work, M. I. Daion and G. S. Akopyan for presenting materials, and A. P. Shmelev for discussing the results.

LITERATURE CITED

1. V. N. Bolotov and M. I. Devishev, Nuclear Instr. Meth. (In press).
2. V. N. Bolotov and M. I. Devishev, Nucl. Instr. Meth., Vol. 31 (1964).
3. G. S. Akopyan, M. I. Daion, V. M. Knyazev, and I. G. Solodovnikov, Pribory i Tekhn. Eksp., No. 2 (1964).

4.   Van der Varden, Mathematical Statistics. [Russian translation], IL (1960).
5.   J. W. Cronin, E. Engels, M. Pyka, and R. Roth, Rev. Scient. Instr., 33:946 (1963).
6.   R. Kajikawa, J. Phys. Soc. Japan, 28:1365 (1963).
7.   M. I. Daion and L. F. Klimanova, Zh. Eks. i Teor. Fiz., 45:2078 (1963).
8.   Y. Matsukawa, J. Appl. Phys., 2:239 (1963).

A LIQUID-HYDROGEN TARGET

Yu. M. Aleksandrov, A. N. Zinevich, and E. M. Leikin

Liquid-hydrogen targets are frequently used in research work carried out in accelerators. The main requirements laid upon such targets are: economy, represented by a low evaporation rate of the liquid hydrogen, a small thickness of constructional materials in the "in" and "out" paths of the beam and reaction products (ensuring the recording of charged products over a wide range of angles and energies and also a low level of background from the "empty" target), simplicity in manufacture, and convenience and reliability in operation.

As regards constructional design and type of thermal insulation, the majority of targets may be divided into metal targets with high-vacuum thermal insulation and foam-plastic targets in which materials of the foam-polystyrene type are used for constructional purposes and thermal insulation.

In addition to this there are targets in which the irradiated part (the appendix) is formed by a "hood" on the Dewar used for prolonged storage of the liquid hydrogen [1,2], and targets in which liquid helium is used in order to keep a constant level of liquid hydrogen [3, 4].

Foam-plastic targets [5, 6]* are, as a rule, comparatively simple in manufacture and use, but they are very uneconomical. The principal mechanism of heat transfer in such targets is due to the thermal conduction of the foam plastic, the inner cavities of which are evacuated on fairly intensive cooling; the rate of heat inflow depends on the thickness of the thermally insulating layer and the temperature drop at its boundaries, and even in the case of targets furnished with a "nitrogen jacket" the consumption of liquid hydrogen reaches some 1 liter/h. In addition to this, the mechanical properties of foam plastic targets is often extremely limited.† In such targets it is hard to ensure the extraction of the charged reaction products.

Metal targets with high-vacuum thermal insulation are free from these defects [7-14]. At a pressure of 10^{-5} mm Hg or lower, the thermal conductivity of the residual gas is practically negligible. One of the main sources of heat transfer in such targets is the transfer of heat by radiation through the vacuum space; this may be considerably weakened by the use of thermal and radiation screens. Hence, targets with high-vacuum insulation are good at retaining their liquid hydrogen, the consumption of which, as a rule, is an order lower than in foam-plastic targets, usually not exceeding 50-100 cm^3/h. A target of this kind has been described in [15]; in this, the hydrogen requirements were no greater than 1.5 liters/week.

However, the use of high-vacuum thermal insulation greatly complicates the construction of the target and makes it very cumbersome in manufacture. In order to reduce the radiative heat transfer, all the inner surfaces of the target, including the screens, must have a high reflection coefficient, i.e., must be carefully polished. In the majority of cases the high vacuum is maintained by continuous pumping of the target during operation. Since the pumping of a target filled with liquid hydrogen requires certain minimum safety precautions, the vacuum system must include emergency devices to cut off the pumps from the main body of the tar-

---

*See bibliography in [6].

†We have experienced cases in which foam-polystyrene targets withstood only one or two fillings.

Fig. 1. A liquid-hydrogen target.

get whenever necessary. All this makes the whole apparatus extremely cumbersome and also complicates its use.

It should be noted that the advantages of the two types may to some extent be combined by using metal targets with vacuum-powder insulation made of Mipor (microporous rubber). Such targets require no careful polishing and only require preliminary pumping to backing-pump pressure; they nevertheless involve a higher consumption of hydrogen than in targets with high-vacuum insulation. No description of such targets has been published.

In constructing a liquid-hydrogen target, we attempted to combine convenience of operation and economy with simplicity of manufacture and use, in a metal target with high-vacuum insulation. In construction, our target was similar to the metal—helium Dewar of [16].

Figure 1 shows a photograph of a target placed in the γ-ray beam of the Lebedev Physical Institute synchrotron; Fig. 2 shows the construction in vertical and horizontal sections.

The outer casing of the target, ensuring thermal insulation, is made in the form of a cylinder 400 mm in diameter and 750 mm high, constructed of copper 2 mm thick. The upper part of the casing is connected to the lid and to the lower part by means of steel flanges. Six rectangular flanges, a vacuum valve, and a large tube for introducing the beam are soldered to the lower part. The rectangular flanges are intended for taking the reaction products out of the target; they make it possible to work over a wide angular range. The tops of these flanges are furnished with stainless steel windows 60 × 110 mm in size and 0.1 mm thick. All the vacuum sealing in the outer casing is effected with rubber gaskets.

Inside the casing are the nitrogen reservoir with its heat screen and the hydrogen store with its appendix. The nitrogen reservoir, 9 liters in volume, consists of two halves soldered together, pressed from copper sheet 1.5 mm thick. The reservoir is fixed to the lid with six adjustable stays. Nitrogen is poured in through an opening in the upper part of the reservoir, into which a stainless steel bellows with a flange is soldered, and through an opening in the main flange of the target lid.

A heat screen is fixed to the nitrogen reservoir; inside this are the hydrogen-containing spaces. In the operational condition, the screen is held at the temperature of boiling nitrogen and reduces radiative heat transfer from the outer casing to the inner cavity. The heat screen has the form of a cylinder with a flat bottom, 350 mm in diameter and 500 mm high; it is made of copper 1 mm thick. At the level of the hydrogen appendix and the rectangular flanges of the outer casing, openings are made in the screen; these are covered with copper foil 0.05 mm thick, and at the points of entrance and exit of the beam by aluminum foil 0.01 mm thick.

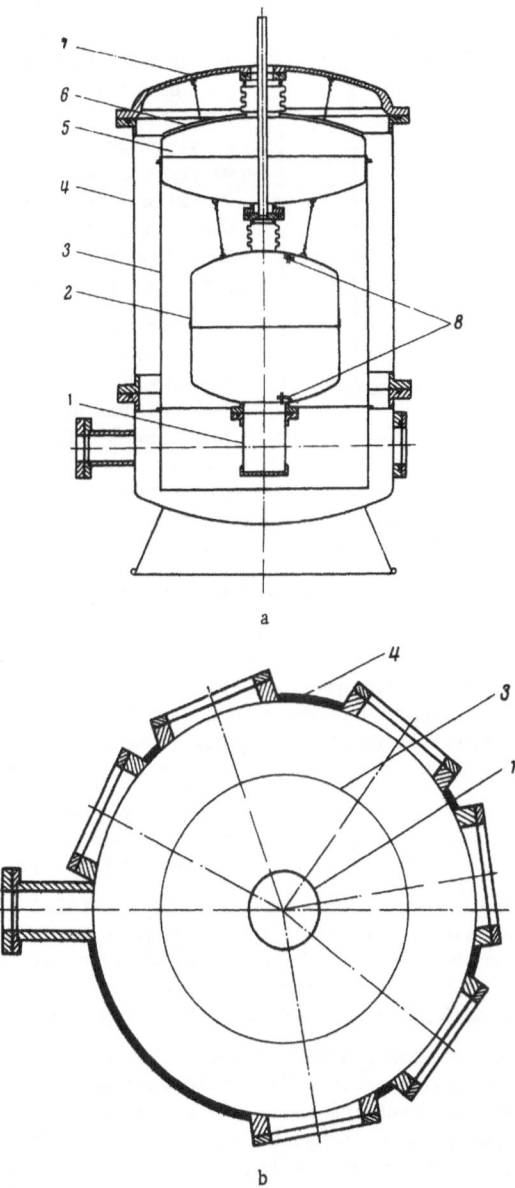

Fig. 2. Construction of a liquid-hydrogen target. (a) Vertical section; (b) horizontal section; (1) appendix; (2) hydrogen store; (3) heat screen; (4) outer casing; (5) nitrogen reservoir; (6) activated charcoal; (7) lid; (8) capacitive level detectors.

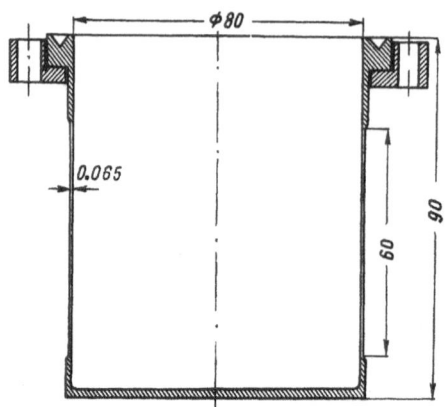

Fig. 3. Construction of the hydrogen appendix of the target.

The hydrogen store consists of two halves soldered together, pressed from copper sheet 1 mm thick. The store is connected to the flange in the bottom of the nitrogen reservoir by means of a stainless steel bellows. The store is fixed to the nitrogen reservoir by means of steel suspensions. The hydrogen vapor is passed in and out through a thin-walled stainless steel tube 17 mm in diameter soldered into the bottom of the nitrogen reservoir and coming out through an opening in the flange of the outer-casing lid. A stainless steel flange is soldered to the lower part of the store for connecting with the appendix.

The appendix (Fig. 3) is a cylinder 80 mm in diameter and 90 mm long; the thickness of the cylinder walls in the irradiated part (60 mm long) is 0.065 mm. The appendix is machined entirely from a Dural block (D-16T). The inner and outer surfaces of the appendix are carefully polished. The vacuum sealing at the junction of the appendix with the hydrogen store is effected by means of an indium gasket, which ensures reliable operation at the temperature of liquid hydrogen. The total volume of the store and appendix is 10.6 liters.

The inner surface of the casing and lids, the heat screen, and the outer surfaces of the nitrogen reservoir and hydrogen store are carefully polished.

The amount of liquid hydrogen in the target is controlled by means of two capacitive detectors, one of which is fixed to the upper part of the hydrogen store and the other at the level of the appendix flange. The detectors are plane condensers forming part of the oscillatory circuit in an electronic generator; their capacity changes when there is any liquid hydrogen between the plates. A change in the capacity of the detectors cuts off the generator oscillations and this is indicated by pointers.

The thermal condition of the target in the course of preliminary cooling by liquid nitrogen (in preparation for filling with liquid hydrogen) is regulated in accordance with the thermal emf generated by two copper—constantan thermocouples, one of which is fixed to the appendix and the other to the heat screen.

The target is first pumped out to a vacuum of 10^{-6} mm Hg by means of a TsVL-100 diffusion pump. Before filling with hydrogen, the target is disconnected from the pumps and remains in that state during the whole cycle of operations involving hydrogen. The high vacuum inside the target casing is maintained by means of activated charcoal in thermal contact with the nitrogen reservoir. For continuous working with hydrogen over a period of two weeks, the pressure at the end of the working cycle never rises above 10^{-5} mm Hg.

Data obtained in work carried out with this target since 1962 are presented below. The average consumption of liquid nitrogen is no greater than 10 liters/day. On filling the target with liquid para-hydrogen, the consumption is approximately 60 cm^3 of liquid per hour. On using liquid hydrogen comprising 75% of ortho- and 25% of para-hydrogen, the period of working with a single filling reaches 120 h. On first pouring liquid hydrogen into a target cooled to the temperature of boiling nitrogen, 15 liters of liquid hydrogen are used; 4 liters of these are used in additional cooling of the target.

The time dependence of the quantity of liquid hydrogen present when using a mixture of ortho- and para-hydrogen may be calculated by solving the thermal-balance equation, allowing for the inflow of heat from outside and for heat evolution associated with the ortho—para conversion. The variation in the concentration of ortho-hydrogen α with time t is given by the equation [17]:

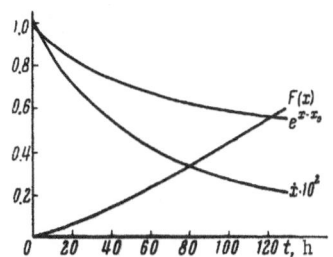

Fig. 4. Time dependence of $F(x)$, e^{x-x_0}, and $\dot{x}$.

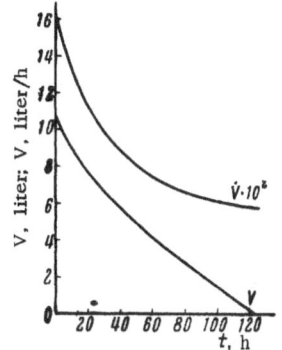

Fig. 5. Time dependence of V and $\dot{V}$ for B = 57 cm³/h.

$$\alpha(t) = \left[\frac{1}{\alpha_0} + kt\right]^{-1},$$

where α_0 is the initial concentration at t = 0, and k is the conversion constant. The change in the volume of liquid hydrogen with time is described by the equation

$$-\lambda \frac{dV(t)}{dt} = u \frac{d\alpha}{dt} V(t) + A, \tag{1}$$

where $V(t)$ is the volume of liquid hydrogen at the instant t, λ is the latent heat of para-formation, and u is the heat of conversion. The term A on the right-hand side of the equation describes the access of heat from outside, resulting from the thermal conductivity of the residual gas, thermal radiation through the vacuum space, thermal conduction through the suspensions, etc. The value of A remains constant in time.

Let us introduce the following notation: $u/\lambda = f$, $f/(1/\alpha_0 + kt) = x$. In this case, the solution of Eq. (1) satisfying the initial condition $V(0) = V_0$ at t = 0 has the form

$$V(t) = e^{x-x_0} \times$$

$$\times \left\{ V_0 - B \frac{e^{x_0}}{k} f \left[E(-x) - E(-x_0) + \frac{e^{-x}}{x} - \frac{e^{-x_0}}{x_0} \right] \right\},$$

where $x_0 = f\alpha_0$ and $E(-x)$ is the integral exponent [18]. Substituting into this formula the values $f = 1.43$, $\alpha_0 = 0.75$, and k = 0.0126 h^{-1}, we obtain

$$V(t) = 0.34 e^x [V_0 - 330BF(x)]. \tag{2}$$

Here, F(x) denotes the dimensionless function $E(-x) + (e^{-x}/x) -$ 0.12. The time dependence of F is shown in Fig. 4. The same figure gives the time dependence of e^{x-x_0}. The quantity B = A/λ is measured in cm³/h and characterizes the rate of evaporation of liquid hydrogen resulting from the influx of heat from outside; it may be obtained by analyzing the thermal balance of the target. However, the calculation requires exact values of the reflection coefficients of the various surfaces, the thermal conductivities of the suspensions, etc., these are usually unknown. By using formula (2) we can find the value of B by measuring the time for which the liquid hydrogen remains in the target. If x_{max} is the corresponding value of x, then, according to (2),

$$B = \frac{V_0}{330F(x_{max})}. \tag{3}$$

The rate of evaporation of liquid hydrogen $\dot{V}(t)$, is described by the formula

$$\dot{V}(t) = \dot{x}(t) V(t) - B, \tag{4}$$

in which $\dot{V}(t)$ is expressed in cm³/h. The value of $\dot{x}(t)$ is also shown in Fig. 4.

The value of B for the given target found from the total hydrogen-retaining time (120 h) was 57 cm³/h. The experimentally measured time dependence of the amount of liquid hydrogen in the target is accurately described by formula (2). The time dependence of V and $\dot{V}$ for B = 57 cm³/h is shown in Fig. 5.

The authors offer their sincere thanks to A. B. Fradkov for a number of useful discussions, I. V. Pintelin for making the appendices, and V. F. Grushin and K. I. Yablonin for assistance in the work.

LITERATURE CITED

1. R. Littauer, Rev. Scient. Instr., 29:178 (1958).
2. R. R. Wilson, Rev. Scient. Instr., 29:732 (1958).
3. G. S. Janes et al., Rev. Scient. Instr., 27:527 (1956).
4. C. A. Swenson et al., Rev. Scient. Instr., 25:608 (1954).
5. L. Marshall, Rev. Scient. Instr., 26:614 (1955).
6. V. I. Petrukhin et al., Pribory i Tekhn. Eksp., No. 2, p. 22 (1964).
7. L. Cook, Rev. Scient. Instr., 22:1006 (1951).
8. E. A. Whalin et al., Rev. Scient. Instr., 26:59 (1955).
9. J. K. Walker et al., Nucl. Instr., 22:138 (1963).
10. A. V. Bogomolov et al., Joint Institute for Nuclear Research Preprint R-396 (1959).
11. V. T. Cocconi et al., Nuovo Cim., 22:494 (1961).
12. E. H Bellamy et al., Nucl. Instr., 7:293 (1960).
13. D. Bodansky et al., Phys. Rev., 93:1367 (1954).
14. D. E. Nagle, Phys. Rev., 97:480 (1955).
15. G. M. Lewis et al., Nuovo Cim., 27:384 (1963).
16. A. B. Fradkov, Pribory i Tekhn. Eksp., No. 4 (1958).
17. A. H. Larsen et al., Rev. Scient. Instr., 19:266 (1948).
18. E. Jahnke and F. Emde, Tables of Functions. Russian edition published by Gostekhizdat, Moscow-Leningrad (1949). Fourth edition published in English by Dover, New York (1951).

APPARATUS FOR INVESTIGATING π^+-MESON PHOTOPRODUCTION

Yu. M. Aleksandrov, V. F. Grushin, V. A. Zapevalov, and E. M. Leikin

The investigation of reactions of the type $\gamma + p \rightarrow \pi^+ + n$ due to bremsstrahlung from electron accelerators necessitates determination of the energy of the γ ray responsible for the event of interest. Since there is a kinematic relationship between the γ-quantum energy and the energy and angle of emission of the π meson formed, this problem can be solved by an apparatus which detects only π^+ mesons emitted at a given angle with a particular energy. The use of scintillation counters with a small entrance aperture as a π^+-meson detector ensures sufficiently accurate angular discrimination.

The energy of π^+ mesons in the near-threshold region, i.e., for $E_\gamma < 250$ MeV, can be determined by magnetic analysis and measurement of the range in matter. Range measurement is particularly attractive, since, in this case, the detector is as simple as possible and the stopped π^+ mesons can be reliably identified by the $\pi \rightarrow \mu$ decay, which has a characteristic time $\tau_\pi = 2.55 \cdot 10^{-8}$ sec. The operation of the corresponding experimental setups [1-7] is based on registration of the delayed coincidences between the gate pulse produced by the π^+ meson and the pulse due to the decay μ meson. The choice of a sufficiently long gate time ($\approx 4\tau_\pi$) prevents additional counting losses and the uncertainty in their length does not introduce any error into the detection efficiency. Our apparatus, which was designed for an investigation of the reactions $\gamma + p \rightarrow \pi^+ + n$ at γ-ray energy 230 MeV [8] was based on the use of a π^+-meson detector in which the mesons were identified by $\pi \rightarrow \mu$ decay and their energy was determined from their range in matter.

For operation in high-background conditions we required an apparatus which ensured more reliable identification of π^+ mesons. Figure 1 shows a block diagram of the version of the apparatus designed for the detection of π^+ mesons at angles close to the direction of the γ-ray beam.

Before entering the detector, the particles are subjected to magnetic analysis, so that the detector receives, in addition to π^+ mesons, only positrons with the same momenta. More than 10^5 positrons traverse the detector at an angle of 0° for every π^+ meson stopped in it. Owing to the low resolving time of the delayed coincidence circuit for the detection of $\pi \rightarrow \mu$ decay ($\approx 10^{-7}$ sec; see above), we employed special measures to reduce the number of openings of the gate by positrons. The incorporation of an anticoincidence counter (4) in the detector enabled us to use the difference in the ranges in matter of π^+ mesons and positrons with the same momenta. Stoppage of a π^+ meson of the required energy in counter 3 produces a pulse at the output of the triple coincidence circuit CC_1 and this pulse is then delivered to the gate-forming circuit. When a positron enters the detector, a pulse from the coincidence circuit CC_2 is produced in addition to the pulse from CC_1, and this second pulse stops the first pulse from opening the gate. This arrangement reduced the number of openings of the gate by positrons by a factor of several tens. Positron opening pulses could not be completely suppressed owing to cases of radiation slowing down of positrons in counter 3, where the π^+ mesons stopped. For a further reduction of the number of such opening pulses, particle discrimination was effected in the channels of counters 2 and 3 by the amounts of energy left in these counters. The minimum energy loss in counter 2 for a π^+ meson which stops in counter 3 is about 6 MeV, i.e., approximately three times greater than the probable energy loss of a fast positron. Using the Landau distribution [9], it is easy to show that the detec-

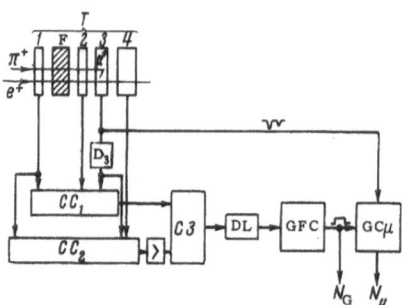

Fig. 1. Block diagram of first version of apparatus. T — Telescope of scintillation counters 1, 2, 3, and 4; F — copper filter; CC_1, CC_2 — triple coincidence circuit; SC — suppression circuit; DL — delay line; GFC — gate-forming circuit; $GC\mu$ — gate for μ-meson pulse; D_3 — fast discriminator in channel of counter 3.

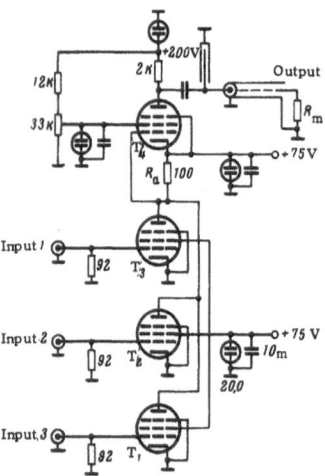

Fig. 2. Triple coincidence circuit.

tion efficiency of counter 2 for positrons can be reduced to 0.03, while the π^+-meson detection efficiency of this counter can be maintained at about 1. In the channel of counter 3 we introduced a discriminator with a bias corresponding to an energy release of 6 MeV, which was 1.5 times greater than the probable energy loss of fast positrons in this counter. With this bias, the positron detection efficiency was 0.1,* while the $\pi \rightarrow \mu$ decay count was reduced by not more than 5%.

Displacement of the gate pulse in time was effected by a delay line connected between the suppression circuit and the gate-forming circuit. The apparatus provided for registration of the number of gate pulses (N_G) and the number of delayed coincidences (N_μ).

In the second version of the apparatus, designed for operation in ordinary conditions, the counter 4, the CC_2 coincidence circuit, and the suppression circuit were omitted. The triple coincidence circuit was replaced by a double coincidence circuit in which the counters 1 and 2 were connected. The pulse from counter 3 went only to the μ-meson gate.

In the construction of the individual components of the first version of the apparatus, the high repetition rate of the input pulses was taken into account. Figure 2 shows the triple coincidence circuit. Its operation is based on linear addition of the currents in the common anode load of the coincidence tubes T_1, T_2, and T_3 (all 6Zh9P tubes). Normally, the tubes are conducting and the voltage on their control grids is zero. The tubes are cut off by negative pulses from FEU-36 photomultipliers operating on long matched RK-2 cables. Tubes with the same anode current and cutoff voltage were chosen for the circuit. In view of the low value of R_a, the length of the pulse at the output of the circuit does not exceed the lengths of the pulses produced by the photomultiplier. The amplitude of the output pulse reaches 3-4 V and the discrimination coefficient for triple coincidences relative to double coincidences is 1.5. To prevent a shift of the level at large loads, the circuit has no coupling capacitors, and the coincidence circuits have a direct-current connection with the coincidence selection element (CSE). The CSE consists of a fast discriminator T_4 (6Zh9P tube), the input impedance of which is R_a. The cutoff of T_4 is regulated by the screen voltage and is chosen so that the discriminator conducts only when all the tubes in the coincidence circuit are simultaneously closed.

---

*Owing to the nonlinear dependence of the light yield in the scintillators on the meson energy in the low-energy region [10], this value was increased by a factor of about 1.5.

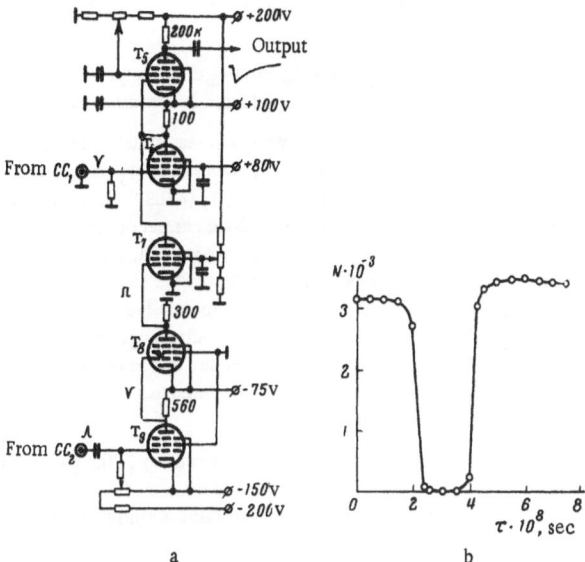

Fig. 3. Suppression circuit (a) and self-suppression curve (b).

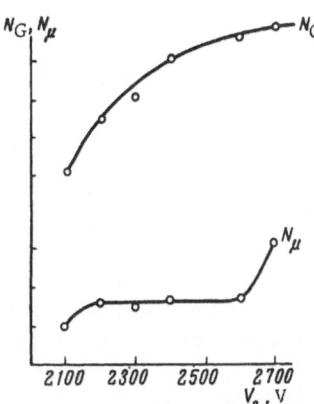

Fig. 4. N_G and N_μ (in arbitrary units) as functions of voltage V_2 on photomultiplier of counter 2.

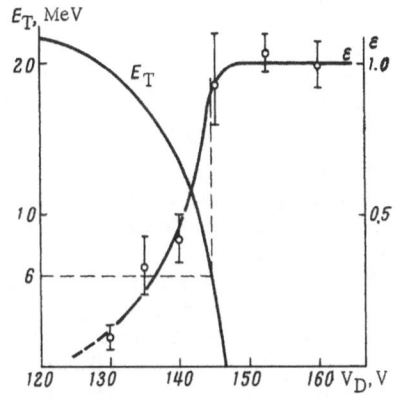

Fig. 5. Threshold of energy loss E_T and relative efficiency ε of counter 3 in relation to voltage V_D on discriminator D_3.

The discriminator T_4 operates on an RK-50 coaxial cable terminating in a matched load. Owing to the small anode load and high transconductance of T_4, short pulses of fairly large amplitude are produced at the discriminator output. In addition, further shaping of the output pulse is effected at the anode of T_4 by a 0.5-m length of short-circuited RK-50 cable.

The resolving time of the circuit is determined by the length of the pulses from the FEU-36. It is usually 6-8 nsec. The second triple coincidence circuit is constructed in a similar way.

Figure 3a shows the suppression circuit. This circuit is a linear anticoincidence circuit similar in its operation to the above-described coincidence circuits. The stages in it also have direct-current coupling with the high load. The suppressing pulse is formed at the anode of T_8 (6Zh9P). This pulse causes the normally cutoff anticoincidence tube T_7 (6Zh9P) to conduct and the anode current of this tube shifts the working point of the discriminator T_5 (6Zh9P) toward a higher voltage bias. As a result, a triple coincidence pulse arriving at T_6 (6Zh9P) in the "shadow" of a suppressing pulse does not make the discriminator T_5 conduct. The width of the suppression curve can be regulated by altering the amplification factor of T_9 (6Zh5P). The large anode load of the discriminator T_5 provides a long pulse with a short leading edge at the output of this circuit and from this pulse the gate is subsequently formed (see below). Figure 3b shows the self-suppression curve recorded when counter 4 in CC_2 was replaced by counter 2. This demonstrates the high efficiency of electronic suppression and the short dead time of the circuit.

In view of the presence of the large load in the channel of counter 3, the gate circuit for the μ-meson pulse was based on the same principle. This circuit is a linear double coincidence circuit with direct-current coupling to the CSE, which consisted of the already mentioned fast discriminator based on a 6Zh9P tube. A similar circuit was used as the discriminator D3.

The discrimination of particles by the energy left in counter 2 is effected by altering the amplification factor of the photomultiplier. As Fig. 4 shows, reduction of the photomultiplier supply voltage leads to a monotonic decrease in N_G, whereas the dependence of N_μ on the photomultiplier voltage has a plateau. The selected photomultiplier working voltage corresponded to the maximum of the ratio N_μ /N_G within the N_μ plateau.

A change in the D3 discrimination level alters the threshold of the energy loss E_T above which π^+ mesons stopped in counter 3 are detected by the apparatus. Figure 5 shows the experimental dependence of the relative efficiency ε of counter on the voltage V_D on the discriminator, obtained on measurement of the differential cross section of the process $\gamma + p \to \pi^+ + n$ at angle $\theta_\pi = 74°$ in the laboratory system. The same figure shows the curve connecting E_T and V_D, obtained from the dependence of ε on E_T, calculated for the working conditions of the experiment.

Figure 6 shows a diagram of the second version of the apparatus. Pulses from counters 1 and 2 arrive at the phase-inverting stages T_{10} and T_{11} (6E5P tubes). The height of the pulses is clipped and reaches 4-5 V on the anode load of the 6E5P. The circuit provides for shaping of the pulses by lengths of short-circuited coaxial cables. The double coincidence circuit is based on a 6A3P tube (T_{12}), which is cut off by the two control grids. When the tube is made to conduct by coincident pulses from counters 1 and 2, its anode potential, owing to the large anode load, is reduced by the discharge of the parasitic capacitance of the anode circuit by the current through the tube. The time constant for restoration of the anode potential after the end of the coincidence pulses is the same as that of the anode circuit. Thus, at the output of the coincidence circuit there appears a pulse with a short leading edge, proportional to the length of overlap of the input pulses, and a long trailing edge (several microseconds). This pulse is then transmitted through the two amplifying stages T_{13} and T_{14} (6Zh9P tubes), between which a variable delay line is introduced. For the transmission of the leading edge of the pulse without distortion the amplification channel operates at fairly high frequency. Where necessary, a UR-3 amplifier is also incorporated in it.

The gate pulse of amplitude $\simeq 6$ V and length $\simeq 120$ nsec is formed in the anode of a 6E5P tube (T_{15}) by an artificial delay line. Figure 7 shows the prompt coincidence curve, which characterizes the shape of the gate in the absence of shaping of the photomultiplier pulses. The operating principle of the gate-forming circuit in the first version of the apparatus is similar to that described.

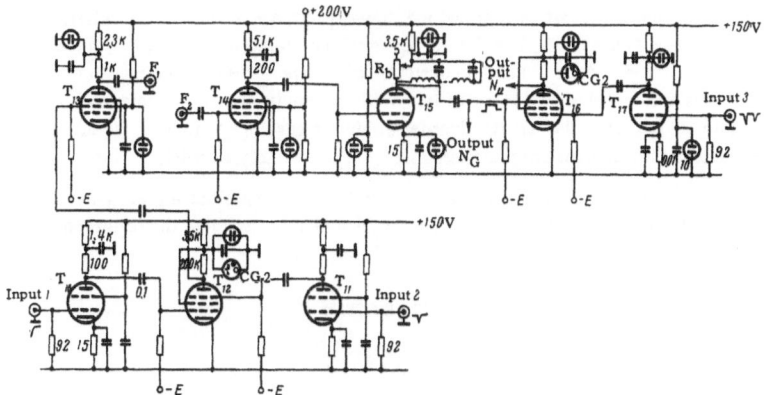

Fig. 6. Diagram of second version of apparatus. E indicates variable negative voltages.

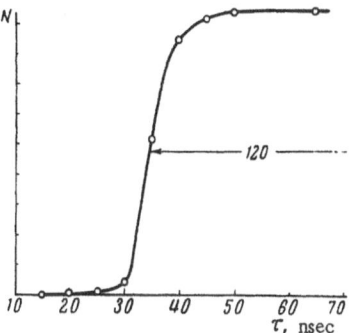

Fig. 7. Shape of gate pulse.

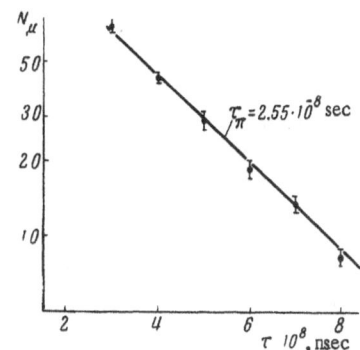

Fig. 8. Typical curve of $\pi - \mu$ decay.

The gating circuit for the μ-meson pulse is a double coincidence circuit based on a 6A3P tube (T_{16}), similar to the above-described double coincidence circuit. The μ-meson pulse from counter 3 before arriving at the circuit passes through a phase-inverting stage incorporating a 6E5P tube (T_{17}). In this version of the circuit, the number of gate pulses N_G and the number of delayed coincidences N_μ are registered.

One method of checking the operation of the apparatus is to determine the relationship between the number of counts N_μ and the time shift of the gate pulse, i.e., to record the $\pi \rightarrow \mu$ decay curve. Figure 8 shows the results of such measurements. The experimental data, plotted on the graph in a semilogarithmic scale, lie satisfactorily on a straight line with a slope which characterizes the lifetime of the π^+ meson.

The detection efficiency for stopped π^+ mesons depends significantly on the time interval after stoppage within which the circuit can detect the decay μ meson. Since pulses from the stopped π meson and from the decay μ meson pass through the same channel, the registration of delayed coincidences cannot begin until the prompt coincidences cease to be registered. This necessitates the introduction of a shift of the onset of registration of the μ meson relative to the instant of stoppage of the π meson by at least the length of the pulse from the photomultiplier. The home-produced photomultiplier which produces output pulses of the smallest length is the FEU-36. However, owing to the insufficient width of the transmission band of the output channel of this photomultiplier, the main pulse at the output is usually followed by a group of afterpulses of comparable amplitude. The appearance of these afterpulses is due to the action of the resonance circuit formed by the inductance of the conductor of the last emitter and the stray capacitance of this emitter relative to the ground. This made it very difficult to use the FEU-36 for the registration of delayed coincidences, since the after-pulses resembled cases of $\pi \to \mu$ decay. Reliable operation of the apparatus with the registration of π^+ mesons from $\pi \to \mu$ decay was accomplished by the use of improved photomultipliers in which decoupling capacitors were introduced into the circuit of the last two dynodes within the envelope to remove the afterpulses [11]. These measures reduced the amplitude of the afterpulses by approximately an order.

The authors express their sincere gratitude to V. G. Pol'skii, who contributed greatly to the successful completion of this work, and to K. I. Yablonin, who put a great deal of work into the construction of the whole apparatus.

LITERATURE CITED

1. W. L. Kraushaar et al., Phys. Rev., 78:486 (1950).
2. O. Chamberlain et al., Phys. Rev., 79:394 (1950).
3. M. Jakobson, A. Schulz, and J. Steinberger, Phys. Rev., 81:895 (1951).
4. C. E. Wiegand, Phys. Rev., 83:1085 (1951).
5. W. Kraushaar, Phys. Rev., 86:513 (1952).
6. W. Imhof, R. Kalibjian, and V. Perez-Mendez, Rev. Scient. Instr., 29:476 (1958).
7. G. M. Lewis et al., Nuovo Cim., 27:384 (1963).
8. Yu. M. Aleksandrov, V. F. Grushin, V. A. Zapevalov, and E. M. Leikin, Dokl. Akad. Nauk SSSR, 160(2) (1964).
9. L. D. Landau, J. Phys. (USSR), 8:204 (1944).
10. J. Birks, Scintillation Counters [Russian translation], IL (1955).
11. V. G. Pol'skii, Dissertation, MÉI (1961).

USE OF STATIC CHARACTERISTICS OF GAS-DISCHARGE GAPS
IN THE DESIGN OF DECATRON COUNTING CIRCUITS

V. A. Zapevalov

The decatron is a gas-discharge decimal counter. Transfer of the discharge from one cathode to the next one in a two-pulse decatron is accomplished by means of two pulses shifted in time, one with respect to the other. The first pulse is applied to the first guides, while a second, delayed pulse is applied to the second guides (when the discharge moves in the forward direction).

The problem of stable motion of the discharge in the decatron is associated with the transfer of the discharge from the first to the second-guide electrodes and then to the following cathode. Satisfaction of the conditions for full transfer of the discharge from the first guide to the second leads to an increase in the ignition voltage in the gas-discharge gap of the glowing cathode and, thus, to a deterioration of the conditions for the discharge returning to a formerly glowing cathode. Full transfer of the discharge to the second guide also leads to a reduction of the ignition voltage in the discharge gap of the following cathode to a magnitude approximately equal to discharge glow voltage in the decatron and, thus, to a reliable transfer of the discharge to this cathode. A complete analysis of the current pulses at the cathodes and guides is quite cumbersome. The use of static characteristics of the decatron discharge gaps, and graphical analysis of the discharge transfer process in the decatron, provides much valuable information for the design of decatron control circuits and also for the selection of suitable parameters of auxiliary circuits.

We will make use of two kinds of volt−ampere characteristics: (1) the ignition characteristic (Fig. 1a); and, (2) the discharge extinction characteristic (Fig. 1b). Let us note once more that ignition of the discharge at electrodes as it moves in the decatron takes place at a voltage nearly equal to the operating voltage of the discharge in the gas-discharge gaps of the decatron.

The volt−ampere characteristic of a circuit consisting of a gas-discharge gap and a resistance R whose source voltage is varied from zero to U (ignition characteristic) can be plotted from the points of intersection of the straight line corresponding to R with a line passing through point A so that $U_A = U_O$ (Fig. 1a). Assuming that extinction of the discharge gap takes place at zero current, the volt−ampere characteristic resulting from varying the voltage from U to zero (extinction characteristic) is shown in Fig. 1b. In this figure, U_O is the operating voltage of the gas-discharge gap, and U_{ig} is the ignition voltage of the discharge gap while the discharge glows at the adjacent electrode ($U_{ig} \simeq U_O$). The volt−ampere characteristic $I_\Sigma = f(U)$ is shown in Fig. 2.

1. TRANSFER OF DISCHARGE FROM CATHODE TO FIRST GUIDE

In normal operation of the two-pulse decatron, a voltage positive with respect to the cathode − the guide bias voltage − is applied to all first- and second-guide electrodes. For a graphic analysis of discharge transfer from a glowing cathode to the following first guide, the guide voltage must be varied from a voltage equal to

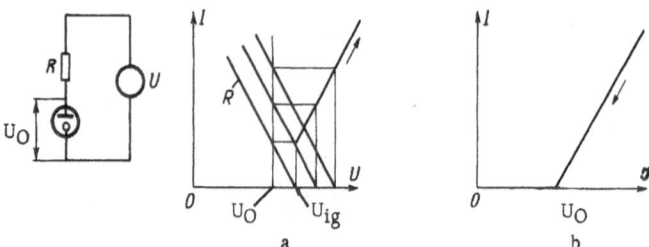

Fig. 1. Volt—ampere characteristic of a circuit containing a gas-discharge
gap: (a) ignition characteristic; (b) extinction characteristic.

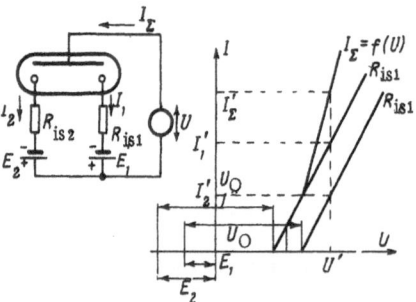

Fig. 2. Volt—ampere characteristic $I_\Sigma = f(U)$. The discharge glows simultaneously in two discharge gaps.

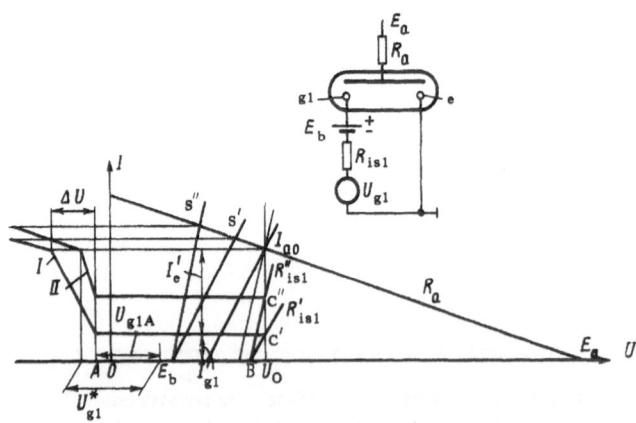

Fig. 3. First-guide current as a function of voltage across it. Curves I and
II correspond, respectively, to $R'_{is1} > R''_{is1}$. The discharge passes from the
cathode to the first guide.

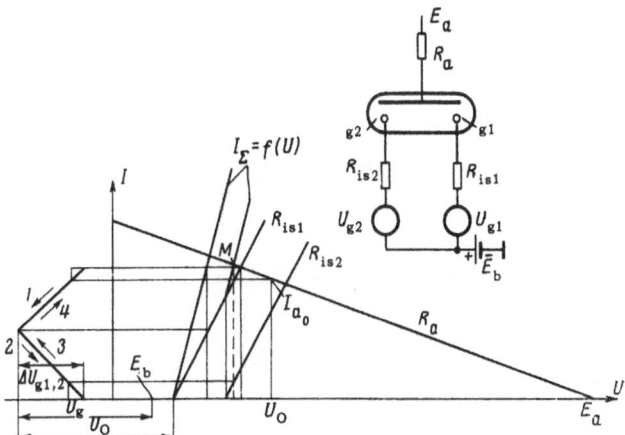

Fig. 4. First- and second-guide currents as a function of the voltages across them during the transfer of the discharge from the first to second guide. (1) and (3) correspond to the variation of currents for $U_{g1} > U_{g2}$; (2) and (4) to $-U_{g1} < U_{g2}$.

the guide bias voltage to a certain other voltage with a polarity opposite with respect to the cathode voltage. The function $I_{g1} = f(U_{g1})_{U_{g2}=0}$ can be plotted with the help of the circuit shown in Fig. 3 and of the volt–ampere characteristics of Fig. 1.

At a certain value of the voltage at the first guides, equal to U_{g1A} with respect to E_b, the first guide will ignite, and the first-guide current can in this case be determined from the point of intersection of the volt–ampere characteristic ABC (Fig. 3) with the characteristic $I_c = f(U)$ of the glowing anode–cathode gap, points B' and B" (the latter characteristic is a vertical line drawn from the point on the voltage axis which corresponds to the operating voltage U_O). The voltage U_{g1}^*, which corresponds to full transfer of the current from the cathode to the first guide, is equal to $U_{g1}^* = I_{a_0} R_{is1} + E_b$. Here, I_{a_0} is the zero-state decatron current when the discharge glows at the cathode, and R_{is1} is the source resistance in the first-guide circuit.

The following conclusions can be drawn from the analysis of discharge transfer from cathode to the first guide:

1. Beginning at $U_{g1} = U_{g1}^*$, which corresponds to the voltage of full transfer of the discharge current to the first guide ($I_C = 0$, $I_{g1} = I_{a_0}$), the guide current is almost independent of the voltage U_{g1}. In order to find the operating conditions of the decatron at $U_{g1} > U_{g1}^*$, it is necessary to find the point of intersection of the volt–ampere characteristic ABC with the load line R_a (Fig. 3, points S' and S").

2. The voltage ΔU, which corresponds to the transfer of the current from the cathode to the first guide, is the higher the larger the source resistance of the control generator.

3. The transfer of current from the cathode to the first guide can be abrupt, provided that

$$\frac{|U_{g1A}| - |E_b|}{R_{is1}} \geq I_{c0}. \tag{1}$$

4. ΔU increases as the anode current of the decatron increases.

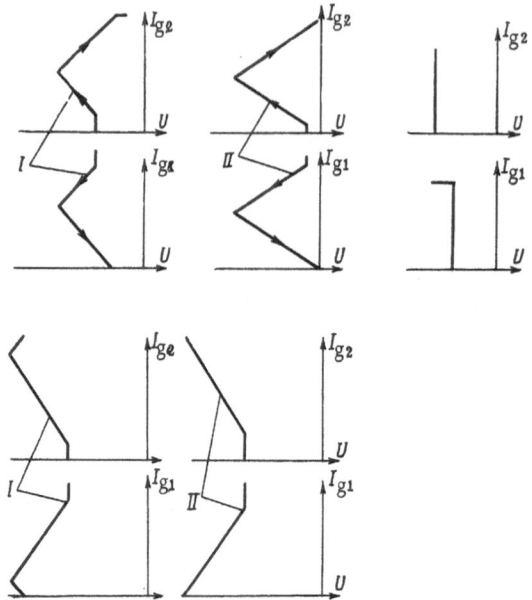

Fig. 5. First- and second-guide current curves for various values and ratios of control-generator source resistances: (a) Curve I: $R_{is}' = R_{is1}' = R_{is2}'$; curve II: $R_{is}'' = R_{is1}'' = R_{is2}''$; $R_{is}' < R_{is}''$. (b) Curves I and II plotted for the ratio $R_{is1}/R_{is2} = 5$ and $R_{is1}/R_{is2} = \infty$, respectively. (c) $R_{is1} = R_{is2} = 0$.

2. TRANSFER OF DISCHARGE FROM FIRST TO SECOND GUIDE

According to [1], the minimum time during which the current exists at the first or second guide can be determined from

$$\tau_{min} = \frac{1}{3f_{max}}, \qquad (2)$$

where f_{max} is the maximum trigger pulse repetition rate for the given decatron type. From this it follows that reliable operation of the decatron requires that the currents exist at the guides for a time longer or equal to τ_{min}. The generation of two pulses shifted in time for transferring the discharge in the decatron is not diffi-cult. It is much more important to determine if these pulses ensure normal operation of the decatron, which is, in particular, essential if a single triode section is used per decatron. For a graphic representation of the dis-charge transfer from the first to the second guide, let us temporarily hold constant the first-guide voltage at the level U_{g1} (with respect to the bias voltage), and vary the potential of the second guides from E_b to $U_{g2} = U_{g1}$. Then, keeping U_{g2} constant, we shall decrease U_{g1}.

The second guide will ignite when the potential between it and the anode reaches the ignition voltage. The operating conditions of the decatron will now be determined by the currents of two glowing discharge gaps and can be found from the point of intersection of R_a with the volt–ampere characteristic $I_\Sigma = f(U)$ plotted for any fixed values of U_{g1} and U_{g2} (Fig. 4, point M). Graphically constructed curves for discharge transfer from the first to second guide, plotted for various different source resistances, are shown in Fig. 5. The figure indicates that as the control-generator source resistance decreases, the value of $\Delta U_{g1,2}$ necessary for full trans-fer of the discharge can take place at $U_{g1} = U_{g2}$ provided that $R_{is1}/R_{is2} \gg 1$ (if the conditions for ignition

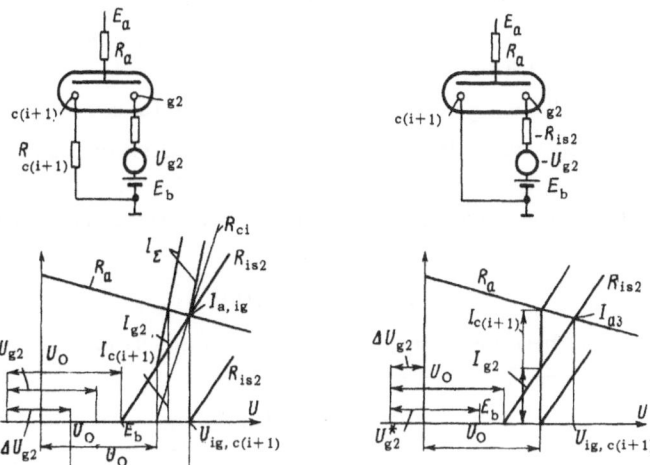

Fig. 6. To the transfer of discharge from the second guide to the following
cathode: (a) resistance in $(i + 1)$-th cathode circuit $[R_{c(i+1)}$ not equal to
zero]; (b) resistance in $(i + 1)$-th cathode circuit equal to zero.

are satisfied at the second guide). The last circumstance enables discharge transfer with equal control-pulse
amplitudes at both guides so that the demand of the time difference of the trailing edges of control pulses becomes much less stringent. This is particularly important in control circuits using the decatron—control
circuit—decatron arrangement. All this refers to the case of equal maximum values of the voltages U_{g1} and
U_{g2}.

When the control-pulse amplitudes are equal, reduction of R_{is} to zero make the transfer of the discharge
to the second guide possible only when U_{g1} is less than U_{g2}. This will demand from the control circuit a
greater difference in control-pulse fall times in order to ensure full transfer of the discharge current from the
first to second guide. If the discharge transfer is to begin at $U_{g1} > U_{g2}$, R_{is1} must be greater than a certain
$R_{is\,1min}$ given by

$$R_{is\,1min} = \frac{\Delta U_{g2} R_a}{E_a - (U_O + \Delta U_{g2} + \vartheta)} ,$$ (3)

$$U_{1,\,g2} = U_O + \Delta U_{g2}.$$

Here, ΔU_{g2} is the maximum spread in the second-guide ignition voltage when the discharge is transferred to it
from the first guide, and $\vartheta = E_b - |U_{g1}|$.

3. TRANSFER OF DISCHARGE FROM THE SECOND GUIDE TO FOLLOWING CATHODE

When the discharge is present at the second guide the ignition voltage of the following cathode, as well
as that of the preceding cathode, are considerably lower (approximately equal to U_O). When the absolute value
of the pulse amplitude at the second guides is reduced, the discharge will transfer to the following cathode. Reliable transfer of the discharge is governed by the following factors:

1. Difference in fall times of the control pulses: $dU_{g1}/dt > dU_{g2}/dt$.

2. The first-guide current should be either zero or of a value at which the ignition voltage of the previously glowing cathode is considerably higher: $U_{ig, ci} > U_{ig, c(i + 1)}$.

3. $R_{c(i +1)} < [E_bR_a]/[E_a - (U_O + E_b)]$, where $R_{c(i +1)}$ is the resistance in the $(i + 1)$-th cathode circuit.

4. The rate of change of the anode and second-guide circuit voltages should satisfy the inequality $dU_{g2}/dt < dU_a/dt$ [2].

The value of $U_{g2}^{\bullet}$, with respect to E_b, at which the current transfer to the cathode begins can be calculated from Fig. 6a:

$$U_{g2}^{\bullet} = I_{a, ig} R_{is2} + E_b. \tag{4}$$

Here $I_{a, ig}$ is the decatron current at the moment the anode voltage reaches the potential $U_a = U_{ig, c(i+1)}$, [where $U_{ig, c(i +1)}$ is the ignition voltage of the $(i + 1)$-th cathode]:

$$I_{a, ig} = \frac{E_a - U_{ig, c(i+1)}}{R_a} \simeq \frac{E_a - U_O}{R_a}.$$

If $R_{c(i +1)} = 0$, the region of U_{g2} variations corresponding to full transfer of the current from the second guide to following cathode is equal to (Fig. 6b):

$$\Delta U_{g2} \simeq \frac{E_a - U_O}{R_a} R_{is2} \tag{5}$$

If, however, $R_{c(i +1)} \neq 0$, then, according to Fig. 6a,

$$\Delta U_{g2} \simeq \frac{(E_a - U_O) R_{c(i+1)}}{R_{c(i+1)} + R_a} + \frac{E_a - U_O}{R_a} R_{is2}. \tag{5'}$$

4. PLOTTING THE SHAPE OF CURRENT PULSES AT THE DECATRON ELECTRODES

The above graphic description of discharge transfer in the decatron makes it possible to plot the shape of current pulses at the decatron electrodes. Figure 7 shows the shapes of current pulses corresponding to various control generator pulses. No scale is given along the time axis. This means that all figures are valid for time scales for which transient processes due to the gas discharge and to parasitic circuit parameters can be neglected. Figures 7a and 7b show the shape of current pulses (I_{g1} and I_{g2}) plotted for an OG-4 decatron when $R_{is1} = R_{is2} = 50$ kΩ, and $I_{a_0} = 0.5$ mA. Figure 7c corresponds to $I_{a_0} = 0.5$ mA, $R_{is1} = 50$ kΩ, $R_{is2} = 10$ kΩ, and equal fall times of the first- and second-guide pulses. Figures 7d and 7e, and Fig. 8a show current pulses for various shapes of control pulses, obtained experimentally for different ratios between control generator source resistances. The current pulses shown in Fig. 8b have been calculated graphically.

A comparison of Fig. 8a and Fig. 8b indicates that the calculated and experimentally obtained pulse shapes agree quite well. The pulse shapes shown in Figs. 7a and 8 are most suitable for reliable transfer of the discharge. This is due to the fact that the current pulses corresponding to control pulses are separated in time at each guide (there is only a small amount of overlapping) which allows the decatron to operate with maximum allowable electrode currents (under the given operating conditions) and considerably reduces the possibility of return of the discharge to a previously glowing cathode.

The performance of the decatron cannot be judged on the basis of the shape of current and control pulses shown in Figs. 7a and 7b. This problem can be solved only experimentally.

In order to investigate the possibility of operating the decatron with guide pulses having equal fall rates and with $R_{is1}/R_{is2} \gg 1$, we built the circuit shown in Fig. 9, which delivers control pulses starting at different times but ending at one and the same moment (Fig. 10).

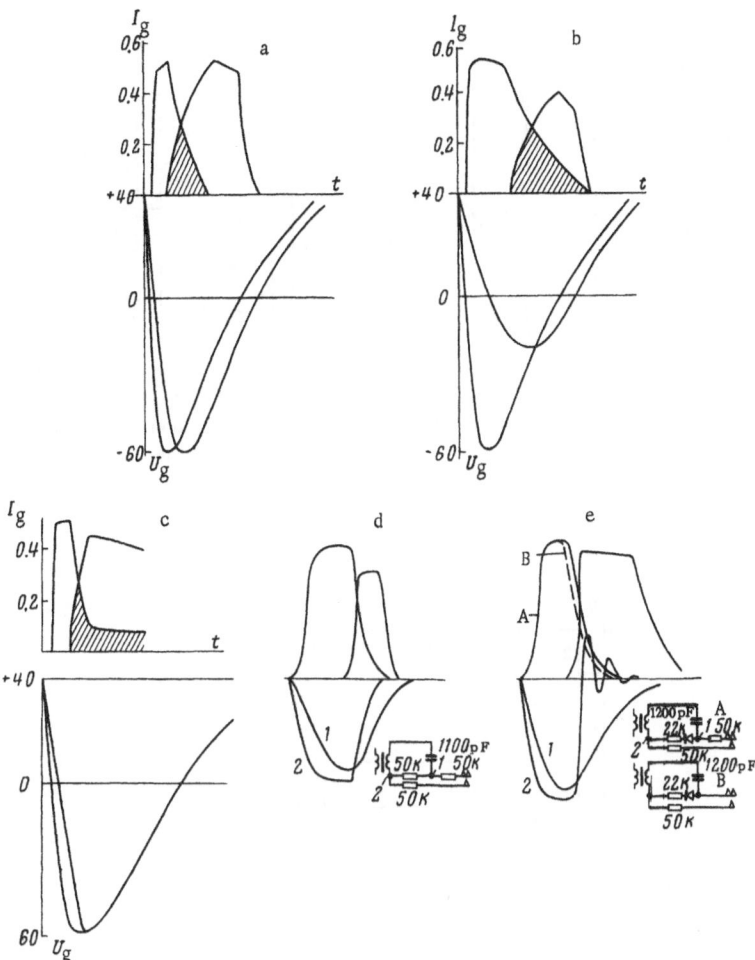

Fig. 7. First- and second-guide current pulses for various shapes, proportion,
and time shift of control pulses.

Different control-generator source resistances were set by means of the R'_{is1} and R'_{is2} potentiometers. We
have determined the dependence of the source resistance of the second-guide control generator at which the
OG-5 decatron ceased to operate on the decatron current (at $U_{g1} = U_{g2}$). This resistance has been called the
critical resistance and denoted by R_{cr}. The source resistance of the first-guide control generator remained constant during the entire experiment and was equal to about 63 kΩ.

The function $R_{cr} = f(R_a)$ is shown in Fig. 11a. The currents I_{g1} and I_{g2} were found from the experimental point M on curve $R_{cr} = f(R_a)$ shown in Fig. 11a. The current I_{g2} has been assumed as the minimum operating second-guide current. The $R_{cr} = f(R_a)$ curve calculated graphically by using the reference point M is
shown in the same figure (broken curve).

The function $\eta_{cr} = I_{g2}/I_{g1} = f(R_a)$ corresponding to $R_{cr} = f(R_a)$ is shown in Fig. 11b. From the shown
curves follows that the second-guide current and the ratio between the first- and second-guide currents are important for reliable operation of the decatron. Normal operation of the decatron with low anode currents requires that $I_{g2} \geq I_{g2, \min}$. In order to ensure normal operation of the decatron with sufficiently large anode

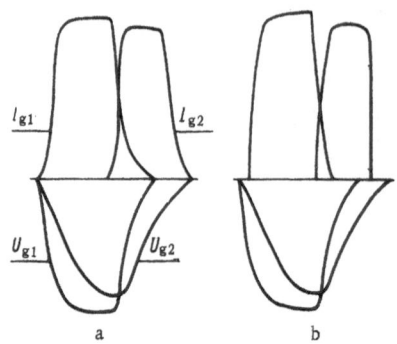

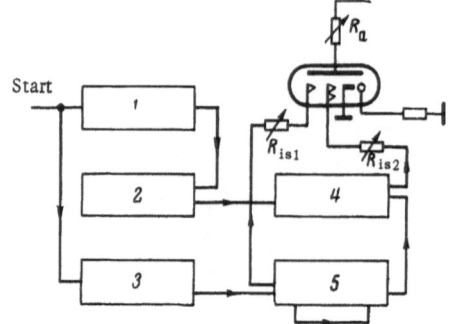

Fig. 8. First- and second-guide current pulses: (a) experimental; (b) calculated graphically; R_{is1} = 50 kΩ, R_{is2} = 0, OG-4 decatron.

Fig. 9. Block diagram of circuit used for determining the domain of $R_{cr} = f(R_a)$ and $\eta_{cr} = f(R_a)$ ensuring stable operation of the decatron: (1), (2), (3) univibrators; (4) mixer; (5) shaper.

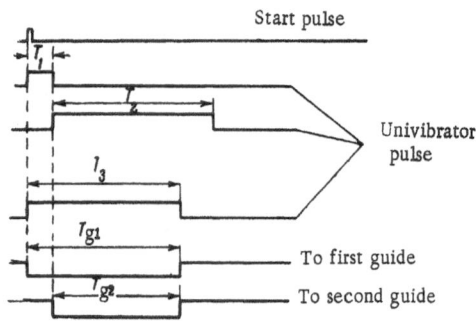

Fig. 10. Time diagram of circuit operation.

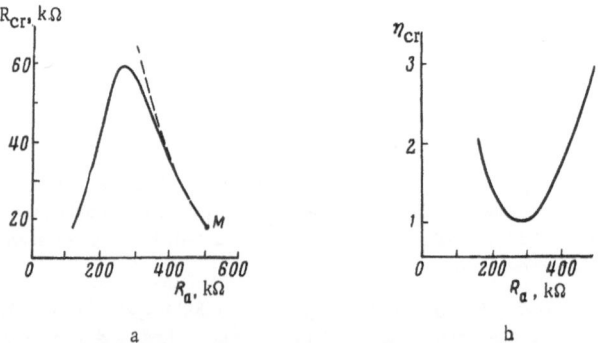

Fig. 11. Functions $R_{cr} = f(R_a)$ and $\eta_{cr} = f(R_a)$ for the OG-5 decatron.

currents $I_{a_0} > I_{a_0}$, min, it is necessary that $I_{g2}/I_{g1} > 1$. The last inequality makes it possible to provide good conditions for the transfer of the discharge from second guide to the $(i + 1)$-th cathode, and prevents the return of the discharge to the former i-th cathode. Here, I_{a_0}, min is the minimum current at which the decatron is still operating.

Thus, the OG-5 decatron will operate reliably with generator source resistances $R_{is2} = 18$ kΩ and $R_{is1} = 63$ kΩ, provided R_a is more than 160 kΩ but less than 500 kΩ (with $E_a = 450$ V). It can now be stated that the decatron operation is quite stable with first- and second-guide current pulses of the shape shown in Figs. 7b and 7c within a definite current range, typical of the given decatron type.

5. RESETTING THE DECATRON TO ZERO

Selection of Guide Bias Voltage

The guide bias voltage should, in our opinion, be such that it ensures reliable resetting of the decatron to zero and reliable transfer of the discharge in azimuthal direction. Resetting of the decatron can be accomplished by several methods: by opening the common circuit of all cathodes, by opening the cathode and guide circuits, or by the pulse method (the pulse is applied to the common cathode circuit or to the zero cathode). When the common cathode circuit is opened, the best conditions for receiving the discharge are present at the guides adjacent to the glowing cathode (first and second guides); the discharge transfers then to one of these guides. The steady-state anode voltage of the decatron can, in this case, be found as the point of intersection of the volt—ampere characteristic of the circuit consisting of the gas-discharge gap R_{cp} and the bias source, with the load line R_a. If $U_a \geq U_{ig,c_0}$ (where U_{ig,c_0} is the firing voltage of the "cold" anode—zero-cathode gap), this gap will ignite and the discharge can pass from the guide to the zero cathode (Fig. 12). The guide bias voltage satisfying the conditions for this method of zero reset can be found from

$$E_b \geqslant (U_{ig,\,c_0} - U_O) - (E_a - U_{ig,\,c_0})\frac{R_{cp}}{R_a}. \qquad (6)$$

Here, R_{cp} is the coupling resistance between the bias source and the guides. Guide bias voltages found experimentally and calculated from (6) for various R_{cp} are given in the table.

Selection of Zero-Cathode Resistance

The zero-cathode resistance (R_{c_0}) should satisfy the conditions for reliable transfer of the discharge to the zero cathode when the decatron is reset to zero, and also for stable advance of the discharge from the ninth or first cathode to the zero cathode. When the conditions for discharge ignition at the zero cathode are satisfied, $R'_{c_0,\,max}$ at which the discharge is present at the zero cathode only can be found from

$$R'_{c_0,\,max} = \frac{E_b\,R_a}{E_a - (E_b + U_O)}. \qquad (7)$$

If $R_{c_0} > R_{c_0,\,max}$ the discharge can be present simultaneously at the zero cathode and the guide adjacent to the glowing cathode. In such a case there is no guarantee that the decatron will be reset to zero. The value of $R_{c_0,\,max}$ which ensures stable forward as well as backward motion of the discharge through the zero cathode can be found, in accordance with Fig. 13, from

$$R''_{c_0,\,max} = \frac{E_b\,R_a}{E_a - (E_b + U_O)}. \qquad (8)$$

Thus, the zero-cathode resistance which ensures reliable operation of the decatron should satisfy the inequality

$$R_{c_0} \leqslant \frac{E_b\,R_a}{E_a - (E_b + U_O)}. \qquad (9)$$

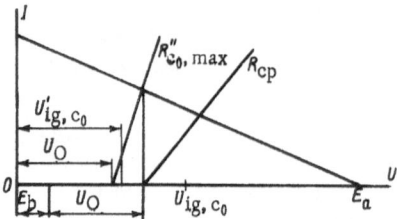

Fig. 12. To the determination of maximum zero-cathode resistance for resetting the decatron to zero by opening the common cathode circuit.

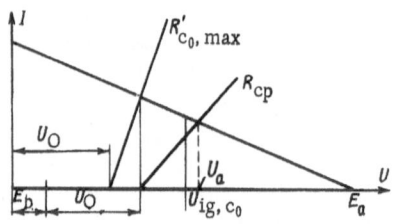

Fig. 13. To the determination of maximum zero-cathode resistance ensuring reliable advance of discharge in azimuthal direction.

Type OG-4 Decatron, $U_O = 125$ V, $E_a = 450$ V, $R_a = 820$ kΩ, $U_{ig,c_0} = 210$ V

R_{cp}, kΩ	E_b, V (experimental)	E_b, V (calculated)
3	85	85
50	72	71
100	55	57
200	25	29

Pulsed Zero Reset

Pulsed zero reset of a decatron can be accomplished by either positive or negative pulses. Positive pulses are applied to the cathodes, while negative pulses are applied to the zero electrode.

Positive-Pulse Reset. When a positive reset pulse is applied to the decatron cathodes, the anode potential can be found from the point of intersection of the characteristic corresponding to the glowing discharge gap and the reset-pulse-generator source resistance with the load line R_a. When the decatron anode voltage reaches a value equal to $U_{ig,g} + E_b$, the discharge will ignite at one of the guides adjacent to the glowing cathode. The anode voltage is in this case determined by the point of intersection of the volt−ampere characteristic $I_\Sigma = f(U)$ with the load line R_a (Fig. 14a, point M). When the anode voltage rises to U_{ig,c_0}, the discharge will pass to the zero cathode provided R_{c_0} and E_b satisfy the conditions (9) and (6), respectively. The reset pulse amplitude can be found from

$$U_{res} = E_b + \frac{E_a - (E_b + U_O)}{R_{cp} + R_a} R_{cp}. \tag{10}$$

Negative-Pulse Reset. The reset pulse is applied to the zero cathode. As soon as the voltage across the anode−zero cathode gap rises to U_{ig,c_0} the discharge will start to pass from the glowing cathode to the zero electrode (Fig. 14b). If $R_{c_0} + R_{is,res}$ is such that

$$\frac{U_{ig,c_0} - U_O}{R_{c_0} + R_{ig,res}} \geqslant I_{a_0}, \tag{11}$$

the discharge will pass abruptly to the zero cathode (here, I_{a_0} is the zero-state decatron current and $R_{is,res}$ is the source resistance of the reset pulse generator).

From expression (11) follows that

$$(R_{ig,res} + R_{c_0})_{max} = \frac{U_{ig,c_0} - U_O}{E_a - U_O} R_a = \frac{E_{b,max} R_a}{E_a - U_O}. \tag{12}$$

Here $E_{b,max}$ is the bias voltage satisfying zero reset by the method of opening the common cathode circuit provided $R_{cp} = 0$. The maximum value of $R_{c_0,max}$ satisfying expression (9) is given by

$$R_{c_0,max} = \frac{E_{b,max} R_a}{E_a - (E_{b,max} + U_O)}. \tag{13}$$

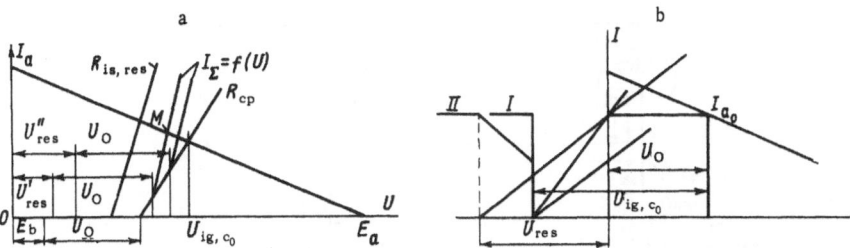

Fig. 14. Pulsed zero reset: (a) positive-pulse reset; (b) negative-pulse reset; (I) $R_{is,res} + R_{C_0} \leq (U_{ig,c_0} - U_O/I_{a_0}$; (II) $R_{is,res} + R_{C_0} > (U_{ig,c_0} - U_O/I_{a_0}$.

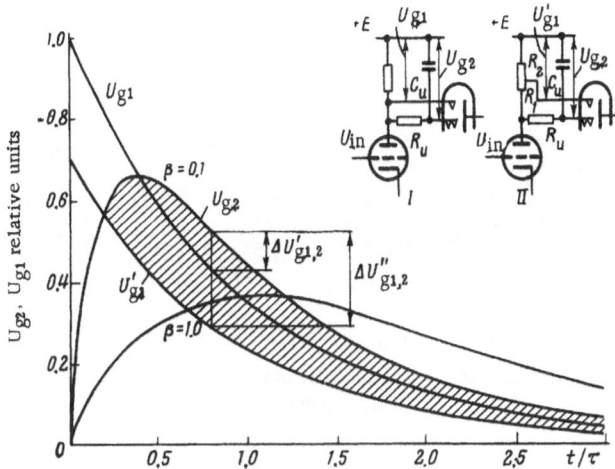

Fig. 15. Shape of decatron guide pulses in normalized coordinates.

It is evident by comparing (9) and (12) that if $R_{is,res} + R_{C_0} < (R_{is,res} + R_{C_0})max$, R_{C_0} certainly satisfies the condition for zero reset by opening the common cathode circuit. When the inequality (11) is fulfilled, the reset pulse amplitude should satisfy

$$U_{res} \geqslant U_{ig,c_0} - U_O. \tag{14}$$

If $R_{is,res} + R_{C_0} > (R_{is,res} + R_{C_0})max$, the reset generator voltage corresponding to full transfer of the discharge to the zero cathode is equal to

$$U_{res} = I_{a_s}(R_{is,res} + R_{C_0}). \tag{15}$$

6. THE PRINCIPLE OF AMPLITUDE MATCHING

One of the earliest discharge-advance circuits which provides $\Delta U_{g1,2}$ sufficiently high for reliable opera-tion of the decatron is based on the principle of "matching the shift-pulse amplitudes" [3]. The circuit was designed for use as a data acquisition unit of a 40-channel pulse height analyzer and this application predeter-mined the selection of power supplies and of the type of coupling between the individual circuit elements so that a minimum number of components could be used.

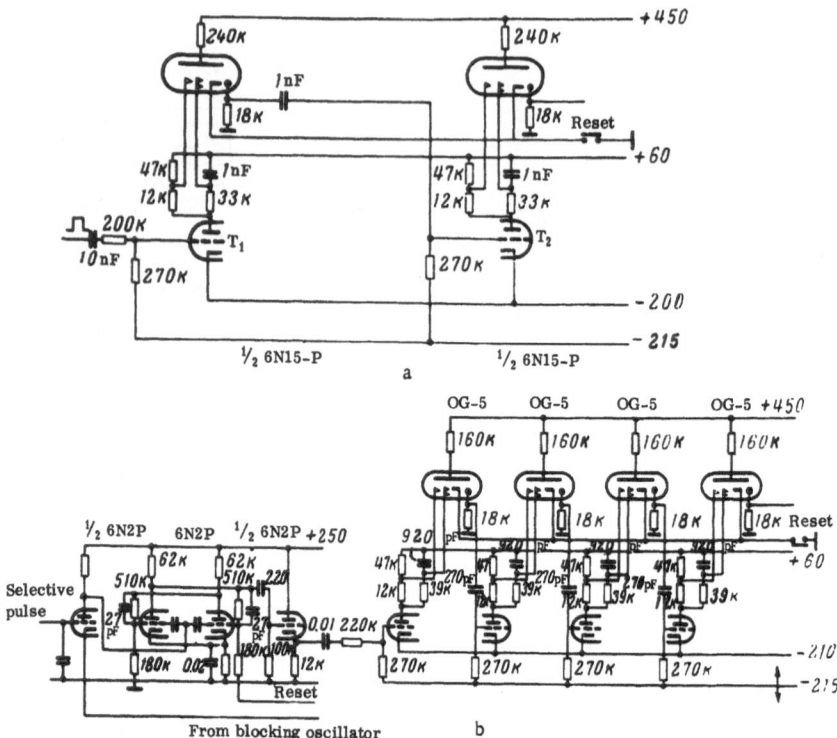

Fig. 16. Decatron counter circuits using the principle of matching the amplitudes of discharge-shifting pulses.

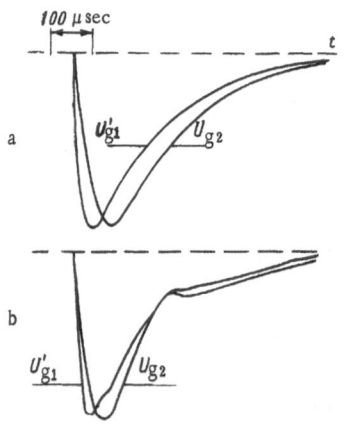

a

b

Fig. 17. Oscillogram of discharge-shifting pulses at the decatron guides: (a) without decatron; (b) with decatron.

Figure 15 shows in normalized coordinates the shape of U_{g1} and U_{g2} pulses (plate circuit of tube T_2 in Fig. 16a) as functions of $\beta = Rc/\tau$ and t/τ, where τ is the time constant corresponding to the U_{g1} pulse. The figure indicates that obtaining $\Delta U_{g1,2}$ adequate for reliable operation of the circuit in Fig. 15 I is possible only with quite high β. The increase of β results in a loss of U_{g2} amplitude which can considerably impair the decatron operation. A voltage difference $\Delta U_{g1,2}$ which ensures reliable operation of the decatron can be easily obtained with the circuit shown in Fig. 15 II (curve U'_{g1}). In this circuit, the pulse applied to the first guide is matched to the second-guide pulse obtained by integration.

The curves shown in Fig. 15 give only a qualitative picture, since they do not take into account the finite rise time of the zero-cathode pulse, the possibility of grid currents in T_2, the effect of the tube plate circuits on the decatron, and the fact that $R_{i2} = 0$ (R_{i2} is the ac plate resistance of tube T_2).

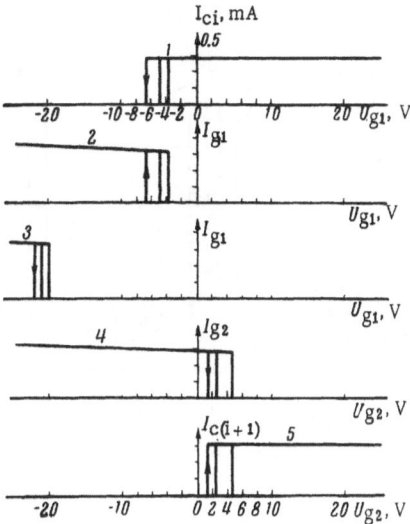

Fig. 18. Currents through OG-4 decatron electrodes as a function of the voltage across them for $R_{is1} = R_{is2} = 3$ kΩ and $R_{c_0} = 0$: (1) $I_{ci} = f(U_{g1})$, $U_{g2} = +25$ V; (2) $I_{g1} = f(U_{g1})$, $U_{g2} = +25$ V; (3) $I_{g1} = f(U_{g1})$, $U_{g2} = -25$ V; (4) $I_{g2} = f(U_{g2})$, $U_{g1} = +25$ V; (5) $I_{c(i+1)} = f(U_{g2})$, $U_{g1} = +25$ V.

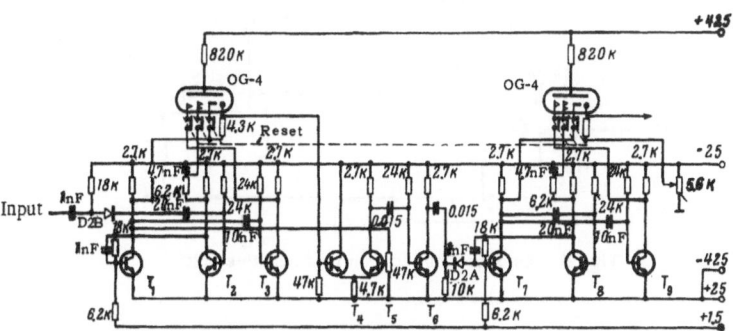

Fig. 19. Transistorized decatron scaler.

The above-described method of matching the discharge-shifting pulses was used in the design of the circuits shown in Fig. 16a (5 kHz), Fig. 16b (11.5 kHz), and also in the design of several other circuits using OG-4 decatrons. For OG-4 and OG-5 decatrons, $\Delta U_{g1,2} \simeq 20$ V, when $U'_{g1} = U_{g2} = 80$-100 V. Oscillograms of the shifting pulses in the circuit of Fig. 16a are shown in Fig. 17.

7. TRANSISTORIZED DECATRON SCALERS

Many transistorized discharge-shifting circuits have been described in the literature. All these devices use either high-voltage transistors or transistor—transformer circuits.

The results of the analysis of discharge transfer from electrode to electrode in the OG-4 decatron (Fig. 18) enabled the design of a simple transformerless discharge-shifting circuit (Fig. 19) using P13 transistors with a

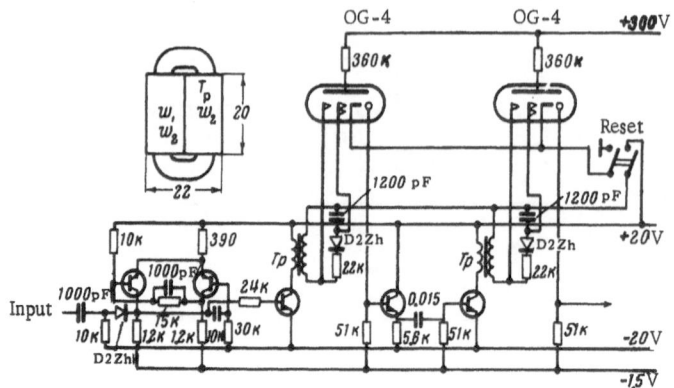

Fig. 20. Transistor−transformer decatron scaler. Transformer parameters: KhVP steel core, S_{st} = 0.75 cm^2, w_1 = 1500 turns, w_2 = 6000 turns, PE-0.05 wire.

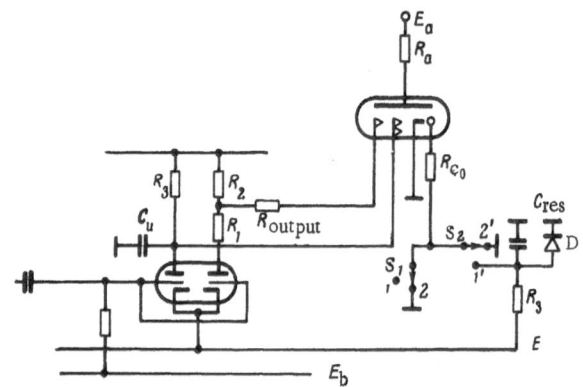

Fig. 21. Zero reset of decatron by a negative pulse.

−25-V supply. In the circuit shown in Fig. 19, the second-guide pulse, which has negative polarity and a duration of 300 μsec, is taken from the integrating circuit capacitance in the collector of T_2, which, together with T_1, forms a univibrator.

The first-guide pulse with a duration of 150 μsec is taken from the collector of transistor T_3, the differentiated T_1 collector pulse being applied to the base of T_3. Simultaneously with the first- and second-guide pulses, a positive pulse from the univibrator (T_1) is applied to all cathodes (including the zero cathode).

This method of discharge shifting made it possible to achieve the maximum rated operating speed of the OG-4 decatron of 2-2.1 kHz. It has been found experimentally that at low operating speeds − up to 300-400 Hz − it is possible to find a quite extensive range of zero-cathode voltages in which the decatron operates reliably. This range extends for the OG-4 decatron from −5 to −12 V with respect to the common negative bus; this fact has been made use of in the discharge-shifting circuit of the second decatron. The zero cathode is not coupled to the other electrodes, and the voltage applied to it is taken from the potentiometer P. This simplifies triggering of the shifting circuit of the second decatron.

The first decatron is coupled with the discharge-shifting circuit of the second decatron through a differential amplifier (T_4, T_5) and amplifier T_6. A disadvantage of this circuit is the difficulty of resetting the decatrons to zero. The table (p. 206) lists the coupling resistances R_{cp} between the guides and the bias source, and their corresponding bias voltages suitable for the OG-4 decatron, which ensure reliable resetting of the decatrons to zero. It is clear from the table that zero reset cannot be accomplished by opening the cathode circuits when the guide bias voltage is +25 V. The decatrons are for this reason reset to zero by disconnecting the first guides, second guides, and cathodes of each decatron separately. In addition to the above-described circuit we have designed a decatron scaler version using low-voltage transistors and step-up transformers (Fig. 20). The design was made in order to see if such circuits can be used with low-frequency decatrons of the type OG-4, A101, etc. It has been found that transistorized, decatron drive circuits with step-up transformers are suitable for decatrons whose maximum operating speed is at least more than 10 kHz.

8. SELECTION OF COUPLING AND CONTROL CIRCUITS ENSURING RELIABLE DECATRON OPERATION

Reliable operation of decatron circuits depends on the decatron quality, and on the correct selection of components of the decatron coupling circuit as well as of the components included in the circuit which shapes the shifting pulses. Reliable azimuthal motion of the decatron discharge requires that the decatron control circuit satisfies the following conditions: the stages shaping the guide pulses should have a low source resistance, the source resistance of generators (we have in mind generators equivalent to the pulse-shaping stages) should be such that their ratio is $R_{is1}/R_{is2} \gg 1$, the operating speed of the circuit is improved by using direct coupling between the generator and decatron guides.

Circuits meeting these demands can be realized by two methods. In the first case, the tube used for shaping the shifting pulses is specified in advance. In this case, the source resistance of the generator equivalent to the pulse-shaping stage can be reduced only by decreasing the anode load resistance which, as was mentioned above, results in an increased guide bias voltage necessary for reliable zero reset of the decatron (we mean zero reset by opening the cathode circuits). The increase of guide bias voltage requires a higher amplitude of shifting pulses at the guides, while reduction of the anode load results in their decrease. From the above follows that the first two demands are best satisfied by a circuit in which R_{is1} can be significantly reduced with a simultaneous reduction of the guide bias voltage. In such a case, the decatron cannot be reset to zero by opening the cathode circuit. Thus, the decatron will operate more reliably if zero reset is accomplished by some other method (Fig. 21), and if the guide bias voltage is selected close to its minimum value.

The necessity of satisfying the above-mentioned conditions follows from the previously described analysis of decatron operation with the help of steady-state characteristics of gas-discharge gaps. The decatron is, however, a pulse-operated device. Thus, the stated demands must be amended by allowing for pulsed operation of the decatron.

The reliability of charge transfer from the first to second guide is improved by using the ratio $R_{is1}/R_{is2} \gg 1$, and also by generating two control pulses shifted, one, with respect to the other by a sufficient amount. From all the methods of generating such pulses in decatron-shaping stage—decatron circuits known to us, the differentiating and amplitude matching methods are, in our opinion, most suitable [3]. For this reason, the decatron control circuit should be based on the principle of control-pulse amplitude matching.

This method makes it possible to generate control pulses which satisfy the conditions for the best transfer of the discharge from the first to second cathode as well as the conditions for reducing the ignition potential of the cathode following the second guide (Figs. 22 and 26). The variation of the ignition voltage of the cathode following an operating second guide can be illustrated by the curves shown in Figs. 22-26 in [2].

The use of control stages having low $R_{is1,2}$ and low bias voltage results in a reduction of shifting pulse amplitudes and, consequently (Fig. 23), to a reduction of the ignition voltage of the following cathode. The ignition voltage of the cathode which follows an active second guide depends to a considerable degree on the rate of change of the anode and second-guide voltages (Figs. 24 and 25).

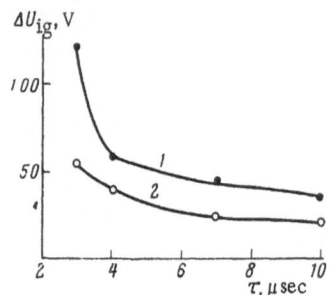

Fig. 22. Ignition voltage as a function of pulse duration τ at the second guide: (1) $U_{ig} = f(\tau)$, I_{g2} = 1.5 mA; (2) $U_{ig} = f(\tau)$, I_{g2} = 2 mA.

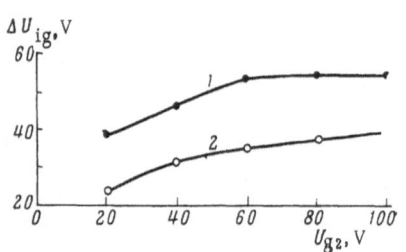

Fig. 23. Ignition voltage as a function of pulse amplitude at the second guide: (1) $U_{ig} = f(U_{g2})$, τ = 7 µsec; (2) $U_{ig} = f(U_{g2})$, τ = 10 µsec.

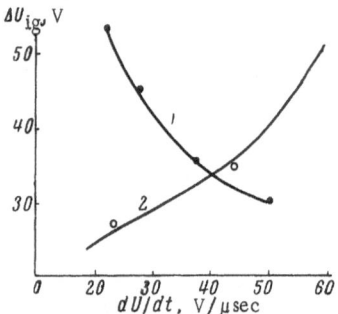

Fig. 24. Ignition voltage as a function of the rate of rise of the anode and second-guide voltage: (1) $U_{ig} = f(dU_a/dt)$; (2) $U_{ig} = f(dU_{g2}/dt)$.

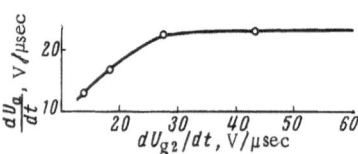

Fig. 25. Rate of rise of the anode voltage as a function of the rate of fall of the second-guide voltage.

The function $dU_a/dt = f(dU_{g2}/dt)$ is shown in Fig. 25. The curve is composed of two sections: (1) when dU_{g2}/dt < 28 V/µsec, dU_a/dT increases with an increasing rate of change of the second-guide voltage; (2) when dU_{g2}/dt > 28 V/µsec, dU_a/dt = const. The region dU_{g2}/dt > 28 V/µsec corresponds to the most reliable transfer of the discharge from guide to cathode so that, in this region, the transfer takes place without cutoff of the second-guide current. Cutoff of the second-guide current when dU_{g2}/dt > dU_a/dt leads to an increase of ignition voltage of the cathode which follows an active second guide. Thus, correct design of the circuit requires that the condition $dU_a/dt \geq dU_{g2}/dt$ is satisfied. An increase of dU_a/dt is possible only at the expense of reducing the resistance R_a; the reduction of decatron anode load increases the decatron current

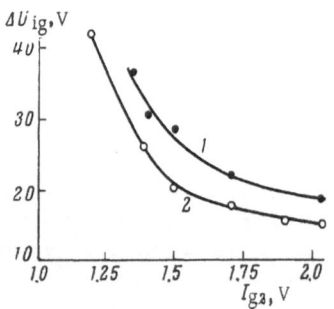

Fig. 26. Ignition voltage as a function of anode current: (1) τ = 7 µsec; (2) τ = 10 µsec.

which, in turn, reduces the ignition potential of the following cathode (Fig. 26) and improves the reliability of decatron operation. The decatron current should be thus held at the maximum allowable value of the given type.

$R_{is1,2}$ can be reduced by employing tubes with low R_i. The decatron can then be reliably reset to zero by opening the cathode circuits because of the large values of R_2 and R_3 (coupling resistances between the guides (Fig. 21). It is our opinion that preference should be given to circuits in which $R_{is\,1,2}$ are reduced by decreasing the resistance in the plate circuits of the control tubes (with dc coupling between the decatron and control tube). The latter is due to the fact that after the pulse at the control-tube grids ends, the time constant of the plate circuits changes sharply from $\tau_a' = [R_i R_3 /(R_i + R_3)] C_{int}$ to $\tau_a'' = R_3 C_{int}$. This leads to a shift of the dc component level at higher operating speeds and to a possible cutoff of the decatron.

Finally, reliable operation of the decatron requires that $R_{c_0} < R_{c_0},max = [E_b R_a/E_a - (E_b + U_O)]$. The realization of the demands mentioned above makes it impossible to reset the decatron to zero by opening the cathode circuit. This led to the development of the zero reset circuit shown in Fig. 21. In this circuit the capacitor C_{res} is charged to a voltage sufficient for resetting the decatron by a negative pulse (14). At the moment of reset this capacitance is discharged to zero through the zero cathodes (the diode D prevents the capacitor voltage from becoming positive). The switch S_1 should be set to position 1 before switching S_2 to position 1'. After the capacitor discharges S_1 is returned to position 2 and the switch S_2 can then be set back to position 2'. This reset procedure is necessitated by the fact that manipulation of the switch S_2 only is liable to transfer of the discharge from the zero to the ninth or first cathode when the switch is thrown from one position to the other. The above method was successfully used for resetting to zero 40 OG-4 decatrons by employing a circuit with the following parameters: $C_{res} = 5 \mu F$, $R_{ch} = 200 k\Omega$, D7Zh diode, $R_{c_0} = 51 k\Omega$, and $R_a = 750 k\Omega$.

LITERATURE CITED

1. F. M. Yablonskii, Radiotekhn. i Elektron., 5 : 338 (1960).
2. F. M. Yablonskii, Radiotekhn. i Elektron., 6 : 1941 (1961).
3. V. A. Zapevalov, Pribory i Tekhn. Eksp., 2 : 186 (1961).

CONTRIBUTION TO THE DESIGN OF THREE-MESH
BANDPASS FILTERS

V. S. Voronin and S. S. Semenov

Bandpass filters are required to operate with transformers between their unmatched terminal impedances in some applications in physics research. One of the cases of practical interest here is the use of three-mesh band filters as converting four-terminal networks in high-power wideband microwave amplifiers. The three-mesh bandpass filter used in this application must be so designed that power transfer from the anode circuit of the tube in the preceding stage to the cathode-grid circuit in the next stage will be maximized with allowable frequency response nonuniformity over a specified frequency band, while the input and output capacitances of the tubes and the inductances of the connections must appear as component elements in the converting network. This article presents a simple method for the design of such three-mesh bandpass filters. Consider a balanced three-mesh bandpass filter (Fig. 1). We introduce the notation ω_1 and ω_2 for the cutoff frequencies of the frequency passband, and write

$$\omega_0 = \sqrt{\omega_1\omega_2} = \frac{1}{\sqrt{LC}} = \frac{1}{\sqrt{L_1C_1}}, \quad Q = \frac{R}{\sqrt{\frac{L}{C}}}, \quad Q_1 = \frac{2R}{\sqrt{\frac{L_1}{C_1}}},$$

$$\Omega = \left(\frac{\omega}{\omega_0} - \frac{\omega_0}{\omega}\right)Q, \quad \beta = \frac{1}{QQ_1}. \tag{1}$$

Then

$$Y = \frac{1}{R}(1 + j\Omega), \quad Z = j2R\beta\Omega \tag{2}$$

and the normalized transfer coefficient of this balanced three-mesh filter will be

$$K = \frac{2\dot{U}_2}{\dot{I}_1R} = \frac{1}{(1 + j\Omega)(1 + j\beta\Omega - \beta\Omega^2)}. \tag{3}$$

It is readily seen that the minima of the transfer coefficient (Fig. 2) correspond to the frequencies

$$\Omega_{\min} = \pm\sqrt{\frac{1-\beta}{3\beta}} \tag{4}$$

and both are equal to

$$|K_{\min}|^{-2} = 1 + \frac{4}{27}\frac{(1-\beta)^3}{\beta}. \tag{5}$$

215

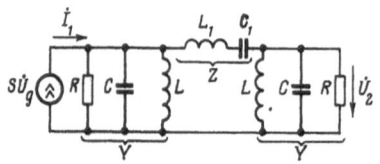

Fig. 1. Circuit of balanced three-mesh filter.

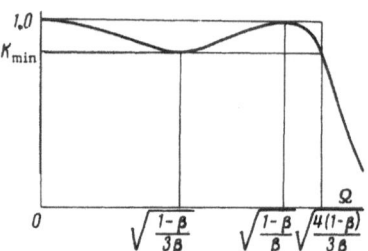

Fig. 2. Normalized frequency response of balanced three-mesh filter.

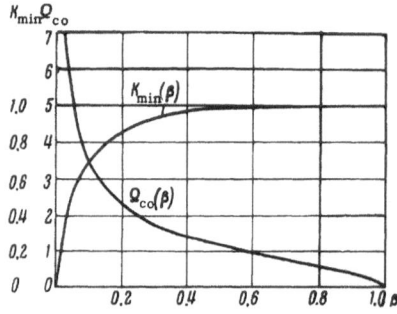

Fig. 3. Variation of normalized transfer coefficient $\dot{K}$ and band cutoff frequency Ω_{co} with β.

The maxima correspond to the frequencies

$$\Omega_{max} = \begin{cases} 0, \\ \pm \sqrt{\dfrac{1-\beta}{\beta}} \end{cases} \qquad (6)$$

and all three are equal to

$$|K_{max}| = 1. \qquad (7)$$

The bandpass cutoffs are found from the equation

$$\Omega_{co} = 2\Omega_{min} = \pm 2\sqrt{\dfrac{1-\beta}{3\beta}}. \qquad (8)$$

Once the allowable nonuniformity in the frequency response has been specified, $|K_{min}|$, we can proceed to use the band cutoff frequencies ω_1 and ω_2 and the capacitance C to find β and Ω_{co} with the aid of Eqs. (5) and (8) or with the aid of the graphs (Fig. 3), and then all parameters of components of the balanced three-mesh filter can be found using Eq. (1).

The second phase in design is to convert the balanced three-mesh bandpass filter so obtained to an unbalanced three-mesh filter exhibiting the same frequency response and providing maximum power transfer between unmatched terminal impedances with no intervening transformers.

If the problem is not one of maximizing power transfer, then the simplest approach would be to resort to the method proposed by Pawsey [1] so as not to alter the frequency response in the conversion. In this method, the balanced filter is divided on its axis into two identical parts (Fig. 4). One part is increased in its impedance level by using a positive real multiplicative factor a and this part is then reconnected to the part left unaltered. As a result, we obtain an unbalanced circuit with a frequency response between unequal impedances which is identical to the frequency response of the original balanced network. But the power transferred to the resistor R in the right-hand half of the filter will now be $(a + 1)/2$ times less than the optimum case where an ideal transformer is employed.

If power transfer between unequal impedances must be maximized, we can proceed as follows. The inductance L_1 is now represented as two series-connected inductances L_1' and L_1''

$$L_1 = L_1' + L_1'' \qquad (9)$$

and two ideal transformers with transformation ratios n_1 and n_2 are included in the circuit:

$$n_1 = 1 + \dfrac{L_1'}{L}, \quad n_2 = 1 + \dfrac{L_1''}{L} \qquad (10)$$

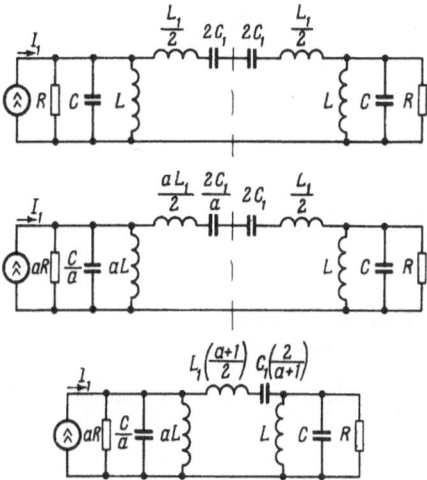

Fig. 4. Conversion of balanced three-mesh filter to unbalanced
filter, while retaining maximum power transfer feature.

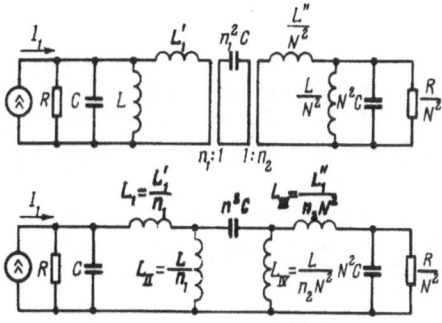

Fig. 5. Conversion to balanced three-mesh filter to unbalanced
filter, with maximum power transfer retained.

(see Fig. 5), so that we end up with the required transformation ratio N:

$$N = n_1/n_2. \tag{11}$$

By changing the order in which the inductances L_1', L, and L_1''/N^2, and L/N^2 are included [2, 3], we can produce a circuit using no ideal transformers. Expressing the parameters of the resulting unbalanced filter in terms of the parameters of the original balanced network [Eqs. (9)-(11)], we find

$$n_1 = \frac{2 + L_1/L}{1 + 1/N}, \quad n_2 = \frac{2 + L_1/L}{1 + N}; \tag{12}$$

$$L_1' = (n_1 - 1)L, \quad L_1'' = (n_2 - 1)L; \tag{13}$$

$$L_I + L_{II} = L, \quad L_I = \frac{L_1'}{n_1} = \left(1 - \frac{1}{n_1}\right)L, \quad L_{II} = \frac{L}{n_1};$$

$$L_{III} + L_{IV} = \frac{L}{N^2}, \quad L_{III} = \frac{L_1''}{n_2 N^2} = \left(1 - \frac{1}{n_2}\right)\frac{L}{N^2}, \quad L_{IV} = \frac{L}{n_2 N^2};$$

(14)

$$C_0 = C_1 n_1^2. \tag{15}$$

The circuit described is realizable only when

$$N \leqslant 1 + \frac{L_1}{L}. \tag{16}$$

This three-mesh filter can be tuned by successively tuning to the geometric-average frequency ω_0 of the plate circuit and grid circuit while leaving the circuit branch with the capacitor C_0 open:

$$\omega_0 = \frac{1}{\sqrt{(L_I + L_{II})C}}, \quad \omega_0 = \frac{1}{\sqrt{(L_{III} + L_{IV})CN^2}} \tag{17}$$

and likewise the circuit obtained by shorting capacitances C and CN^2, and consisting of the capacitor C_0 and series-connected parallel-pair inductances $L_I L_{II}$ and $L_{III} L_{IV}$

$$\omega_0 = \frac{1}{\sqrt{\left(\dfrac{L_I L_{II}}{L_I + L_{II}} + \dfrac{L_{III} L_{IV}}{L_{III} + L_{IV}}\right)C_0}}. \tag{17a}$$

The method described has been used in the design of the radio-frequency system of the ring phasotron (synchrocyclotron). Two coaxial lines with output branches at distances $1/n_1$ and $1/n_2$ of their length from their shorted terminations were employed as the filter inductances. The characteristic impedance of the transmission lines was set so that the lines would perform satisfactorily when the input and output capacitances of the tubes were shorted [4].

This type of three-mesh bandpass filters on lines in high-power wideband microwave amplifiers made it possible to devise load components for the tubes with fairly high input impedance combined with convenient design features in the connections between filter and oscillator tubes.

LITERATURE CITED

1.   D. C. Pawsey, IEEE Trans. on Circuit Theory, Vol. CT-10, No. 4 (1963), pp. 521-523.
2.   J. Stewart, Circuit Theory and Design. Wiley, New York (1956), pp. 56-62. [Russian translation (1962).]
3.   G. V. Zeveke, P. A. Ionkin, A. V. Netushil, and S. V. Strakhov, Fundamentals of Circuit Theory. Gosenergoizdat, Moscow-Leningrad (1963). p. 165.
4.   D. P. Linde, Fundamentals of the Design of Microwave Tube Oscillators. Gosenergoizdat, Moscow-Leningrad (1959), pp. 262-280.

SYSTEM FOR STABILIZING THE SUPPLY CURRENT
OF AN ELECTROMAGNET

V. S. Voronin and S. S. Semenov

A system for feeding the distributed windings of an electromagnet used in the first starting stage of the 30-MeV radial-sector annular phasotron of FIAN (Physical Institute of the Academy of Sciences of the USSR) is described. The electromagnet is supplied from dc generators of the P101 type rated at 65 kW, 230 V, and 282 A. In order to stabilize the direct component of current and suppress voltage pulsations at the generator outputs to a level of 0.03%, control units of 6S18S tubes are included; some of these are connected in parallel with the excitation winding of the generator and others in parallel with the load. A voltage for suppressing pulsations and stabilizing the steady component reaches the control grids of these tubes from the output of a dc amplifier. The latter is based on the principle of the parallel operation of a narrow-band dc amplifier with regeneration, taken from a BT-4 system and having a low zero drift, and a wideband ac amplifier. A feature of the amplifier is the use of intensive negative feedback for stabilizing the operating conditions of the tubes and uniformly matching the narrow- and wide-band frequency characteristics. The stabilizing system described makes it possible to use dc generators as sources of high-stability current or high-stability voltage.

1. REQUIREMENTS LAID ON THE SUPPLY SYSTEM

The characteristics of the supply system for the magnet of the annular phasotron are principally governed by the specific construction of the accelerator magnet and the necessity of preserving strict tolerances on the stability of the magnetic-field characteristics.

It is well known that the magnetic field of the annular phasotron [1,2] has a strong spatial inhomogeneity, varying over a narrow ring from extremely low values of the order of 20 G to saturation values of 7000 to 10,000 G.

In the case in question, the required law of magnetic-field variation in the various units of the electromagnet is secured by means of four distributed windings laid in channels of the pole-pieces, rather like the windings of electrical machines, and a single concentrated winding arranged on the yoke [3] (Fig. 1). The corresponding windings of the positive units are connected in series and are supplied from individual stabilized-current sources having a common reference voltage. The negative units are supplied in an analogous manner.

Since the annular phasotron belongs to the family of strong-focusing accelerators, all the tolerances relating to the accuracy and stability of the magnetic-field characteristics required to keep the working point $\nu_r\nu_z$ well away from resonances in such accelerators must naturally be satisfied. In addition to this, the annular phasotron imposes additional requirements associated with the condition of orbit similarity; these reduce to a tolerance on the field index n in each element of periodicity. Since the field of the annular phasotron is extremely dependent on radius, the azimuthal symmetry is very sensitive to distortions of n depending on the azimuth, and the tolerance thus laid upon n is in practice one of the most rigid.

219

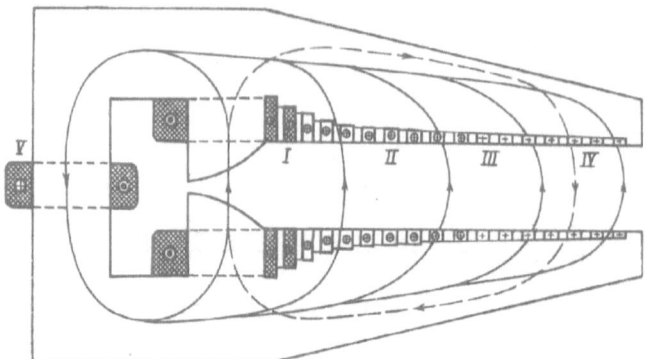

Fig. 1. Schematic diagram of one of the forty units of the electromagnet in the
annular phasotron. (I)-(V) Principal windings of the electromagnet.

For lower frequencies, the field instability is approximately equal to the instability of the ampere-turns; hence, the currents in the windings of the electromagnet must be stabilized to an accuracy of the order of a few hundredths of a percent. An important feature in this is the fact that any instability of the currents in the windings of the electromagnet is almost entirely reflected in pulsations and instabilities of the supply voltage, over a fairly wide frequency band, since the inductive reactance of the distributed turns is small, and falls off far less with increasing pulsation frequency than is the case in ordinary, widely used constant electromagnets.

The region most sensitive to instabilities of the supply currents is the region of weak magnetic field at the beginning of the working range, where the field consists of the difference between the field set up by the winding on the yoke (compensating the magnetic potential drop in the iron of the magnetic circuit) and the leakage fields of the distributed windings situated at large radii. The total leakage field approximately equals -180 G. In order to obtain a working field of $+20$ G, a field of $+200$ G must be created by the fifth winding. The current in the principal fifth winding is regulated by a signal from a magnetic-field detector situated in the region of weak magnetic field.

2. VOLTAGE-STABILIZING SYSTEM FOR THE DC GENERATORS

By way of high-stability current sources for feeding the first principal excitation windings of the magnet, we use a power plant comprising two dc generators of the P101 type rated at 65 kW, 230 V, and 282 A, and a synchronous motor. Feeding of the second, third, and fifth principal windings of the electromagnet is effected by an arrangement analogous to the generator-excitation circuit (Fig. 2). Regulation in the generator-excitation circuit is achieved by means of motor-driven rheostats (RVM) controlled automatically (AP) when the current in the excitation winding of the electromagnet deviates by more than 5% from the value set by the reference voltage. Smooth control (within 5%) is effected by means of groups of control tubes (GCT) of the 6S18S type (six tubes) connected in parallel with the generator-excitation winding. In view of the fact that the generator-excitation winding has considerable inductance (about 40 H), the depth of control with respect to excitation falls with increasing frequency. For effective suppression of current pulses in the load, we therefore need a second control link so as to vary the voltage on the load at a rate corresponding to a frequency band on several tens of kc/sec. For this purpose, eighteen 6S18S tubes are connected in parallel with the first principal windings of the electromagnet, the loads of these being 0.01-H chokes with resistances of 0.2 Ω. The 100-Ω automatic-bias resistances in the cathode leads of the 6S18S tubes stabilize the position of the working point for different anode voltages and give a uniform distribution of the current over the tubes; the negative current feedback arising from this improves the linearity of control and reduces the effect of scatter in the tube parameters. The resistances in the grid circuits of the tubes make the operation of the tubes independent of each other and prevent interruptions to the operation of the whole unit on breaking the contacts in the panel or on short-circuiting the electrodes of one of the tubes.

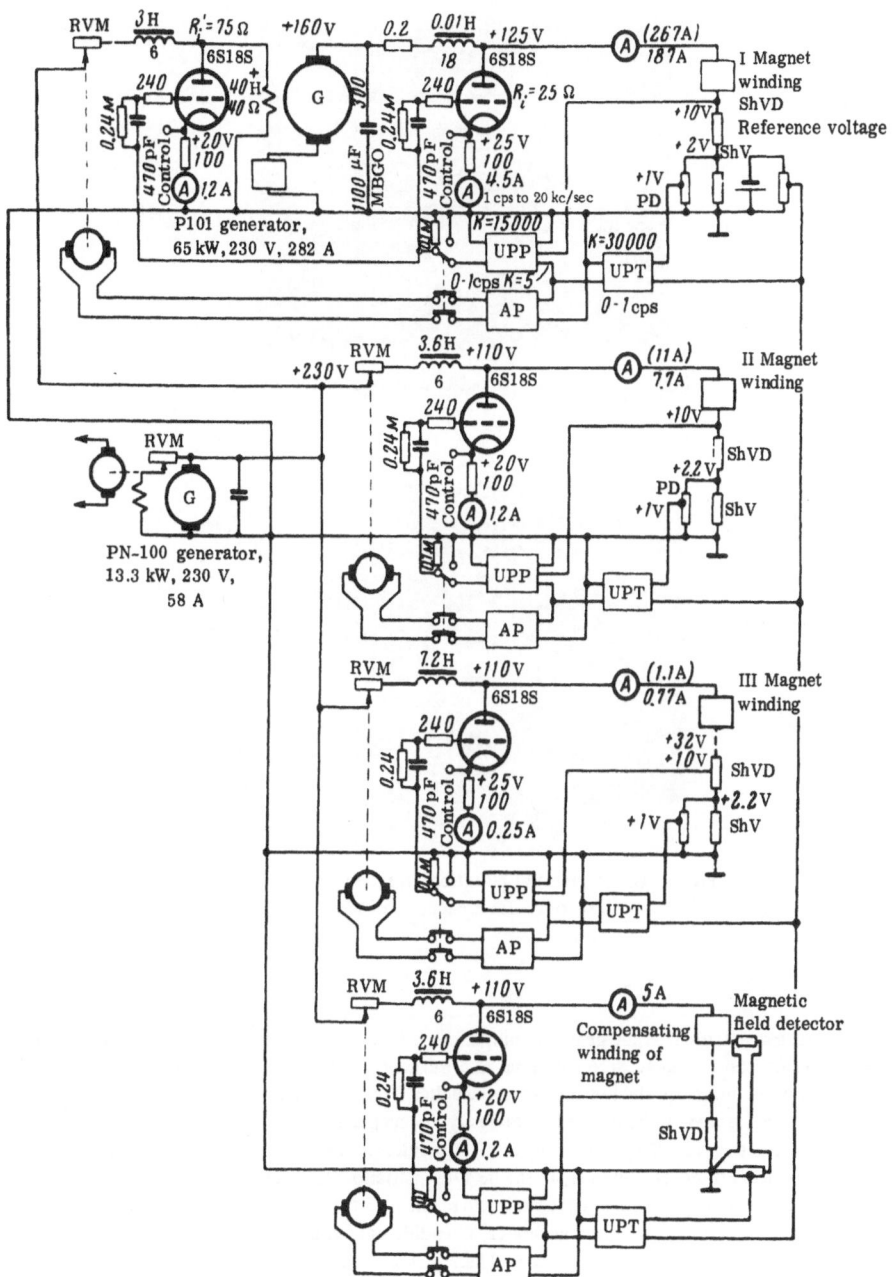

Fig. 2. Supply system for the electromagnet of the annular phasotron.

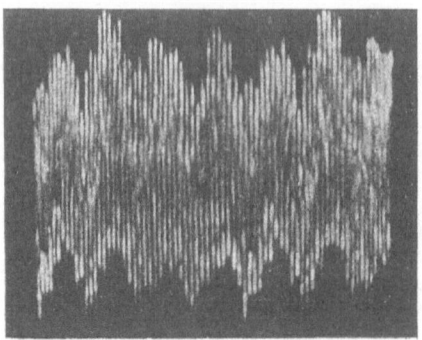

Fig. 3. Oscillogram of pulses at the generator output.

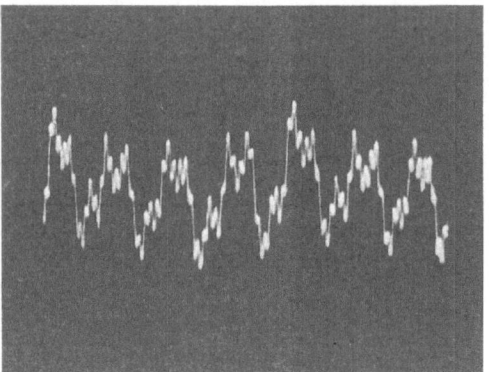

Fig. 4. Oscillograms of pulsations in the generator output after
shunting with a capacity of 1200 μF.

The greatest pulsations at the generator output came from the collector pulses (frequency 700 cps, amplitude 4 V); superimposed on these were pulsations at a frequency of $(^4/_3) \cdot 50 = 66.7$ cps and amplitude 3 V, due to asymmetry of the magnetic field of the generator pulses (Fig. 3). The relative high frequency of the 700-cps pulsations enabled these to be reduced quite easily to acceptable values by shunting the generator output with a condenser battery of 1200 μF, made up of 40 condensers of the MBGO type, each rated at 30 μF for 300 V. This made little difference, however, to the 66.7-cps pulsations (Fig. 4). Considerably less reflected in pulsations of generator output voltage were the 100% to 150% voltage pulsations appearing at a frequency of 150 cps in the excitation winding of the generator, when this was fed from a controlled thyratron rectifier in the original version of the stabilization system. These pulsations were unacceptably high, however, when the second, third, and fifth excitation windings of the magnet were fed from controlled thyratron rectifiers. The inclusion of passive filters for suppressing the pulsations was economically undesirable for the large currents required. In addition to this, such filters would be included in the negative-feedback circuit, which would reduce rapidity of action in the system and would require the introduction of corresponding corrective circuits. For these reasons it was decided to dispense with controlled thyratron rectifiers, simply using the rectifiers in the first stage as nonadjustable current sources, and securing control entirely from the groups of 6S18S tubes.

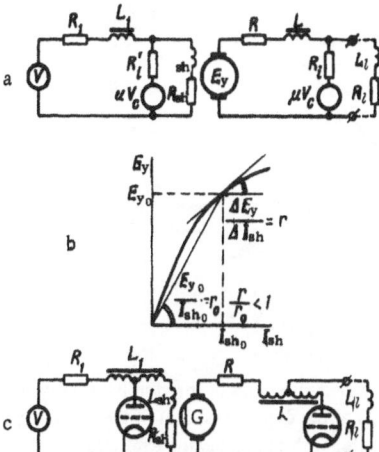

Fig. 5. To illustrate the calculation of the stabilization system.
(a) Equivalent circuit of the stabilization system; (b) characteristics of generator excitation; (c) autotransformer arrangement
for connecting the control tubes.

Assuming that the internal resistance of the control tubes is the same for ac and dc (which is acceptable for triodes) and also that the tubes operate on the linear parts of their characteristics, without cutting off with respect to current and anode voltage, we can use the active two-pole theorem for Fig. 5a and write

$$I_{sh} = \frac{I_{s.c.}}{1 + \dfrac{Z_{sh}}{Z_{int}}} = \frac{\dfrac{V}{R_1 + j\omega L_1} - \dfrac{\mu V_c}{R_i'}}{1 + \dfrac{(R_{sh} + j\omega L_{sh})(R_i' + R_1 + j\omega L_1)}{R_i'(R_1 + j\omega L_1)}} = \frac{R_i'}{R_i' R_{sh} + R_i' R_1 + R_{sh} R_1} \times$$

$$\times \frac{V - \mu V_c \dfrac{R_1}{R_i'}\left(1 + \dfrac{L_1}{R_1}\right)}{1 + j\omega \dfrac{L_{sh}(R_i' + R_1) + L_1(R_i' + R_{sh})}{R_i' R_{sh} + R_i' R_1 + R_{sh} R_1} + (j\omega)^2 \dfrac{L_1 L_{sh}}{R_i' R_{sh} + R_i' R_1 + R_{sh} R_1}} = \frac{1}{r_0} \frac{V - \mu V_c \dfrac{R_1}{R_i'}(1 + j\omega\tau_1)}{1 + j\omega(a + b) + (j\omega)^2 ab}, \qquad (1)$$

where $r_0 = (V/I_{sh})|_{\omega = 0}$.

For independent excitation, when $V = V_0$, by putting $\Delta E_y = r\Delta I_{sh}$ (Fig. 5b), we obtain

$$\frac{\Delta E_y}{-\mu V_c} = \frac{r}{r_0} \frac{R_1}{R_i'} \frac{1 + j\omega\tau_1}{1 + j\omega(a + b) + (j\omega)^2 ab}. \qquad (2)$$

In the self-excitation condition, when $V = E_y$,

$$\frac{\Delta E_y}{-\mu V_c} = \frac{r}{r_0} \frac{R_1}{R_i'} \frac{1 + j\omega\tau_1}{\left(1 - \dfrac{r}{r_0}\right) + j\omega(a + b) + (j\omega)^2 ab} \qquad (3)$$

and for the no-load voltage at the load terminals,

$$\frac{\Delta U_{n.l.}}{-\mu V_c} = \frac{\Delta E_y}{-\mu V_c} \frac{R_i}{(R_i + R)} \frac{1}{\left(1 + j\omega \dfrac{L}{R_1 + R}\right)}. \qquad (4)$$

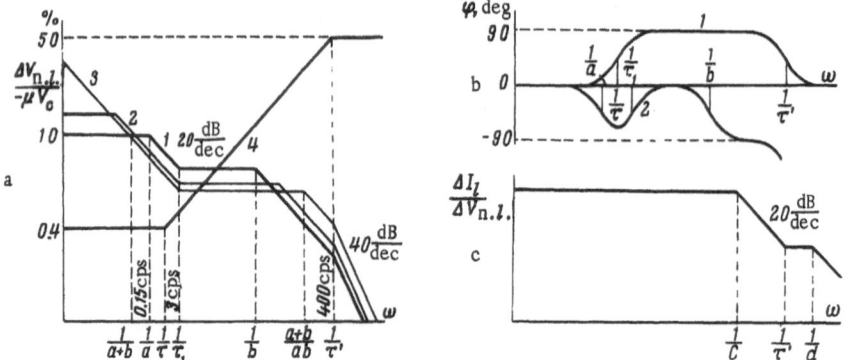

Fig. 6. Asymptotic frequency characteristics of the stabilization system (log–log scale). (a) Asymptotic frequency characteristics of the depth of control of the no-load voltage at the load terminals, using tubes connected in parallel with the generator-excitation winding: (1) with independent excitation; (2) with self-excitation $r/r_0 = 0.5$; (3) limiting case for self-excitation $r/r_0 \lesssim 1.0$; (4) asymptotic frequency characteristic for the depth of control of the no-load voltage at the load terminals, using tubes connected in parallel with the load. (b) Phase characteristics of the depth of control of the no-load voltage at the load terminals: (1) using tubes connected in parallel with the load; (2) using tubes connected in parallel with the generator-excitation windings. (c) Asymptotic frequency characteristic of the depth of current control in a resistive–inductive load $R_l L_l$.

The frequency characteristic for the control of no-load voltage by means of tubes connected in parallel with the load has the form

$$\frac{\Delta U_{n.l.}}{-\mu V_c} = \frac{R}{R_i + R} \frac{1 + j\omega \dfrac{L}{R}}{1 + j\omega \dfrac{L}{R_i + R}} = \frac{R}{R_i + R} \frac{1 + j\omega\tau}{1 + j\omega\tau'}. \tag{5}$$

Asymptotic frequency characteristics corresponding to Eqs. (2)-(5) are given in Fig. 6a.

By using the superposition method, with due choice of the quantities $R_l L_l R$ and L, we can obtain the overall frequency characteristic for the depth of no-load voltage control with a given degree of nonuniformity.

We can easily see that the choke in the excitation circuit is required, in addition to R, not only to increase the depth of control at high frequencies, but also to eliminate the bridge effect, in which control signals with the same amplitude add together in antiphase at the point of intersection of two characteristics (Fig. 6b). In order to ensure that the control tubes should be fully used with respect to both current and voltage, "auto-transformer" connection of the tubes (Fig. 5c) may be appropriate in certain cases. In this case, however, the stray inductance of the autotransformer/choke must be taken into account, since this will reflect on the operation of the system at high frequencies.

Considering that $Z_l = R_l + j\omega L_l$, and using the active two-pole theorem, we obtain for the depth of current control in the load

$$\Delta I_l = \frac{\Delta U_{n.l.}}{R_l + j\omega L_l + \dfrac{(R + j\omega L) R_i}{R_i + R + j\omega L}} = \Delta U_{n.l.} \cdot \frac{R_i + R}{R_l R_i + R_l R + R R_i} \times$$

$$\times \frac{1 + j\omega \dfrac{L}{R_i + R}}{1 + j\omega \dfrac{L_l (R_i + R) + L (R_l + R_i)}{R_l R_i + R_l R + R R_i} + (j\omega)^2 \dfrac{L_l L}{R_l R_i + R_l R + R R_i}} = g\Delta U_{n.l.} \cdot \frac{1 + j\omega\tau'}{1 + j\omega (c+d) + (j\omega)^2 c\, d}. \tag{6}$$

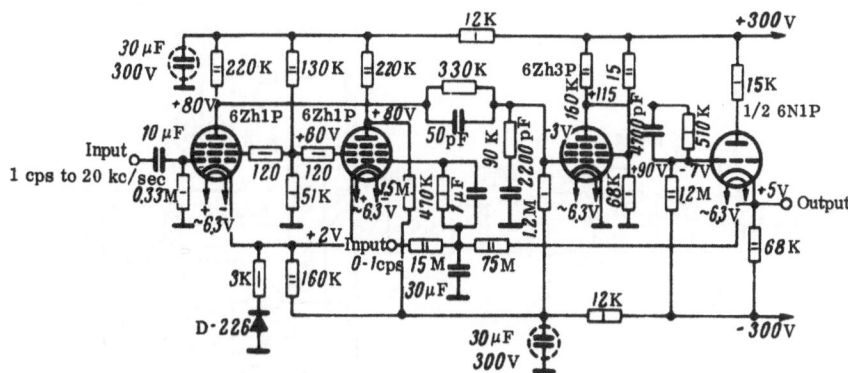

Fig. 7. Amplifier used for suppressing pulsations (UPP).

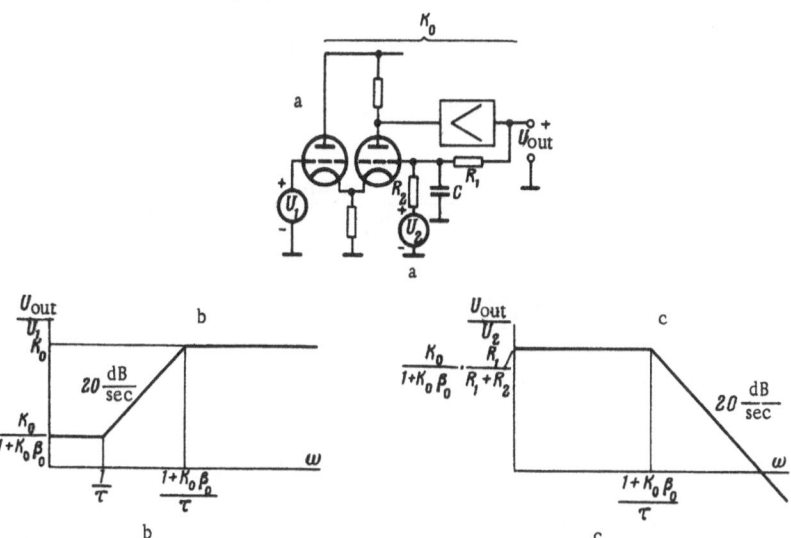

Fig. 8. Arrangement of amplifier. (a) Equivalent circuit of the UPP; (b) asymptotic characteristic of UPP amplifier with respect to input U_1; (c) asymptotic frequency characteristic of UPP amplifier with respect to input U_2.

Supposing that the frequency characteristic of the depth of control of the no-load voltage is made fairly uniform, we obtain the asymptotic frequency characteristic of the depth of current control in the load shown in Fig. 6c. Since the fall at high frequencies takes place at a rate of 20 dB per decade, on completing the frequency independent negative current-feedback circuit this system should be stable without requiring any additional correcting circuits. On applying this stabilization circuit to the supply system of the annular-phasotron electromagnet, however, it proved necessary to introduce a number of correcting devices in order to adjust the frequency characteristic in the high-frequency region, at which the cable capacities and the more complex nature of the load began to have an effect.

3. DC AMPLIFIER FOR THE COMPARISON CIRCUIT AND THE SUPPRESSION OF PULSATIONS

The supply system for the magnet of the annular phasotron employs dc amplifiers based on the principle of the parallel operation of a narrow-band dc amplifier with regeneration and a wide-band ac amplifier.

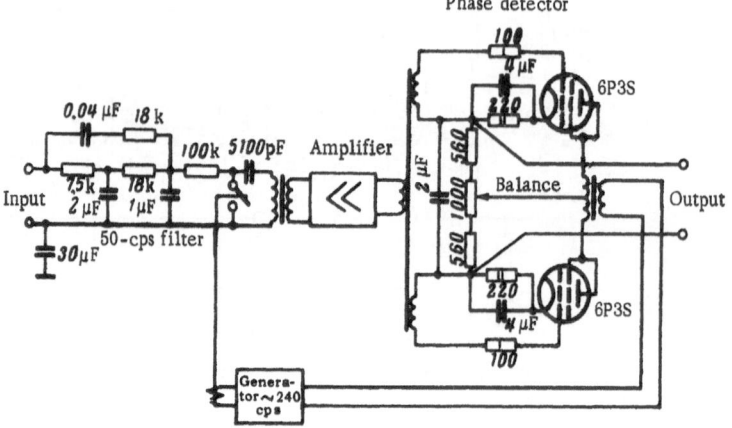

Fig. 9. DC amplifier from BT-4, with 50-cps filter.

The arrangement of the amplifier appears in Fig. 7. For the transmission factor of this kind of amplifier with respect to input U_1 (Fig. 8), we may write

$$U_{\text{out}} = U_1 \frac{K_0}{1 + K_0\beta_0} \frac{1 + j\omega\tau}{1 + \dfrac{j\omega\tau}{1 + K_0\beta_0}}, \tag{7}$$

where

$$\tau = C \frac{R_1 R_2}{R_1 + R_2} \quad \text{and} \quad \beta_0 = \frac{R_2}{R_1 + R_2}.$$

For the transmission factor with respect to input U_2 we obtain

$$U_{\text{out}} = -U_2 \frac{R_1}{R_1 + R_2} \frac{K_0}{1 + K_0\beta_0} \frac{1}{1 + \dfrac{j\omega\tau}{1 + K_0\beta_0}}. \tag{8}$$

The overall transmission factor is obtained from the expression

$$U_{\text{out}} = \frac{K_0}{1 + K_0\beta_0} \frac{U_1(1 + j\omega\tau) - U_2 \dfrac{R_1}{R_2}\beta_0}{1 + \dfrac{j\omega\tau}{1 + K_0\beta_0}}. \tag{9}$$

It is not difficult to see that, if

$$U_2 = -K_0 \frac{R_2}{R_1} U_1, \tag{10}$$

the transmission factor will not depend on the frequency:

$$U_{\text{out}} = K_0 U_1. \tag{11}$$

In the stabilization system considered (Fig. 2), the input is connected through C_{in} = 10 µF and leak resistance R_{in} = 0.33 MΩ, where

$$R_{in} C_{in} \gg \frac{\tau}{1 + k_0 \beta_0} \tag{12}$$

to a high-resistance shunt, the stability of which may be quite low, from which the voltage of the current pulsations in the magnet windings passes to the amplifier. Input U_2 receives a signal from the narrow-band $[\omega_u \gg (1 + K_0 \beta_0)/\tau]$ dc amplifier (from the BT-4) with low zero drift (Fig. 9), which amplifies the difference signal between the reference voltage and the voltage from an accurate low-resistance shunt. The maximum voltage on the output of the amplifier (see Fig. 7) is ± 80 V and is fed to the grids of the 6S18S control tubes. A D-226 diode keeps the charge of condenser C = 30 µF from the source of negative voltage at the instant at which the amplifier is switched on and while it is warming up. Generally speaking, if the amplifier inputs U_1 and U_2 are respectively connected to an ac amplifier with a lower limiting frequency of $\omega_l = 2\pi f_l$, and a dc amplifier with an upper limiting frequency of $\omega_u = 2\pi f_u$, the limiting frequencies being given by a phase shift of approximately 45°, and

$$\omega_l < \frac{1 + K_0 \beta_0}{\tau} < \omega_u \quad \text{and} \quad U_2 = -K_0 \frac{R_2}{R_1} U_1, \tag{13}$$

then the maximum nonuniformity of the frequency characteristic will be given by one of the expressions

$$\frac{\Delta U_{out}}{U_{out}} < \frac{\omega_l \tau}{1 + K_0 \beta_0} \quad \text{or} \quad \frac{\Delta U_{out}}{U_{out}} < \frac{1 + K_0 \beta_0}{\omega_u \tau}. \tag{14}$$

The magnet-supply system described has proved extremely stable and reliable in service.

LITERATURE CITED

1.   A. A. Kolomenskii et al., in the book: Proceedings of the Internat. Conf. on High-Energy Accelerators and Instrumentation. CERN (1959), pp. 89-99.
2.   A. A. Kolomenskii et al., in the book: Proceedings of the Internat. Conf. on High-Energy Accelerators, Dubna, 1963. Atomizdat (1964), pp. 653-657.
3.   V. N. Kanunnikov, Zh. Tekhn. Fiz., 33(5):592-602 (1963).